Today's conservation literature emphasizes land[illegible]
population genetics without addressing the behav[illegible]
the long-term survival of populations. This book [illegible]
practical arguments for considering behavio[illegible]
conserve biodiversity.

Prominent scientists and wildlife mana[illegible]
volume to address a number of issues, i[illegible]
of behavioral research to conserva[illegible]
variation as a component of biodi[illegible]
to solve conservation problems [illegible]
and management practices [illegible]

The book is unique in [illegible] empha[illegible]
opposed to captive a[illegible]
research has concent[illegible] this
volume demonstrates that the complete ec[illegible]logical fra[illegible]ork, not just
behavioral ecology, provides valuable techniques and knowledge for
conserving biodiversity.

BEHAVIORAL APPROACHES TO CONSERVATION IN THE WILD

BEHAVIORAL APPROACHES TO CONSERVATION IN THE WILD

edited by

JANINE R. CLEMMONS
Department of Zoology, University of Wisconsin

and

RICHARD BUCHHOLZ
Department of Biology, Northeast Louisiana University

PUBLISHED BY THE PRESS SYNDICATE OF THE UNIVERSITY OF CAMBRIDGE
The Pitt Building, Trumpington Street, Cambridge CB2 1RP, United Kingdom

CAMBRIDGE UNIVERSITY PRESS
The Edinburgh Building, Cambridge CB2 2RU, UK http://www.cup.cam.ac.uk
40 West 20th Street, New York, NY 10011–4211, USA http://www.cup.org
10 Stamford Road, Oakleigh, Melbourne 3166, Australia

First published 1997
Reprinted 1998

Printed in the United Kingdom at the University Press, Cambridge

Typeset in Linotron Times 10/13pt

A catalogue record for this book is available from the British Library

Library of Congress Cataloguing in Publication data

Behavioral approaches to conservation in the wild / edited by Janine
R. Clemmons and Richard Buchholz.
p. cm.
Based on papers from a symposium held at the Animal Behavior
Society Annual Meetings in Lincoln, Neb., 1995.
Includes index.
ISBN 0 521 58054 4 (hc.) ISBN 0 521 58960 6 (pbk).
1. Wildlife conservation–Congresses. 2. Animal behavior–
Congresses. I. Clemmons, Janine R. (Janine Rhea), 1957– .
II. Buchholz, Richard, 1964– . III. Animal Behavioral Society.
Meeting (1995: Lincoln, Neb.)
QH76.B44 1997
639.9–dc20 96-31559 CIP

ISBN 0 521 58054 4 hardback
ISBN 0 521 58960 6 paperback

Contents

Contributors

Peter Arcese Department of Wildlife Ecology, 1630 Linden Drive, University of Wisconsin, Madison, WI 53706, USA

Luis F. Baptista Department of Birds and Mammals, California Academy of Science, Golden Gate Park, San Francisco, CA 94118, USA

Steven R. Beissinger Ecosystem Sciences Division, 151 Hilgard Hall #3110, University of California, Berkeley, CA 94720-3110, USA

Creagh Breuner Department of Zoology, Box 351800, University of Washington, Seattle, WA 98195, USA

Richard Buchholz Department of Biology, Northeast Louisiana University, Monroe, LA 71209–0520, USA

Scott P. Carroll Department of Entomology and Center for Population Biology, University of California, Davis, CA 95616, USA

John R. Cary Department of Wildlife Ecology, 1630 Linden Drive, University of Wisconsin, Madison, WI 53706, USA

Janine R. Clemmons Department of Zoology, Birge Hall, 430 Lincoln Drive, University of Wisconsin, Madison, WI 53706, USA

Charlotte Deerenberg Zoological Laboratory, University of Groningen, PO Box 14, 9750 AA, Haren, The Netherlands

Hugh Dingle Department of Entomology and Center for Population Biology, University of California, Davis, CA 95616, USA

Kent Dunlap Department of Zoology, University of Texas, Austin, TX 78712, USA

John A. Endler Department of Ecology, Evolution, and Marine Biology, University of California, Santa Barbara, CA 93106, USA

Gene S. Fowler Department of Biology, Pomona College, Claremont, CA 91711, USA

Leonard Freed Department of Zoology, University of Hawaii at Manoa, Hawaii HI 96822, USA

Sandra L. L. Gaunt Borror Laboratory, Department of Zoology, 1735 Neil Avenue, Ohio State University, Columbus, OH 43210, USA

Kathleen Hunt Department of Zoology, Box 351800, University of Washington, Seattle, WA 98195, USA

Lukas F. Keller Department of Wildlife Ecology, 1630 Linden Drive, University of Wisconsin, Madison, WI 53706, USA

Jan Komdeur (1) Zoological Laboratory, University of Groningen, PO Box 14, 9750 AA, Haren, The Netherlands; (2) National Environmental Research Institute, Department of Wildlife Ecology, Kaloe, Grenavej 12, DK– Roende, Denmark

Jaan Lepson Department of Zoology, University of Hawaii at Manoa, Hawaii HI 96822, USA

Dennis T. Logan Senior Scientist, Coastal Environmental Services, Inc., 1099 Winterson Road, Suite 130, Linthicum, MD 21090, USA

Jenella E. Loye Department of Entomology and Center for Population Biology, University of California, Davis, CA 95616, USA

Ian G. McLean Department of Zoology, University of Canterbury, Private Bag 4800, Christchurch, New Zealand

Patricia G. Parker Department of Zoology, 1735 Neil Avenue, The Ohio State University, Columbus, OH 43210–1293, USA

Katherine Ralls National Zoological Park, Smithsonian Institution, Washington, DC 20008, USA

Edmund H. Smith Adjunct Professor of Zoology, Sonoma State University. Present address: 4090 Harrison Grade Road, Sebastopol, CA 95472–9774, USA

Scott H. Stoleson School of Forestry and Environmental Studies, Yale University, New Haven, CT 06511, USA

Thomas A. Waite School of Forestry, Michigan Technological University, Houghton, MI 49931, USA

John C. Wingfield Department of Zoology, Box 351800, University of Washington, Seattle, WA 98195, USA

Blair E. Witherington Florida Marine Research Institute, Tequesta Field Station, 9100 SE Federal Highway, Tequesta, FL 33469, USA

Preface

We had two concerns that spurred us into assembling the chapters in this book. We suspected that animal behaviorists could play a larger, more productive role in conservation biology than they have so far, and we knew that conservation biologists could adopt some powerful tools and perspectives from ethology that would be valuable in reducing the alarming loss of Earth's biological diversity. As we explored these issues with our colleagues in both fields, we were surprised by the clichés that seemed to serve as intellectual blinders. Behavioral biologists rarely recognized behavioral contributions to conservation other than in captive breeding and reintroduction programs of endangered species. The conservation biologists typically claimed that behavioral research was mostly "descriptive natural history."

To counter this disciplinary stagnation, and with the financial help of the US National Science Foundation and the Animal Behavior Society, we individually and jointly explored the role of behavioral study in conservation efforts. This was accomplished through the organization of and attendance at national and international symposia and special paper sessions on the common goals of conservation biology and behavioral studies. In addition we met with conservation or behavior discussion groups to be sure that we would entertain and include a variety of perspectives from the two disciplines. Along the way we collected an international assemblage of stimulating and insightful authors engaged in fieldwork at the interface of ethology and conservation biology.

We would like to share with the reader the reasons why this volume is shaped as it is. This book has two major objectives. The first is to stimulate behavioral researchers to think about how their specific areas of study can contribute to the conservation of biological diversity. The second is to show conservationists the relevance of behavioral research in solving a variety of conservation problems. While this book is far from inclusive at representing the diverse work of all behavioral scientists having this goal in mind, it is an attempt to clarify and emphasize the point that behavioral research has already been of conservation value, and expansion of conservation-oriented behavioral research is essential if we are to achieve our highest potential at conserving biological diversity.

This book is intended in part for conservation biologists for several reasons. The first is that the relevance of behavioral research to conservation biology most likely has not been made clear to most conservation biologists, except perhaps in the realm of captive breeding and reintroduction programs. The reintroduction of animals, for example, poses a variety of problems pertaining to such behavioral phenomena as conspecific imprinting and the learning of food finding or predator avoidance skills. These kinds of problems faced by captive breeders seem to be somewhat better recognized, although nonetheless in need of research, whereas the direct relevance of behavioral research to the preservation, monitoring, and management of wild populations may be less clear to many biologists (including many behavioral biologists). Common sense would suggest that animal behavior most certainly should be a vital consideration in conservation or management plans and implementation. The survival and reproductive success of all animal species impinges on behavior – how individuals, social groups, or populations respond to and interact with the environment. The realm of influence is not restricted to the Animal Kingdom: the survival of many species of plants also depends directly or indirectly on behavioral processes of animals that in turn affect a variety of other processes, including pollination, seed dispersal, and nutrient cycling. Behavioral biology, however, is more than a subdiscipline of ecology. This book provides numerous examples showing what behavioral biology can offer uniquely in the way of theories, methods, technology, factual knowledge, and insight into solving conservation problems.

A second consideration is that behavioral diversity itself is an important but often overlooked component of biological diversity. Failure to identify and preserve endangered behavior will undermine species' potential for not only survival but evolutionary change, and hence, long-term persistence. Efforts aimed solely at conserving genetic diversity do not automatically subsume the survival of behavioral diversity.

Thirdly, conservation biology has been dubbed the 'crisis science.' How to preserve as much of the world's biological diversity as possible is a multifaceted task to which a multitude of disciplines can contribute.

Conservationists may not have the time to conduct the necessary research to gain answers using traditional scientific approaches, which are costly in terms of time and funding. In many cases, conservationists need 'best guess' answers to urgent problems. Behavioral biology can offer a tremendous data base and unique perspective to increase the chance that a given conservation strategy will be successful.

Viewed another way, conservation biology is not just a crisis science. The problems, approaches, and emphasis characterizing conservation biology today will change over the next several decades. Whereas in the concerned eyes of many, our time and resources would be better devoted simply to halting environmental destruction, conservationists predict that the activity of preserving land and ecosystems will subside within the next few decades. The major tasks of conservationists then will be to manage species within those preserves, restore destroyed or altered areas, and use or enjoy biological resources sustainably both inside and outside preserved areas. To these major tasks of the present and especially the future, behavioral research will have much to contribute.

With those words to the conservation biologist, we believe that the onus is ultimately on behavioral biologists to show how their discipline has and can contribute to solving conservation problems. This book barely scratches the surface. Why behavioral biologists as a whole have not taken a lead in conservation issues is for anyone to speculate, and a few of our authors do this. Rest assured that one reason cannot be that basic research has no place in applied science. Most of the contributors to this book would not consider themselves applied scientists. While conservationists can provide some of the questions to help guide research directions, pure science generates new questions and answers in unexpected ways. Theory directs lines of questioning and provides guidelines for answering. To the degree that theory depends on unique insight, there are no prescriptions for where to look.

Behavioral biologists are both personally and professionally dependent on conservation. As pointed out by Niko Tinbergen, behavioral biology depends perhaps even more than other biological fields on the comparison of traits among extant species for elucidating evolutionary history and process. The reason is that behavior does not fossilize, so that the comparative method is our main recourse for studying evolution. The extinction of each animal species, therefore, is the deletion of another data point in the potential information base on which our profession is dependent. Thus, anyone who wishes to understand the evolutionary basis of behavior has much to gain by the conservation of biological diversity.

This book limits its focus on behavioral research that applies to the conservation of wild populations as opposed to captive-bred and reintroduced populations, only because we feel that others are already working together to further research in that area. Although captive breeding is and will probably always remain an important aspect of conservation biology, much more thinking is needed about how behavioral research can be applied to the conservation of wild populations. This book is also strongly oriented toward non-human behavioral research, although human behavioral research is badly needed. Tinbergen wrote repeatedly about the accelerating pace of human-induced alterations of the environment, calling for comparative research that would allow us to determine not only how non-human species but also how humans could cope with the changing environment. Humans, like other species, are biologically constrained and have only limited behavioral options under a given set of conditions.

I realise full well that I am advocating a swing of animal ethology towards a more applied course. But it must by now be clear to many scientists that 'applied' science can be intellectually just as stimulating as 'pure' exploratory research. Even if this were not the case, the seriousness of our own situation is such that, unless we address ourselves to these vital problems, our very survival, and certainly our welfare, and with this our own science, will be in serious danger. In a sick, greatly impoverished and damaged society there will be no place for, no resources to spare and no interest in attending to, those 'growing points' of our science that have been discussed in such persuasive and stimulating ways in many of the other contributions to this volume. There is no lack of plans for future research, nor is there among the younger generation a shortage of keen and gifted recruits. If our science is to flourish, it must be seen not only to plan future ethological exploration, but also to work towards convincing our fellow-men of the necessity for its further growth. As part of this it will be the task of all the sciences concerned to work towards further integration and towards a more general awareness that the behavioural sciences are not a dispensable luxury, but an essential part of our overall effort to ensure a healthier future for society, in which alone man's highest mental potential can be fully realised.
(Tinbergen 1976, in *Growing Points in Ethology*, ed. P. Bateson & R. Hinde, Cambridge University Press)

To this eloquent plea from one of the 'fathers' of ethology, we add that the efforts and tools of the ethologist and conservationist will be for naught unless the rampant growth of the human population and the profligate ways in which we waste our natural resources are brought under control. The approaches and techniques described in this book represent important advances in maintaining biodiversity as functional ecosystems, but alone they will fall short of our common goal. Saving what we study and cherish requires personal and political action on the part of our readers, whose efforts will be contagious only if they are sincerely and persistently demonstrative of their love for the natural world.

Janine R. Clemmons
Richard Buchholz

General acknowledgments

A great many people aided us in the initial effort to explore this topic and then produce this book. Jane Brockmann and Chuck Snowdon first suggested to us and continuously encouraged the organization of a symposium entitled "Conservation and Behavior in the Wild" at the Animal Behavior Society annual meetings in Lincoln, Nebraska, 1995, where many of the ideas in this book were set forth. Lee Drickamer was additionally helpful and supportive of our interest in getting more animal behaviorists active in conservation. Jack P. Hailman assisted us in the administrative arena to raise funding for the symposium. We owe much to the discussion of ideas with students and faculty at the University of Wisconsin, Madison, and the University of Florida, Gainesville, and to all of the contributors and attendees of the symposium and first conservation paper session at the ABS meetings. We are grateful for the logistical support provided to J.R.C. by the Department of Zoology at the University of Wisconsin, Madison, and to R.B. first by the Department of Zoology at the University of Florida, and then the Department of Biology at Northeast Louisiana University. Tracey Sanderson, our editor at Cambridge University Press, from the beginning expressed a keen and patient interest in our book and was tirelessly instructive about the editorial aspects of book production. The symposium and part of this book were financially assisted by the US National Science Foundation and the Animal Behavior Society.

Special thanks go to the many reviewers of the chapters in this book. All of the chapter manuscripts were peer-reviewed by at least one animal behaviorist and one conservation biologist, and typically reviews were solicited from several of each. We appreciate the considerable effort of these reviewers, many of whom provided valuable perspectives for our authors.

We are especially grateful to each of our authors for their allegiance to the goals of this book. Given that this book would be the first on this topic, we felt a duty to focus the chapters, more so than editors of a multi-authored volume might normally. As a result, our authors patiently complied with our requests for more detailed outlines and considerable rewrites. We are impressed by and grateful for their efforts and proud of their final products. Finally, we have no less to thank our family and friends for listening sympathetically to the day-to-day updates on the progress of this book.

Part I:
Problems and issues

In the first four chapters our authors consider general issues and controversies surrounding the applicability of behavior studies to conservation. Currently, it is not clear whether the slow entry of ethology into conservation biology represents real limitations in the ability of behavioral research to contribute to conservation, or whether the absence is due to other factors, such as disciplinary biases. In Chapter 1 we (Clemmons and Buchholz) discuss the traditional approaches used in ecology and ethology, and explore how these fields are likely to find common ground in the interest of conservation. The problem may center around the question of how to integrate different levels of study. We discuss how organismal approaches have a critical role in conservation, whether the objective is to preserve species having special interest or to protect larger systems of ecological complexity.

In Chapter 2, Beissinger outlines the major objectives and methods of conservation biology, and then examines how and where behavioral studies have contributed in the past and may contribute in the future. Beissinger argues that behaviorists need to focus more on species and problems of direct interest to conservation and translate their findings into products more usable to conservationists, especially at the level of larger ecological scales.

Chapter 3 (Arcese, Keller, and Cary) argues that even though conservationists and non-behavioral specialists routinely gather behavioral information, thus seemingly obviating the need for specialists, the unique evolutionary perspective of animal behaviorists can lead to insights and solutions less likely to be discovered by traditional conservationists or wildlife managers. The value of an evolutionary perspective, these authors argue, will become most apparent when there is uncertainty about how animals will respond to a particular environmental change or management strategy.

Finally, Chapter 4 (Dingle, Carroll, and Loye) reminds us that the bias toward the conservation of vertebrates, which comprise less than one percent of the estimated species on Earth, appears to contradict the very mission of conservation biology – to preserve diversity. These authors provide a number of reasons why invertebrate behavior should be given greater emphasis, both as a focus of research and in conservation planning and actions. Invertebrate behavior is at the base of ecosystem functioning, provides critical links among different ecosystems, and is particularly suitable for studying specific problems given the fast generation time and taxonomic diversity of invertebrates.

Chapter 1

Linking conservation and behavior

JANINE R. CLEMMONS AND RICHARD BUCHHOLZ

[Conservation Biology] should attract and penetrate every field that could possibly benefit and protect the diversity of life.
M. E. Soulé, 1986

The discipline of behavioral biology, or ethology, currently is not considered a significant component of conservation biology. The first textbook on conservation biology (Primack 1993), for example, does not mention behavioral biology, even in the list of disciplines of ancillary importance. Many behavioral biologists might find the absence of behavioral research in conservation texts to be a gross oversight, but in truth the lack of ethology in such texts is not historically unwarranted. Aside from significant contributions to conservation by a few select behavioral ecologists (some references in Arcese *et al.* Chapter 3), behavioral papers are uncommon in conservation journals (Dingle *et al.* Chapter 4). In addition, it would not be an exaggeration to say that behavioral researchers, by and large, have not applied themselves to conservation, as evidenced by the scarcity of conservation-related papers published in behavioral journals or, until recently, presented at meetings (Arcese *et al.* Chapter 3; Dingle *et al.* Chapter 4). Also, unlike other professional science organizations (e.g., Ecological Society of America, American Ornithologist's Union), behavioral researchers as a group have not involved themselves in conservation policy-making.

The isolation of behavioral sciences from the interdisciplinary umbrella of conservation biology warrants concern for both disciplines. The animal behaviorist should be concerned with conservation issues because evolutionary studies of animal behavior depend on a diversity of living species occurring in their natural habitats. At the same time the conservationist should be concerned with ethology because the behavior of animals (including humans) characterizes the relationships among species and plays a significant role in ecological processes, including nutrient cycling, pollination, seed dispersal, and in structuring communities (Naiman 1988; Turner *et al.* 1995; Dingle *et al.* Chapter 4).

Conservation biology depends on multiple perspectives and sources of information to solve the variety of threats that are diminishing the world's biological diversity. So, why do the ethologist and conservationist, who would seem to have common goals, find themselves in functionally different orbits that only rarely align to the benefit of wild animals?

It has been proposed that behavioral studies are of limited value to conservation because of a discordance in the levels of focus entertained by behavioral biology and conservation biology (Beissinger, Chapter 2): behavioral research focuses at and below the levels of populations and organisms, whereas many conservation biologists claim that effective conservation must focus on higher levels of ecological organization. The contention is not unreasonable – how useful is an organismal approach when there are millions of species to save? This problem is not unique to behavioral biologists and conservation biologists. The question of how to integrate studies that focus at different levels of biological organization and different temporal and spatial scales has been called the central problem in ecology (Levin 1992). As an introduction to the rest of the chapters in this book, we explore the question, Can study at the organismal level contribute to the preservation of large numbers of species and the ecosystems they inhabit? And, if so, how have historical inertia and philosophical misunderstandings frustrated the recruitment of ethology into conservation biology?

Is the species approach to conservation inadequate?

The individual and population focus of most behavioral studies has often been used as evidence of the irrelevance of ethology to 'real world' conservation. As the dimensions of the extinction crisis grow, it becomes increasingly popular to claim that conservation at the ecosystem or landscape levels is the only solution (Harper 1981; Walker 1995). In part this claim arises from unsuccessful attempts to conserve biodiversity using a species-by-species approach. Despite efforts to protect endangered species under the US Endangered Species Act and various other conservation programs, many target species and countless more inconspicuous or unglamorous types continue to decline (Waller 1996). As a consequence, some conservation biologists have concluded that biological diversity would be more effectively conserved over the long term by protecting the diversity of areas and natural processes under which these millions of species have evolved (Walker 1995; Waller 1996).

Realistically, however, protecting large areas that guarantee the adequacy of space and natural ecological processes for all species without specific consideration of what those needs are will rarely be possible, especially as large undisturbed areas become scarcer. Additionally, ecosystems (or landscapes) are not independent entities: many species require more than one area to complete their life cycles (Dingle *et al.* Chapter 4). What constitutes an area 'large enough' to preserve is dictated not only by acreage, geometrical shape, and proximity to other protected areas, but also ultimately on the particular needs and characteristics of each species that inhabits the area (Allen & Hoekstra 1990). Human behavior must also be considered in the design and management of reserve areas. Humans enter and use wild areas or affect them indirectly in numerous ways (MacKinnon 1974; Chivers 1986; Hutchins & Geist 1987), and the response of animal species to human activities may be just as varied. Greater understanding about how different levels of analyses are necessarily integrated is required to achieve conservation of biological diversity on large scales. We find ample justification, therefore, for a deeper consideration of the role of behavioral studies in the design and management of wildlife areas.

Revalidating behavioral approaches to conservation

Before considering the practical arguments for using behavioral studies in conservation efforts, we briefly review the approaches used to study animal behavior. Consideration of when and how behavioral research can contribute to conservation must necessarily invoke the full spectrum of approaches constituting the framework of ethology. Contrary to popular misconception, not all behavioral biologists spend their days constructing ethograms, nor are behavioral ecologists the only ethologists with something to contribute to conservation.

Behavioral approaches

Behavioral biologists employ four interdependent levels of analysis, known as the Four Whys, or Determinants of Behavior (Tinbergen 1963): (1) Survival Value, or Perpetuation, which includes considerations about the selective advantage of a behavior, although non-adaptive mechanisms may also operate (Hailman 1967); (2) Phylogeny, the evolutionary history of behavior in a population or lineage; (3) Control, the 'immediate' causal mechanisms of a behavior pattern, such as muscular contraction or perception of external stimuli; and (4) Ontogeny, the developmental history of the behavior of individuals as a result of both genetic and environmental factors (Tinbergen 1963; Klopfer & Hailman 1967; Hailman 1967, 1977, 1982). Analyses of control and ontogeny relate to questions about individual organisms, whereas analyses of perpetuation (or survival value) and phylogeny are concerned with populations (Hailman 1977). Organismal-level questions, however, are also relevant to describing populations and species to the extent that organisms, populations, and species share some characteristics (Allee *et al.* 1955).

Although this conceptual framework has been adopted mostly by behavioral biologists, it applies to other biological questions as well, as realized by Tinbergen (1963). In wider biological circles these four levels of analysis are referred to as 'proximate' and 'ultimate' causes (Lack 1954; Mayr 1961). The main advantage of distinguishing the 'four whys' is to avoid misunderstandings about the relative contributions of study at different levels; more than one 'correct' answer to the question 'why?' can occur simultaneously (Sherman 1988).

The four levels of analysis are complementary in their explanations, such that answers from one approach can inform another, and thereby lead to novel insights. For example, in experiments that examined the mechanisms by which lizards disperse to and select new territories, Stamps (1987, 1988) demonstrated that lizards use cues from conspecifics rather than direct estimates of territorial quality (a question about control, or immediate causation). Thus, it would be mistaken to assume that absence of a species from a habitat means that the habitat is somehow unsuitable (a question about adaptation). The concept of conspecific attraction has been applied to metapopulation dynamics (Smith & Peacock 1990) and other problems surrounding habitat selection and re-colonization (Reed & Dobson 1993; Baptista & Gaunt, Chapter 9). Ontogenetic analyses further increase our understanding about habitat selection: animals may continue to return to a site because of imprinting or learning from their parents about migration routes and specific habitats, even after the habitat has been degraded or severely altered (Temple 1978). The mechanisms used to select habitats may also change ontogenetically: inexperienced breeders may select breeding sites using conspecific attraction, but experienced breeders may display site fidelity and return to breeding sites they used in previous years, even if conspecifics are absent (Stamps 1988). Knowledge of control and onto-genetic mechanisms underlying migration and habitat selection (Temple

1978; Hasler & Scholz 1983) as well as a variety of other problems (McLean, Chapter 6) suggests conservation and management strategies. Similarly, comparative evolutionary approaches provide clues about which behavior patterns predispose species to endangerment. Clearly, Tinbergen's four approaches to studying behavior can provide effective and complementary tools for solving conservation problems. We continue this discussion in the next section by exploring how behavioral biology might be applied to typical categories of conservation problems.

Species (or population) approaches in conservation

Although the overall objective in conservation biology is the widespread preservation of biological diversity, there are numerous reasons why individual species or populations become the center of focus. Sometimes a small number of species are singled out for their special value to humans; alternatively, subsets of species may be selected as representatives of larger communities of species. The sustainable use of biological resources, control of exotic species, resolution of human–animal conflicts, and conservation education are examples of conservation issues that focus on a small number of species of special interest (Primack 1993; Meffe & Carroll 1994). Nevertheless, by addressing single-species issues, often times larger numbers of species are benefited indirectly by habitat association or interaction with the target species (but this approach can also backfire: Waller 1996). Once a species is targeted, for whatever reason, knowledge of behavior becomes a critical factor.

Sustainable use

Species having economic or aesthetic value are common victims of overexploitation, habitat alteration, or resource competition with humans. Overexploitation was a major cause of extinction of large vertebrates in recent history (Martin 1967, 1984) and continues to drive populations toward extinction (Redford 1992). As an alternative to either overconsumption of populations or hands-off preservation of habitats, many conservation biologists are advocating the sustainable use of resources (Shaw 1991; Beissinger & Bucher 1992; Beissinger, Chapter 2). Identifying behavioral factors that regulate population growth are often key to the successful management of these species (Stoleson & Beissinger, Chapter 7). These studies employ all aspects of behavioral research, including detailed life history descriptions, hypothesis testing, and implementation of methods and technology.

Human–wildlife conflicts

Biological conservation frequently involves resolving conflicts between wild animals and humans. Conservation programs are often challenged by having to justify the protection of species which threaten human lives or damage agricultural products. Frequently, these problems can be resolved or diminished by understanding the behavior of both the humans and animals in conflict. An example is studies on the pattern of use of wild areas by Grizzly bears (*Ursus horribilis*) and humans, and the circumstances of encounters between these two species (Albert & Bowyer 1991). In addition to the concern for human safety, the behavioral responses of bears to human activities can be disruptive to the bears' abilities to carry out critical biological activities (McLellan & Shackleton 1989). Bears have large home ranges, so that even infrequent encounters with humans, such as with backcountry hikers, roads, or overhead aircraft, can cause bears to deviate off course in attempts to avoid (or pursue) human encounters. The need for behavioral information extends beyond knowing patterns of habitat use and the bears' behavioral reactions to humans. Subtle aspects of learning theory and the cultural transmission of behavior provide greater insight into factors that elicit bear aggression or promote habituation to humans. It is common knowledge among wildlife professionals and even the general public that bears can associate humans with food, and so strategies have been adopted to reduce the availability of human food to bears. The problem of bear attacks, however, persists even in areas where bears have not had the opportunity to learn this association first hand. Hence, while it does not require a behavioral specialist to understand the general concept of classical conditioning, learning and other behavioral theories can contribute to understanding the less intuitive aspects of a problem. Those aspects include the kinds of stimuli that attract bears to humans, stimuli that elicit aggression, the effect of reinforcement schedules on the perpetuation of the bear's association of humans with food, and how the association is transmitted across a population and between generations (McCullough 1982).

Just as our activities bring humans into conflict with wildlife, anthropogenic disturbance can bring wildlife species into conflict with each other. Habitat alteration can bring previously separate species into contact, with the result that one dominant species drives less competitive species or hosts that lack defenses to extinction. Raccoons, opossums, and Brown-headed cowbirds, whose populations have benefited by habitat fragmentation and the concomitant spread of urbanization, have been implicated in the general decline of songbirds across North America (Terborgh 1992; Rothstein & Robinson 1994). Similar 'weedy' species occur on other continents. In many developed countries the extent of forest fragmentation cannot be reversed in the near future; instead, controls of the problem are being sought by elucidating the mechanisms of nest predation and vulnerability to predation – which species are most vulnerable, how human activities help pest species gain access to vulnerable species, how predators or parasites find nests, what kinds of anti-predator tactics are employed by different species, and which are most effective (e.g., Rothstein 1982; Neudorf & Sealy 1992, 1994; Ortega *et al.* 1993; Mark & Stutchbury 1994; Rothstein & Robinson 1994; Dingle *et al.* Chapter 4).

Introduction of exotics

Another source of large-scale ecological devastation arises from the uncontrolled spread of exotic species. Knowledge of behavioral mechanisms can be used to devise strategies for eradicating or reducing harmful exotic species. For example, the accidental introduction of the Brown tree snake (*Boiga irregularis*) on the island of Guam is most likely responsible for the extinction of over half of the native species of birds that were present on the island in the 1940s (Chiszar 1990). Most of the remaining bird species and many other vertebrate populations are in sharp decline. The snake is also responsible for costly power outages. There is the additional concern that the snake could be accidentally introduced to other Pacific islands, repeating the disaster elsewhere. These problems prompted behavioral studies on the snake's mechanisms of locomotion and predation, including studies of sensory modalities used in prey detection, determinants of switching between sensory modalities, plasticity of hunting mode and prey selection, and ontogenetic changes in these behavior patterns (Chiszar 1990; Rodda 1992). These studies have been useful in developing baits for trapping the snakes, for obstructing the snake's access to electrical boxes, for elucidating general mechanisms of colonization, and for making predictions about the fate of other potential prey species on Guam and other islands. Thus, several approaches to studying behavior can be combined for the effective management of destructive exotics.

Indicator species

Single species studies play an important role in monitoring the health of ecosystems. In environmental toxicological analyses, for example, animal behavior can be a more sensitive indicator of stress than more traditional measures, such as mortality or reproductive failure (Smith & Logan, Chapter 12). A currently active area of behavioral research is on behavioral and endocrinological responses to short-term and long-term, predictable and unpredictable changes in the environment. Organisms have evolved mechanisms to cope with natural changes in the environment. The concern in conservation biology is how human-induced changes might stress organisms beyond their capacities to cope. Some environmental changes produce physiological responses before the onset of major shifts in behavior that would disrupt critical biological activities, so that monitoring hormone levels can provide early warning of populations at risk (Wingfield *et al.* Chapter 5).

Disruption of 'umwelten'

Once a species has been targeted as a taxon of concern, conservation biologists must determine what aspects of the environment are critical to its survival, which depends to a large degree on the species' 'umwelt,' or 'self world.' An umwelt describes the attributes of the environment that are perceived and used by an organism. Even subtle alterations of sound, visual, or chemical environments can interfere with an organism's communication or orientation and migration systems. Umwelt studies have proven useful in the protection and management of endangered sea turtles and some bird and insect species whose orientation and migration patterns are disrupted by artificial lights (Witherington, Chapter 13). A broader array of applications, such as how umwelten might be considered in designing buffer zones, corridors, or managing the structure of forests, streams, and other habitats, remains largely unexplored.

Typically, when conservation biologists consider the critical aspects of a habitat, they focus on the type of flora and fauna, on the shape and size of a preserve, or on the characteristics of the surrounding land. By attempting to view the world from a non-human perspective, we become aware of a different set of criteria that define the critical aspects of an environment. The critical characteristics of a forest, for example, include not only certain types of flora and fauna but a structure that results in a variety of light and temperature microenvironments. Many species are 'tuned' to these microenvironments in unexpected ways. A forest canopy gap of a certain size creates a light beam that illuminates the stage for a displaying bird; other light environments in the forest allow the bird to remain concealed from predators (Endler, Chapter 14). Similarly, microclimatic changes may affect thermoregulatory behavior patterns. The decline in populations of Ornate box turtles (*Terrapene ornate*) in North America has been linked to subtle changes in temperature of the microhabitats resulting from changes in human land-use patterns (Curtin & Porter, in press). Turtles along a latitudinal gradient have different preferred temperatures for when they become active. A minimum number of activity hours and days in the breeding season are required to obtain sufficient energy supplies for reproduction, but longer activity periods increase predation risk. Hence, each population is tuned to the local temperature conditions, and alterations of those temperature environments may compromise reproductive success or predator avoidance. An understanding of the ways in which animals sense and react to their environment is of crucial importance to the preservation of viable populations in fragmented and otherwise altered habitats.

Aesthetics and education

Some vertebrate and invertebrate species are favored for how they enrich human lives aesthetically (Buchholz & Clemmons, Chapter 8). Aesthetic value is perhaps less tangible than monetary value but also less vulnerable to the vagaries of economics (Randall 1994). Human emotional attachment to wild animals can be a powerful stimulus for education about larger conservation issues, such as habitat preservation (Cohn 1988). Humans are drawn to some species simply because those species are spectacular, odd, or reminiscent of humans in some fashion. The main focus of our attraction to these species is often their behavior, including their cognitive and physical abilities, emotional expressions, parental care, family roles, play behavior, sexual behavior, social organization, or vocal and visual displays. Obtaining such knowledge requires detailed studies at the organismal level that may demand years in the lives of one or more researchers. Furthermore, behavioral researchers are often responsible for engaging the public with species that are not typically thought of as attractive to people, and thereby promote the conservation of a greater cross-section of biodiversity.

Monitoring and modeling complex systems

Most ecological systems are very complex and must be described in terms of a subset of cause–effect relationships to gain insight into the function of the system as a whole (Moermond 1986; Schoener 1986; Flanagan 1988). Toxicological analyses, for example, may be aimed at determining the effects or potential effects of pollutants on a large number of species of an ecosystem, but because of time, manpower, or financial constraints, actual studies of the effects are carried out only on select species (Smith & Logan, Chapter 12). Once a subset of species has been chosen, detailed knowledge about certain biological aspects is required. In behavioral toxicology studies, for example, in order to demonstrate that a species is seriously affected by a toxin, detailed knowledge is required about what constitutes 'normal' behavior, including knowledge about natural changes and biological rhythms. It then must be demonstrated that the affected behavior has negative consequences for survival and reproduction (Smith & Logan, Chapter 12).

In applied sciences such as conservation, it is not sufficient to describe general ecological patterns and make predictions based on correlation. In order to predict how systems will respond to environmental changes, the mechanisms that produce the observed patterns must be understood (Levin 1992). The challenge is in selecting only the relevant details to supply effective but simplified models of complex systems. Frequently, however, specialists who are familiar with the current body of knowledge and who understand how systems operate at more specific levels of organization are needed to determine which details are most relevant to higher-level models. An apt example is the recent effort to incorporate social behavior in population models in order to determine the minimum population sizes necessary to reduce the risk of extinction (Beissinger, Chapter 2; Arcese *et al.* Chapter 3; Parker & Waite, Chapter 10; Komdeur & Deerenberg, Chapter 11). Social behavior places (or eases) constraints on the number of actual reproducing individuals in a population. More traditional variables, such as adult density or general life history characteristics, can either over- or underestimate a population's productivity and maintenance of genetic variability. Behavioral researchers are cataloguing a lengthy list of behavioral factors that need consideration in population models, including social partitioning of limiting resources (Walters 1991; Komdeur & Deerenberg, Chapter 11), stress responses to environmental changes (Wingfield *et al.* Chapter 5), social determinants of dispersal (Armitage 1991), breeding experience

(Riedman *et al.* 1994), individual differences in life history patterns (Heppell *et al.* 1994), and social inhibition of breeding (Wasser & Barash 1983). Behavioral studies provide information that informs studies at higher scales or levels of organization, thereby improving the reliability of models when applied to conservation problems.

Effective data for political solutions

In depth studies at the organismal level, including studies of individual behavior, are also frequently necessary to provide convincing evidence in policy-making. The demonstration that DDT was linked to the general decline of predatory bird populations was made by investigating causal mechanisms contributing to egg-shell thinning. The inference that concentrations of DDT increased as the chemical passed through the trophic pyramid combined with experimental evidence at the physiological level enabled a policy movement to eradicate the use of the substance (Wilson 1988). Either level of knowledge by itself would probably have been ineffective at the policy interface. Similarly, documenting changes in the mechanisms or development of behavior is necessary to catalyze policy changes that favor the conservation of biodiversity (Smith & Logan, Chapter 12).

Scale, levels of organization, and behavioral links

Conservation at large scales Reauthorization of the Endangered Species Act stirred considerable controversy about what constitutes the most effective strategies for conserving biodiversity – Should conservation be directed at species, ecosystems, or landscapes? (Franklin 1993, and other papers in that volume). The usual reasoning behind ecosystem and landscape approaches is that there are too many species to deal with singly. Unfortunately, these discussions are hindered by confusion between our traditional concept of ecological levels of organization (i.e., organism, population, community, ecosystem, biome) and issues of spatial and temporal scale (Allen & Hoekstra 1990, 1992). Depending on the species in question, any level of organization can be found at most scales of biological interest. Hence, a population of elephants occurs on vastly different spatial and temporal scales than a population of mites (Allen & Hoekstra 1990). A landscape from a mites' perspective could be the surface of one elephant's skin or a few cubic millimeters of soil. The scale at which an organism operates may change depending on the type of behavioral activity (Addicott *et al.* 1987). Ecosystems and biomes, even spatially large ones, are frequently defined by a small set of species that are interlinked by some set of ecological processes (Naiman 1988, and other papers in that volume). In turn, the small set of species or individuals occupying larger scales can create a context in which other species are constrained (e.g., the ecosystem that exists within the rumen of a cow, the cyclic behavior of spruce and budworm populations, or the forest texture created by large numbers of a few ant species; Allen & Hoekstra 1990; Holling 1992; Dingle *et al.* Chapter 4).

When conservation biologists say that the preservation of biological diversity should be aimed at ecosystems or landscapes, what is probably meant is that conservation should be directed at systems occupying larger spatial and temporal scales, thereby encompassing a greater proportion of diversity in an area. We have already discussed reasons why a subset of species may become the central focus, even when large numbers of species are the object of conservation. The following examples illustrate how the behavior of animals can be a major determinant of the size of scale that a species or individual occupies or affects directly.

Large animals typically have large home ranges (Schoener 1968; Harestad & Bunnell 1979). Thus, if the objective is to direct conservation efforts to larger spatial and temporal scales, it makes sense to focus to some degree on large vertebrates (or large plant species, such as trees). The largest animals also happen to be mammals and other vertebrates of high economic or aesthetic value. As a result, justification for their conservation is easier because the public already has greater interest in them. Also, large species are generally at greater risk of extinction because of human activities such as hunting (Redford 1992), and alterations of the landscape, such as with fire and habitat fragmentation, have occurred at the same scale in which large species function (Holling 1992). From an ecological perspective, large species control or provide habitat for many other species that reside within the temporal or spatial context of the larger species. The structure of the African savanna is maintained to a great extent by elephants that forage on and kill large trees and a variety of browsers that keep saplings from reaching fire-resistant size (McNaughton *et al.* 1988). In short, if conservation necessarily must focus on a small number of species, large species have most of the requisite features – habitats of large spatial and temporal scales, vulnerability to extinction, charisma, economic value, keystone characteristics, and they serve as umbrellas for the conservation of many other species.

Body size, however, is not the only factor contributing to large spatial (or temporal) scale. Examples abound to illustrate how behavior can increase the range of scale predicted by size alone. Small-sized birds and insects that migrate may have relatively small home ranges during a given season, but across a year, have ranges that span whole continents or more (Snow 1985; Terborgh 1989). The spatial range of an individual Monarch butterfly (*Danaus plexippus*) varies depending on the geographic location and time of year, but over the span of one spring and summer, a population completes a migration trip from Mexico to Canada and back in four or more consecutive generations (Brower & Malcolm 1991).

Organisms with complex social systems can resemble large organisms in terms of the spatial scales of their effects. Pocket gophers (*Geomys* spp., Huntly & Inouye 1988) live in complex social groups that are spatially clumped across a landscape. Individuals are territorial and show differential use of plant species and vegetation patches. This heterogeneity in behavior at both individual and population levels increases plant diversity. Similarly, prairie dog colonies (*Cynomys ludovicianus*) range in size from tens to hundreds of hectares, with up to 55 individuals per hectare (Whicker & Detling 1988). These colonies exist in a patchwork across a grassland, and ecosystem processes within patches proceed at different rates than those outside of the patches. The processes within patches are interlinked in part by movement of individuals, which is controlled by behavioral mechanisms. As with pocket gophers, the activities of prairie dogs increase species diversity in grasslands by increasing landscape heterogeneity and interspecific competition among other animal and plant species. Social behavior and organization are both the glue that holds individuals together and the fence that spaces them apart to form patterns at different spatial and temporal scales. Without understanding the behavioral actions that determine patterns of organization, we cannot predict the effects of environmental perturbations on the ecological system or determine how to resolve or diminish a problem once the perturbation is in effect.

The need to integrate study at multiple scales For a variety of reasons it is not sufficient to focus entirely on species of large scale. First, it would be reckless to assume that species that share a habitat with a large, keystone species are safely under the preservation umbrella. Consider, for example, a large forest in which the status of a few dominant tree species are monitored. The trees of a tropical forest may still stand, but if human hunting and harvesting of other species persists within the forest, critical processes of seed dispersal and interspecific competition may be interrupted yet go undetected for the life of the trees (Redford 1992). A real world example occurs with the century plants (*Agave* spp.) which require decades to complete a life cycle. These plants may be more endangered than the remaining stands would indicate, because of a sharp decline in populations of their bat pollinators (Howell & Roth 1981). Hence, species that occupy one scale nevertheless interact with species at other scales in the form of competition, predation, mutualism, parasitism, and other behavioral and ecological processes. Conservation practices must consider input from processes operating at multiple interactive scales.

Presently, we do not understand the scale at which most species operate. Most studies relating habitat alteration to conservation currently focus on overall fragmentation of habitats and edge effects, despite the fact that critical problems for declining populations of many species may be associated with changes in microhabitats within the fragments. To illustrate the danger of focusing only at larger scales from the human perspective, consider the decline of ornate box turtle populations in North America discussed previously. Aerial photographs and ground surveys in Green County, Wisconsin, show that the extent of forest area has changed little since the 1950s. The smaller scale vegetation characteristics of the forests, savannas, and prairies, however, have been altered significantly, changing the thermal microenvironments of the turtles and creating effects that would constitute a dramatic climatic change from the turtle's perspective (Curtin & Porter, in press). This is not to say that large-scale habitat destruction, including fragmentation, is unimportant to conservation. Rather, there are other changes occurring in addition to, or even in the absence of, large scale habitat destruction that impact population viability and that can be assessed behaviorally. Moreover, we need to view the world from the organism's perspective to understand what the critical aspects of a habitat are and how they are affected by human land-use.

Conclusion

Conservation biology is concerned with many ecological scales and levels of organization simultaneously, and those scales entertained by behavioral biology are not excepted. The level of detail that behavioral biologists typically engage in is useful for identifying the kinds of information necessary to improve the predictability of population and ecosystem models (Chapters 3 and 10) and to isolate units of biodiversity in need of conservation action (Chapter 8). Also, basic descriptive studies are needed to acquire baseline knowledge on what constitutes normal or adaptive behavior for comparisons with behavior after ecological perturbations have occurred (Chapter 12).

Generally, descriptive approaches such as these can be combined with theoretical analyses to benefit conservation in several ways: to determine species habitat requirements and predict responses of populations to environmental changes (Chapters 2–4, 11, 13 and 14), to help guide conservationists toward unique solutions to specific problems (Chapters 3 and 7), and to discover general principles of behavior through comparative analyses in order to identify taxa or behavioral phenotypes that are susceptible to specific perturbances or that generate critical ecological patterns. Behavioral specialists are also needed to elucidate more subtle aspects of theoretical applications that may be overlooked or misconstrued by non-specialists and to weigh factors according to species differences. Also, the methods and technology employed by behavioral researchers can assist in traditional conservation tasks, such as surveying and monitoring biotic communities (Chapters 5, 6, 9, and 12). In short, the entire spectrum of behavioral biology encompasses and is applicable to the myriad tasks faced by conservation biologists, wildlife managers, and policy-makers.

As conservationists begin to adopt ethological approaches, behavioral biologists will need to increase their awareness of the kinds of research problems that are most crucial to conservation. They must consider the ramifications of their findings to levels of organization that extend beyond their traditional focus (Chapter 2). The behavior of organisms not only generates larger scale ecological patterns, but is constrained by them (Allen & Starr 1982). In considering the larger contexts of their studies, behavioral biologists will not only benefit conservation but advance their own discipline through a greater understanding of the micro- and macroprocesses of evolution.

Of course, there are some conservation tasks that will have little or no functional use for specific behavioral information and theories. Land preservation and reserve design, for example, have and probably will continue to occur largely in the absence of behavioral inputs as well as other scientific principles, not because behavioral information is unhelpful, but because those decisions are guided more often by politics and human societal needs than by scientific rationale, or because there is not sufficient time to construct and execute detailed plans. Nevertheless, those kinds of conservation tasks are short-lived and will probably end shortly after the turn of the century as available sites for preservation are exhausted (Soulé & Wilcox 1980; McNeely *et al.* 1994; Beissinger, Chapter 2). Most of the development of future conservation initiatives will be directed at policy, public education, sustainable use of populations, habitat restoration, and the management and monitoring of populations and ecosystems or landscapes. To these ends, behavioral research has much to contribute and behavioral biologists must prepare themselves for this new role.

While behavioral biologists are making clear signs of their growing desire to 'join the ranks of conservation biology,' successful induction will not rest entirely on their shoulders (Ralls, Chapter 15). The entry of behavioral sciences and other new disciplines into conservation biology will require substantial force to overcome the fierce competition for collegial support and the limited funding that exists among already established conservation biologists and to counter the considerable philosophical and disciplinary biases that exist on both sides of the fence. Scientists, conservationists, and policy makers need to continually reassess the current and potential importance of various disciplines as our understanding of conservation problems evolves and the needs of the world change. Fears that behavioral subdisciplines are merely trying to jump on the conservation band-wagon with little to offer conservation are unfounded and will hinder our conservation goals. Nevertheless, ethologists are forewarned that conservation biology should not be viewed as a convenient and uncritical source of new funding. Just as conservation ecologists must give conservation ethology a fair shake, conservation ethologists must ensure that their peers apply the utmost rigor and ethics to this crisis discipline. The study of animal behavior fits in well with the objectives of conservation biology as a whole. The question, therefore, is not whether behavioral research is useful to conservationists, but how it is or can be. Since answering this question is a relatively new quest, behavioral biologists as a group are only beginning to realize and evince the conservation import of their ideas. The rest of this book is a confident and, we hope, productive first step in that direction.

Literature cited

Addicott, J. F., J. M. Aho, M. F. Antolin, D. K. Padilla, J. S. Richardson, and D. A. Soluk. 1987. Ecological neighborhoods: scaling environmental patterns. *Oikos* **49**: 340–346.

Albert, D. M., and R. T. Bowyer. 1991. Factors related to grizzly bear-human interactions in Denali National Park. *Wildlife Society Bulletin* **19**: 339–349.

Allee, W. C., O. Park, A. E. Emerson, T. Park, and K. P. Schmidt. 1955. *Principles of animal ecology*. W. B. Saunders Company, Philadelphia.
Allen, T. F. H., and T. W. Hoekstra. 1990. The confusion between scale-defined levels and conventional levels of organization in ecology. *Journal of Vegetation Science* **1**: 5–12.
Allen, T. F. H., and T. W. Hoekstra. 1992. *Toward a unified ecology*. Columbia University Press.
Allen, T. F. H., and T. B. Starr. 1982. *Hierarchy: Perspectives for ecological complexity*. University of Chicago Press, Chicago.
Armitage, K. B. 1991. Social and population dynamics of Yellow-bellied Marmots: results from long-term research. *Annual Review of Ecology and Systematics* **22**: 379–407.
Beissinger, S. R., and E. H. Bucher. 1992. Can parrots be conserved through sustainable harvesting? *BioScience* **42**: 164–173.
Brower, L. P., and S. B. Malcolm. 1991. Animal migrations: endangered phenomena. *American Zoologist* **31**: 265–276.
Chiszar, D. A. 1990. The behavior of the brown tree snake: a study in applied comparative psychology. Pages 101–123 in D. A. Dewsbury, editor. *Contemporary issues in comparative psychology*. Sinauer Associates, Sunderland, MA.
Chivers, D. J. 1986. Southeast Asian primates. Pages 127–152 in K. Benirschke, editor. *The road to self-sustaining populations*. Springer, New York.
Cohn, J. P. 1988. Captive breeding for conservation. *BioScience* **38**: 312–316.
Curtin, C. G., and W. P. Porter. Latitudinal gradients in biophysical constraints: geographic variation in box turtle response to shifting land-use and climate. *Ecological Monographs* (in press).
Flanagan, P. W. 1988. Holism and reductionism in microbial ecology. *Oikos* **53**: 274–275.
Franklin, J. F. 1993. Preserving biodiversity: species, ecosystems, or landscapes? *Ecological Applications* **3**: 202–205.
Hailman, J. P. 1967. The ontogeny of an instinct: the pecking response in chicks of the laughing gull (*Larus atricilla* L.) and related species. *Behaviour Supplement*, **15**.
Hailman, J. P. 1977. *Optical signals: Animal communication and light*. Indiana University Press, Bloomington.
Hailman, J. P. 1982. Ontogeny: toward a general theoretical framework for ethology. Pages 133–189 in P. P. G. Bateson and P. H. Klopfer, editors. *Perspectives in ethology*. Plenum Publishing Corporation, New York.
Harestad, A. S., and F. L. Bunnell. 1979. Home range and body weight – a reevaluation. *Ecology* **60**: 389–402.
Harper, J. L. 1981. The meanings of rarity. Pages 89–96 in H. Synge, editor. *The biological aspects of rare plant conservation*. John Wiley & Sons, New York.
Hasler, A. D., and A. T. Scholz. 1983. Olfactory imprinting and homing in salmon: investigations into the mechanism of the imprinting process. D. S. Farner, B. Heinrich, K. Johansen, H. Langer, G. Neuweiler, and D. J. Randall, editors. *Zoophysiology*, Vol. 14. Springer, Berlin.
Heppell, S. S., J. R. Walters, and L. B. Crowder. 1994. Evaluating management alternatives for red-cockaded woodpeckers: a modeling approach. *Journal of Wildlife Management* **58**: 479–487.
Holling, C. S. 1992. Cross-scale morphology, geometry, and dynamics of ecosystems. *Ecological Monographs* **62**: 447–502.

Howell, D. J., and B. S. Roth. 1981. Sexual reproduction in *Agaves*: the benefit of bats; the cost of semelparous advertising. *Ecology* **62**: 1–7.
Huntly, N., and R. Inouye. 1988. Pocket gophers in ecosystems: patterns and mechanisms. *BioScience* **38**: 786–793.
Hutchins, M., and V. Geist. 1987. Behavioural considerations in the management of mountain-dwelling ungulates. *Mountain Research and Development* **7**: 135–144.
Klopfer, P. H., and J. P. Hailman. 1967. *An introduction to animal behavior: Ethology's first century*. Prentice-Hall, Englewood Cliffs, New Jersey.
Lack, D. 1954. *The natural regulation of animal numbers*. Oxford University Press, London.
Levin, S. A. 1992. The problem of pattern and scale in ecology. *Ecology* **73**: 1943–1967.
MacKinnon, J. 1974. The behaviour and ecology of wild orang-utans (*Pongo pygmaeus*). *Animal Behaviour* **22**: 3–74.
Mark, D., and B. J. Stutchbury. 1994. Response of a forest-interior songbird to the threat of cowbird parasitism. *Animal Behaviour* **47**: 275–280.
Martin, P. S. 1967. Prehistoric overkill. Pages 75–120 in P. S. Martin and H. E. Wright, editors. *Pleistocene extinctions*. Yale University Press, New Haven, Connecticut.
Martin, P. S. 1984. Prehistoric overkill: the global model. Pages 354–403 in P. S. Martin and R. G. Klein, editors. *Quarternary extinctions*. University of Arizona Press, Tucson, Arizona.
McCullough, D. R. 1982. Behavior, bears, and humans. *Wildlife Society Bulletin* **10**: 27–33.
McLellan, B. N., and D. M. Shackleton. 1989. Immediate reactions of grizzly bears to human activities. *Wildlife Society Bulletin* **17**: 269–274.
McNaughton, S. J., R. W. Ruess, and S. W. Seagle. 1988. Large mammals and process dynamics in African ecosystems. *BioScience* **38**: 794–800.
McNeely, J. Harrison, and P. Dingwall. 1994. *Protecting nature: Regional reviews of protected areas*. IUCN Gland, Switzerland.
Mayr, E. 1961. Cause and effect in biology. *Science* **134**: 1501–1506.
Meffe, G. K., and C. R. Carroll. 1994. *Principles of conservation biology*. Sinauer Associates, Sunderland, Massachusetts.
Moermond, T. C. 1986. A mechanistic approach to the structure of animal communities: *Anolis* lizards and birds. *American Zoologist* **26**: 23–37.
Naiman, R. J. 1988. Animal influences on ecosystem dynamics. *BioScience* **38**: 750–752.
Neudorf, D. L., and S. G. Sealy. 1992. Reactions of four passerine species to threats of predation and cowbird parasitism: enemy recognition or generalized responses? *Behaviour* **123**: 84–105.
Neudorf, D. L., and S. G. Sealy. 1994. Sunrise nest attentiveness in a cowbird host. *Condor* **96**: 162–169.
Ortega, J. C., C. P. Ortega, and A. Cruz. 1993. Does brown-headed cowbird egg coloration influence red-winged blackbird responses towards nest contents? *Condor* **95**: 217–219.
Primack, R. B. 1993. *A primer of conservation biology*. Sinauer Associates, Sunderland, MA.
Randall, A. 1994. Thinking about the value of biodiversity. Pages 271–285 in K. C. Kim and R. D. Weaver, editors. *Biodiversity and landscapes: A paradox of humanity*. Cambridge University Press, Cambridge.
Redford, K. H. 1992. The empty forest. *BioScience* **42**: 412–422.

Reed, J. M., and A. P. Dobson. 1993. Behavioural constraints and conservation biology: conspecific attraction and recruitment. *Trends in Ecology and Evolution* **8**: 253–255.
Riedman, M. L., J. A. Estes, M. M. Staedler, A. A. Giles, and D. R. Carlson. 1994. Breeding patterns and reproductive success of California Sea Otters. *Journal of Wildlife Management* **58**: 391–399.
Rodda, G. H. 1992. Foraging behaviour of the brown tree snake, *Boiga irregularis*. *Herpetological Journal* 2:110–114.
Rothstein, S. I. 1982. Mechanisms of avian egg recognition: which egg parameters elicit responses by rejector species? *Behavioral Ecology and Sociobiology* **11**: 229–239.
Rothstein, S. I., and S. K. Robinson. 1994. Conservation and coevolutionary implications of brood parasitism by cowbirds. *Trends in Ecology and Evolution* **9**: 162–164.
Schoener, T. W. 1968. Sizes of feeding territories among birds. *Ecology* **49**: 123–141.
Schoener, T. W. 1986. Mechanistic approaches to community ecology: a new reductionism? *American Zoologist* **26**: 81–106.
Shaw, J. H. 1991. The outlook for sustainable harvests of wildlife in Latin America. Pages 24–34 in J. G. Robinson and K. H. Redford, editors. *Neotropical wildlife and conservation*. University of Chicago Press, Chicago.
Sherman, P. W. 1988. The levels of analysis. *Animal Behaviour* **36**: 616–619.
Smith, A. T., and M. M. Peacock. 1990. Conspecific attraction and the determination of metapopulation colonization rates. *Conservation Biology* **4**: 320–323.
Snow, D. 1985. *The web of adaptation: Bird studies in the American tropics*. Comstock Publication Associates, Ithaca, New York.
Soulé, M. E. 1986. Conservation biology and the "real world." Pages 1–12 in M. E. Soulé, editor. *Conservation biology: The science of scarcity and diversity*. Sinauer Associates, Sunderland, Massachusetts.
Soulé, M. E., and B. A. Wilcox, 1980. Conservation biology: its scope and its challenge. Pages 1–8 in M. E. Soulé & B. A. Wilcox, editors. *Conservation biology: An evolutionary-ecological perspective*. Sinauer Associates, Sunderland, Massachusetts.
Stamps, J. A. 1987. Conspecifics as cues to territory quality: a preference of juvenile lizards (*Anolis aeneus*) for previously used territories. *American Naturalist* **129**: 629–642.
Stamps, J. A. 1988. Conspecific attraction and aggregation in territorial species. *American Naturalist* **131**: 329–347.
Temple, S. A. 1978. Manipulating behavioral patterns of endangered birds: a potential management technique. Pages 435–443 in S. A. Temple, editor. *Endangered birds: Management techniques for preserving threatened species*. University of Wisconsin Press, Madison, Wisconsin.
Terborgh, J. 1989. *Where have all the birds gone?* Princeton University Press, Princeton, New Jersey.
Terborgh, J. 1992. Why American songbirds are vanishing. *Scientific American* **266**: 98–104.
Tinbergen, N. 1951. *The study of instinct*. Oxford University Press, New York.
Tinbergen, N. 1963. On aims and methods of ethology. *Zeitschrift für Tierpsychologie* **20**: 410–433.
Turner, M. G., R. H. Gardner, and R. V. O'Neill. 1995. Ecological dynamics at broad scales. *BioScience*, suppl. **1995**: S29–S35.

Walker, B. 1995. Conserving biological diversity through ecosystem resilience. *Conservation Biology* **9**: 747–752.

Waller, D. M. 1996. Biodiversity as a basis for conservation efforts. Pages 16–32 in W. J. Snape III, editor. *Biodiversity and the law*. Island Press, Washington DC.

Walters, J. R. 1991. Application of ecological principles to the management of endangered species: the case of the red-cockaded woodpecker. *Annual Review of Ecology Systematics* **22**: 505–523.

Wasser, S. K. and D. P. Barash, 1983. Reproductive suppression among female mammals: implications for biomedicine and sexual selection theory. *Quarterly Review of Biology* **58**: 513–538.

Whicker, A. D., and J. K. Detling. 1988. Ecological consequences of prairie dog disturbances. *BioScience* **38**: 778–785.

Wilson, D. S. 1988. Holism and reductionism in evolutionary ecology. *Oikos* **53**: 269–273.

Chapter 2

Integrating behavior into conservation biology: Potentials and limitations

STEVEN R. BEISSINGER

I explore the current and potential contributions of behavioral studies to conservation biology, and clarify limitations in applying behavioral ecology and ethology to conserving biological diversity. In my opinion, behavior has played a much more important role at certain scales of conservation problems than at others. Specifically, the behavioral sciences have most influenced conservation at the relatively local scales of populations and species, while the conservation of most of the world's biological diversity will occur not at those scales but at larger scales of ecosystems, landscapes, and biomes. Thus, there appears to be a discordance between the ecological scales at which behavior is most pertinent to conservation and the scales at which conservation efforts will protect the most biological diversity. For behavior to make a larger contribution to conservation, it will have to be translated more often into 'currencies' that can be linked across scales directly to conservation.

To see how I arrived at this opinion, I examine how conservation biology has been formulated as a science and has developed 'tools' to deal with the problem of disappearing diversity. Then I review how behavior fits, or could better fit, into the repertoire of approaches that conservation biology offers for conserving biological diversity, paying particular attention to the currencies needed for translating behavior into conservation and the potential for future behavioral contributions.

The formulation of conservation biology and its tools

Conservation biology emerged as the science of scarcity and diversity (Soulé 1986) in the 1980s in response to the perception that extinction rates had become greatly accelerated (Myers 1979; Wilson 1988; Reid 1992; Smith *et al.* 1993). Conservation biology has the explicit goal of maintaining biological diversity – genetic diversity, species diversity, and ecosystem integrity (Soulé 1985; Beissinger 1990). Biological diversity has grown from the simple concept of species diversity to include genetic diversity, the evolutionary uniqueness of species or taxonomic diversity, the functional processes or interactions of species and physical environments that maintain ecosystems (functional diversity), and landscape diversity or the diversity of ecosystems in a region (Soulé 1985; Beissinger 1990; Noss 1990). This expanded definition of biological diversity emphasizes the processes (e.g., inbreeding, genetic drift, mortality, dispersal, nutrient cycling, succession, and disturbance) and interactions (e.g., coevolution, predation, and competition) that play a crucial role in maintaining or eroding biological diversity. Conservation biology only encompasses a subset of all conservation issues because of its mission to conserve biological diversity. For example, outside of the purview of conservation biology are the more traditional wildlife conservation problems of managing game and pest species, and planning landscapes where humans and wildlife can coexist (e.g., urban ecology), unless these problems directly relate to conserving rare species or ecosystems.

Conservation biology developed from the principles of evolutionary and ecological processes at different spatial and time scales (Beissinger 1990): (1) population genetics and evolutionary biology provide the framework to protect the evolutionary potential of species to adapt to changing environments; (2) demographic processes (mortality and reproduction) drive extinction probabilities and comprise the study of population ecology; (3) underlying demography is the process of individuals making choices (the study of behavior and life history); and (4) community, ecosystem, and landscape ecology provide the foundation for biogeographic distributions of organisms, and determine the potential for ecosystems to be restored or maintained in the face of natural disturbances and human development.

In examining both the sources of conservation principles and the expanded concept of biological diversity, it is clear that behavior should be an explicit part of conservation biology. It is certainly an underlying (implicit) or explicit component of each of the four knowledge areas and is important to many of the processes that are an integral part of maintaining or destroying biological diversity. The linkage between behavior and the application of conservation biology, however, has been weak. Perhaps this has occurred in part because much behavioral research attempts to develop reductionist approaches to elucidate the proximate and ultimate factors responsible for the occurrence and

Table 2.1 *Seven tools that have emerged from the development of conservation biology that can be applied to conserve biological diversity and their dominant context for use*

Dominant context	Tools
Prevent the loss of biological diversity	Reserve and Landscape Design Ecosystem Management Population Viability Analysis
Compromises between conservation and development	Sustainable Development
Recover threatened populations, species or ecosystems	Field Recovery of Endangered Species Captive Breeding and Reintroduction Ecosystem Restoration

evolution of behavioral patterns, while conservation biologists use different kinds of information in their tools designed to solve conservation problems.

Seven 'tools' or knowledge areas, with emerging principles that can be applied to conserve biological diversity at different scales, has marked the development of conservation biology (Table 2.1). These tools can be grouped into those which can be implemented to prevent the loss of biological diversity, to promote compromise between conservation and development, and to recover threatened populations, species, and ecosystems. Most tools are so new that it is too soon to assess their full potentials (Reserve and Landscape Design, Ecosystem Management, Population Viability Analyses, Sustainable Development, and Ecosystem Restoration), whereas others have now established sufficient track records that their potentials to succeed or fail in conserving biological diversity are more certain (Field Recovery of Endangered Species, and Captive Breeding and Reintroduction). Each tool has contexts for which its use is more appropriate than others, although admittedly some of these tools can be used in more than one context. I have placed them according to what I consider each tool's dominant purpose and optimal potential to achieve conservation.

In the remainder of this chapter, I examine these tools individually, assess how these tools explicitly draw upon studies of ethology in their current application, and how behavioral science might be better incorporated into them in the future.

Contribution of behavior to tools that prevent the loss of biological diversity

Reserve and Landscape Design

Reserve and Landscape Design refers specifically to principles derived from biogeography (species–area curves, equilibrium theory of island biogeography, and relaxation rates), metapopulation dynamics, and the ecological and evolutionary consequences of fragmentation and isolation. Typical principles are: (1) larger blocks of habitat are usually better than smaller ones because they protect more species and result in longer population persistence times; (2) blocks of habitat that are closer together are typically better than blocks that are farther apart because this facilitates dispersal and genetic exchange; (3) some connectedness via corridors is often better than none, for the same reasons stated in (2); (4) buffering existing protected areas with zones of appropriate land use can decrease threats from incompatible land use or direct exploitation; and (5) several reserves often help spread the risk of extinction better than a single reserve. Each principle is stated with a modifier (e.g., 'often' or 'usually'), because each conservation situation is unique and needs to be evaluated in its own context. For example, although corridors may facilitate important exchange of individuals between spatially subdivided populations (Harris & Scheck 1991; Soulé & Gilpin 1991), they can assist the spread of disease, fire, and other catastrophes (Simberloff & Cox 1987; Simberloff *et al.* 1992; Hess 1994). Thus, the potential use of corridors must be evaluated on a situation-by-situation basis, considering as criteria the likely use of corridors by target species and the dangers imposed by linking sites.

Most protected areas have been established or managed for biological or other kinds of diversity based on these sorts of principles, if any at all were used. Many parks were established simply to protect a unique ecosystem or a particular species (Diamond 1986), such as breeding areas of elephants in Africa, penguins in Australia, and Elephant seals (*Mirounga angustirostris*) in the United States, or wintering areas of Monarch butterflies (*Danaus plexippus*) in Mexico. Beyond the significance of where a species was breeding or surviving, knowledge of behavior traditionally has not been a particularly important component of the principles of Reserve and Landscape Design, with the important exception of dispersal behavior, which underlies many of them. Two kinds of dispersal data are critical from the perspective of Reserve and

Landscape Design – how far individuals disperse and the willingness of individuals to use corridors if present. A knowledge of dispersal distance, and the effects of different kinds of land uses and barriers on dispersal behavior, is needed to evaluate the current configuration of reserves for many different groups of organisms. In addition, we know little about what species will use corridors for dispersal and what constitutes a minimum or optimal corridor width (Nicholls & Margules 1991; Soulé & Gilpin 1991; Simberloff *et al.* 1992). Thus, behavioral studies that provide correlates of the propensity to disperse across gaps or corridors will be of most value to reserve management design in the future (e.g., Haas 1995) and could help to determine whether it would be useful to make the costly investment in procuring conservation corridors.

Reserves that are specifically created to protect a species are sometimes inadequately designed because the species' life history or behavior was incompletely known. Then, knowledge of behavior can be important in helping to redesign such parks. In many Central and South American countries, for example, most parks and protected areas are found either on the mountain tops or in the lowlands. Yet, nearly one-quarter of the avifauna move annually out of the montane parks partway downslope to search for seasonally fruiting trees or emerging insects in the few small forest fragments which remain mostly on private lands (Guindon 1988; Stiles 1988; Powell & Bork 1995). In this example, a better knowledge of natural history provided by behavioral or ecological studies of the factors likely to be responsible for dispersal behavior would have helped to design the parks properly. Detailed studies of the individual decision rules that triggered dispersal, and variation among individuals in those rules, would probably not be needed.

The immediate need to establish protected areas before ecosystems and species are further destroyed or lost means that delineating park boundaries is usually carried out long before detailed studies have been conducted. Thus, the importance of behavior may only emerge well after protected areas are created. While presently little behavior is explicitly incorporated into the principles of Reserve Design and Landscape Management, in the future a knowledge of behavior may prove to be very useful for redesigning current park boundaries or determining the size of buffer zones around protected areas.

Ecosystem management

Much biological diversity is found on private lands or multiple-use federal lands, thus falling outside of the protection provided by the coarse filter of parks and reserves. Ecosystem management incorporates aspects of both resource conservation and utilization, or, from another perspective, preservation and development (Slocombe 1993; Grumbine 1994). It is emerging as an approach that combines regional planning methods with ecological risk assessment to find compatible land use at a regional level. The objective of Ecosystem Management is to use holistic approaches to manage land and water to provide products and services to meet the needs of human societies, and to conserve biological diversity. Biological principles of Ecosystem Management appear to come mainly from those presented above in Reserve and Landscape Design, but applied to a larger scale to examine the maintenance and juxtaposition of many types of land use or ecosystems. Employing a multispecies perspective, Ecosystem Management may also take the form of simulations that incorporate habitat suitability models, biophysical models of environmental variation such as hydrological or biogeochemical cycles, and Population Viability Analyses.

Individuals rarely are explicitly considered in large and complex dynamic models of ecosystems and land use at the regional scale, and thus behavior beyond natural history tends to be ignored. For example, one of the largest and most recent ecosystem management exercises in the USA, President Clinton's Forest Ecosystem Management Team (FEMAT 1993), developed different scenarios for every parcel of federal land in the Pacific Northwest based mostly on qualitative assessments of habitat affinities for hundreds of species. It was a combination of natural history and risk assessment using the Reserve and Landscape Design principles discussed above. Behavior was sometimes implicitly incorporated in this assessment, such as the unwillingness of Northern spotted owls (*Strix occidentalis*) to cross clear cuts, but only explicitly used in population models for a couple of well-studied species. An exception is the development and incorporation in Ecosystem Management of individually based models, which will be discussed in more detail below. For example, such models are being developed to examine ecosystem management options in the Florida Everglades for a few species of concern (Fleming *et al.* 1994; Wolff 1994).

Given the complexity of Ecosystem Management and the limits on our ability to develop detailed behavioral models of animal movement for a single species, let alone the tens or hundreds that good Ecosystem Management requires, I believe that for the present the influence of behavioral sciences on Ecosystem Management will remain relatively limited. Nevertheless, behavior could become a much more important component of Ecosystem Management as landscape ecology continues to emerge as a credible science (Turner *et al.* 1995*b*). It places importance on examining the interactions of landscape dynamics, ecosystem processes, and biological diversity through the incorporation of spatially explicit models of the occurrence and movement patterns of individuals. One of the key components of landscape ecology is understanding how landscape features influence the movement patterns of individuals. Patterns of movement are especially crucial, because Ecosystem Management requires an understanding of how current and projected spatial configurations of all ecosystems, not just reserves, will affect both biological diversity and the production of goods and services. What kinds of landscape elements are barriers to movement? Why does the same landscape element enhance movement of some species but retard movement of others? How do species perceive landscapes and how does it affect their patterns of movement? Understanding the demographic implications of and factors affecting habitat choice and settlement patterns are as important as quantifying the patterns and correlates of movements. Mechanistic answers to these and other questions are likely only to come from detailed studies of dispersal and habitat choice behavior across a suite of species and ecosystems. They will have broad implications for Ecosystem Management by clarifying the impacts of the current configuration of ecosystems and predicting the impacts of future spatial configurations.

Population Viability Analyses

Population Viability Analyses (PVA) usually involves the development of species-specific models dependent on detailed demographic and environmental information (Shaffer 1981, 1990; Soulé 1987; Burgman *et al.* 1988; Boyce 1992). Demographic models are used to project populations years into the future and evaluate their risk of extinction in relation to environmental variation and various management options (Fig. 2.1). Spatially explicit population viability models range from simple subdivided populations to patch-based metapopulation models to GIS-based (geographic information system) models that are complete spatial arrays of landscapes (Burgman *et al.* 1993). Genetic models that estimate

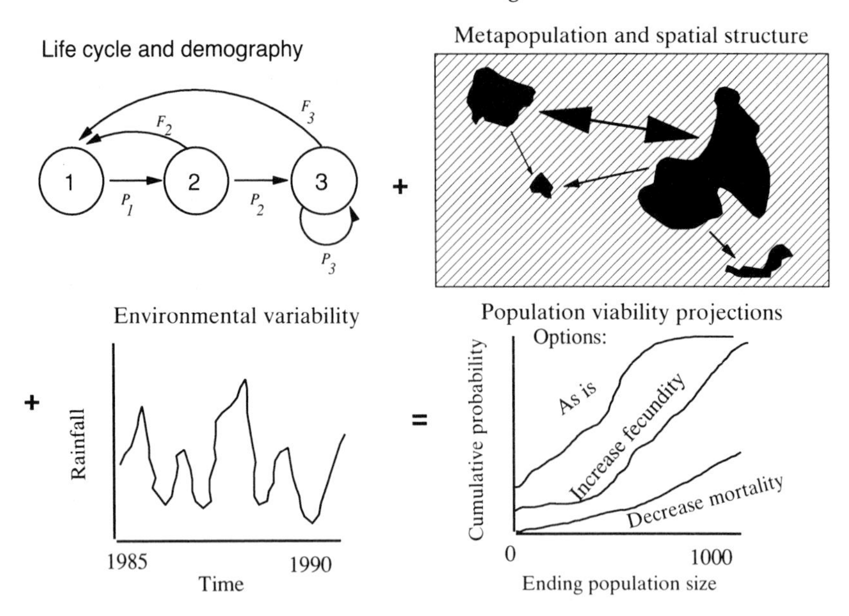

Fig. 2.1. Population viability analysis (PVA) incorporates the life cycle, demography, and spatial structure of the organism under consideration, and the relationship of these factors to environmental variability, to project populations into the future to yield predictions of population size and viability at different time spans under various management options.

effective population size can rarely be directly linked to extinction probabilities (Lande 1988; Reed *et al.* 1988), and are employed less frequently in management and policy decisions. Some approaches try to establish direct links between extinction probabilities and ecosystem management options. Population viability of the Snail kite (*Rostrhamus sociabilis*), for example, was shown to be strongly impacted by Everglades hydroperiod characteristics, especially the interval in years between low or drought water conditions which has been greatly influenced by water management regimens (Beissinger 1995).

Most applications to date use population-based demographic models rather than individual-based models (Menges 1992; Burgman *et al.* 1993). Measurements of demographic rates, of course, implicitly rely upon studies of individuals, but behavioral components may be no more explicit than survivorship, reproductive success, and age. Many PVA applications are based on glorified life table analyses. Thus, behavior is only rarely explicitly treated in these models, with two exceptions. First, behavior can assist in structuring the underlying life cycle diagram (Fig.

2.1) used for demographic modeling (McDonald & Caswell 1993), which may improve model accuracy. Instead of basing the life cycle diagram simply on age classes, behavioral studies led to the formulation of demographic models that incorporated social systems (e.g., breeders, helpers, non-breeders) into the model structure in PVA analyses of threatened woodpeckers (Heppell *et al.* 1994) and gorillas (Harcourt 1995). Behavior was used with demography to partition Everglades water levels into different environmental states in the Snail kite PVA (Beissinger 1995). Incorporation of behavior into PVA analyses should result in substantially improved model prediction, but this needs further investigation given the increased financial investments often required to construct such detailed models. Second, dispersal dynamics are very important to spatial models. Usually detailed data on dispersal rates, mean dispersal distance and the nature of the distribution of dispersal distances (e.g., negative exponential, step function, uniform, etc.) are absent. Thus, mathematical distributions of dispersal distances and rates are often simply assumed based on little information (e.g., Gibbs 1993; Lindenmayer & Lacy 1995), neither is there much information on the role of landscape elements and barriers on dispersal dynamics, as discussed above. The effects of these assumptions on model outcomes are rarely examined. Yet, they probably have an important effect on viability estimates and the conservation conclusions drawn from spatial models. Behavioral studies that supply spatially explicit details of demography and movement patterns would be helpful to PVA models.

There is hope for more incorporation of behavior into PVA models through the use of individual-based spatially explicit models (Dunning *et al.* 1995; Turner *et al.* 1995*a*). For example, Bart (1995) developed such models for the Northern spotted owl: individuals are followed through time and disperse or settle among territories depending upon simple rules based on mating status and territory quality. Thus, behavior is included but it is a far step from a model that incorporates ethology or behavioral ecology – such as the effects of mating system, sexual selection, energetics, psychological influences, or optimality on the pattern of individual movements. Rarely will we have that kind of detailed information for conservation applications! In fact, the demography of most threatened species is not known well enough to construct population-based PVA models, and it is very unusual when there is enough information to construct even simple individual-based models for species of concern. Thus, our ability to develop individual-based models will be severely constrained. Nevertheless, for behavior to become better incorporated into PVA models, it must be able to be translated directly into spatial or demographic consequences that can be used to estimate the probability of extinction in individual- and population-based models.

Contribution of behavior to tools that compromise conservation and development

Sustainable Development

Sustainable Development, like Ecosystem Management, is a buzzword without a single meaning. It has been referred to as the 'use it or lose it' approach to conservation. Sustainable Development can be applied in two contexts: (1) to single-species sustainable harvesting programs such as those with parrots (Beissinger & Bucher 1992; Stoleson & Beissinger, Chapter 7) or Vicuña (Cattan 1989); or (2) to ecosystem-level approaches for sustainable development schemes which identify regional land uses (e.g., certain kinds of agriculture or agroforestry practices) or economic activities (e.g., ecotourism or extractive reserves) that might be compatible with the retention of biological diversity (Reid 1989; Simon 1989; Vincent 1992). For example, development in areas of tropical montane forest might maintain a greater proportion of insects and insectivorous birds (i.e., be sustainable) if farmers planted agroforestry crops, such as shade tree coffee, instead of yucca or banana monocultures (Vannini 1994; Thiollay 1995). Likewise, macaws may be worth more to local people if their populations are protected for ecotourists to view rather than harvested for the pet trade (Munn 1992).

Animal behavior studies seem to have played a small part in examining the impact of land use changes proposed by sustainable development schemes. Like Ecosystem Management, there are just too many species impacted by sustainable development schemes to be considered by behaviorists one at a time. In the ideal case, the response of dozens of species to changes in land use is determined. Thus, it is frequently difficult to gather field data that are more detailed than changes in species occurrence, density, or resource utilization, except in the case of studies of specific target species. Detailed studies of the behavioral ecology of target species, however, could be used to validate the assumptions of coarser approaches. Also, animal behaviorists may be able to provide enlightened estimates of the impacts of development schemes on biological diversity by making comparisons of behavior with species that have been relatively well studied. Just as congeneric species can show unexpectedly different life history patterns or responses to environmental change, so behavioral responses to management can sometimes be surprising. Conservation problems are often site- or situation-dependent, and very different factors can be at work in the field than appear to be acting when viewed from the armchair (Snyder & Snyder 1989).

Behavioral studies could play a significant role in determining the impact of 'sustainable' economic activities on individual target species. Activities like ecotourism can make important contributions to local economies, promote development and provide impetus for habitat protection (Groom *et al.* 1991; Munn 1992; Maille & Mendelsohn 1993; Kangas *et al.* 1995). Ecotourism operations can also impact animal populations or damage ecosystems (Boo 1990; Blane & Jaakson 1994; Jacobson & Lopez 1994; Rinkevich 1995). The ecological impacts of ecotourism operations have received little study, although past experience suggests that wild animals rarely benefit from direct interactions with humans. Behaviorists should conduct more studies of the impacts of ecotourism on particular species.

Demographic models of sustainable harvesting typically have been based on population-level characteristics and, with few exceptions, do not explicitly integrate behavior. Such models have been important for estimating sustainable harvest levels of game populations (i.e., game hunting or fishing) in developed countries, but have been used less frequently in developing countries (Robinson & Redford 1991). Of course, such models have not necessarily prevented overharvesting because of two factors (Ludwig *et al.* 1993). First, the models are often not actually used to manage populations. Market and political forces can cause governments to set harvest quotas without true regard for model results. Second, the models have only been partially successful in predicting population trends. This has occurred in part because many modeling efforts have not incorporated detailed information on mating systems, sex ratio variation, and behavioral characteristics, and environmental variation has made it difficult to distinguish true trend signals. Continued integration of behavior with demographic models may help improve their accuracy, but not necessarily their use by policy makers.

Nevertheless, as the world population is expected to double from 5.7 billion to 11 billion in the next 40 years (UNFPA 1991; Tuckwell & Koziol 1992), humanity is likely to exploit natural ecosystem products at an ever increasing rate and governments will try to find easy ways to apply science to assist in setting animal harvesting quotas. One of the simplest ways to develop harvesting schemes that might truly be sustainable could be to use behavioral and demographic studies to determine what factors limit the population growth of target species, determine if productivity can be increased through simple management options, and then harvest only the extra productivity created by management (Beissinger & Bucher 1992; Stoleson & Beissinger, Chapter 7). This very conservative approach to sustainable harvesting is particularly useful for situations where information needed for quantitative models of harvest rate are lacking or difficult to obtain. As discussed above, such situations seem to be extremely common.

Contribution of behavior to tools that recover threatened populations, species, and ecosystems

Field Recovery of Endangered Species

Field Recovery of Endangered Species depends upon determining what factors limit population growth *in situ* and then reversing those factors. Determining what factors limit population growth is best accomplished by a combination of individual and population level approaches (Caughley 1994). Limiting factors are like specific hypotheses that can be tested only by examining the behavior of individuals in the field – foraging behavior, reproduction, disease infection levels, etc. – and linking behavior to demographic consequences.

Behavioral science has been integrated into recovering endangered species in the wild. Behavioral studies can provide not only the evidence needed to indicate what limits population growth, but can lead to suggestions for creative management practices to reverse those trends. Many chapters in this volume demonstrate this point clearly, so my treatment will be brief. Good behavioral ecology becomes an essential ingredient for the successful recovery of threatened or endangered species, and behavioral ecologists are often quite skilled in hypothesis testing approaches needed to discriminate among potential limiting factors.

Field Recovery of Endangered Species is a tool designed primarily to result in single-species conservation. It has been applied mostly to a handful of lucky terrestrial vertebrates in temperate countries. Nevertheless, recovering endangered species often involves protecting their habitats, and in this way the 'coattails' of these species can be extended to protect whole ecosystems. Endangered species have also acted as 'flagship' species in campaigns to preserve whole ecosystems, and sometimes the success of such programs has been very impressive (Butler 1992; Dietz *et al.* 1994). This requires mounting intensive educational and promotional campaigns that are very time consuming and far removed from the expertise and interests of most behavioral biologists. Such campaigns have sometimes been conducted independent of field recovery efforts and on species that are not the subject of field research (e.g., Butler 1992). Unfortunately, overusing flagship species for conservation gains can unintentionally result in a political backlash that can decrease the effectiveness of conservation programs, as in the use of the Spotted Owl to promote the conservation of old growth forests. To date, the impact of Field Recovery of Endangered Species on conserving whole ecosystems has been smaller than many of the multispecies tools previously discussed which are oriented toward preventing the loss of biological diversity.

Captive Breeding and Reintroduction

Captive breeding of endangered species for conservation has increased tremendously over the past two decades. Techniques for breeding species in captivity have improved, as have techniques for reintroducing captive-bred animals into the wild (Gipps 1991; Wiley *et al.* 1992). Captive breeding has been the difference between survival and extinction in the short term for species like the California condor *(Gymnogyps californianus*), the Guam rail (*Rallus owstoni*), and the Black-footed ferret (*Mustela nigripes*) (Snyder & Snyder 1989; Derrickson & Snyder 1992; Miller *et al.* 1996).

Behavioral considerations are an extremely important component of breeding endangered species in captivity and reintroducing them into the wild for conservation (Lyles & May 1987; Kleiman 1989; Snyder *et al.* 1994). While maintaining genetic diversity has often been emphasized as critical to the success of captive breeding programs (e.g., Foose & Ballou 1988; Allendorf 1993), the real barriers to successful captive breeding are usually behavioral, such as mate choice, social structure, domestication, and disease prevention (Snyder *et al.* 1996). Predator avoidance, habitat choice, and even flocking behaviors are often learned (Kleiman 1989; Miller *et al.* 1990; Snyder *et al.* 1994), and are critical for survival of reintroduced individuals into most wild environments.

Unfortunately few species will be conserved by employing captive breeding and reintroduction. Captive Breeding and Reintroduction for conserving endangered species in the wild is a last-ditch approach that should rarely be invoked because of severe limitations (Snyder *et al.* 1996). (1) Achieving self-sustaining captive populations can be difficult. While some species breed too well in captivity (Lacy 1991; Lindburg 1991), only a handful of taxa have bred in captivity (Conway 1986; Rahbeck 1993) and many species breed poorly in captivity despite extensive efforts (Snyder *et al.* 1996). Predicting which species will breed well in captivity is often difficult, as the breeding success of *Amazona* parrot species has shown (Derrickson & Snyder 1992). (2) Successful reintroduction is rarely achieved. Recent surveys of reintroduction programs of captive-bred animals have shown that few have successfully established wild populations (Griffith *et al.* 1989; Beck *et al.* 1994). Causes of failure ranged from behavioral deficiencies in released animals (e.g., Lyles & May 1987; Kleiman 1989; Fleming & Gross 1993; Snyder *et al.* 1994), especially for species that learn a large portion of their behavioral repertoires, to failure to correct the factors originally causing extirpation. (3) Domestication in captivity is inevitable (Allendorf 1993; Snyder *et al.* 1996), can be quite strong (Belyaev 1979), can proceed rapidly (Moyle 1969; Swain & Ridell 1990), and is difficult to reverse (Knoder 1959; Lyles & May 1987; Derrickson & Snyder 1992). Captive environments differ greatly from wild environments, and species become progressively more adapted to captivity despite comprehensive genetic management. (4) Reintroduction of captive animals risks introducing diseases to wild populations. Disease risks are high for endangered species in captivity owing to enhanced exposure to exotic pathogens (Snyder *et al.* 1996) and perhaps owing to susceptibility from reduced genetic diversity in small populations (O'Brien & Evermann 1988; Thorne & Williams 1988). (5) Financial and space limitations greatly limit how many species can be conserved in captivity and reintroduced to the wild. Costs of captive breeding programs for endangered species run from \$250 000 to \$500 000 per year (Derrickson & Snyder 1992; Balmford *et al.* 1995), and zoological institutions do not have enough space to accommodate viable populations of threatened species (Conway 1986; Sheppard 1995). (6) Captive breeding can divert attention from the problem causing a species' decline (Frazer 1992; Meffe 1992) and preempt investments in better techniques for *in situ* conservation (Snyder *et al.* 1996). Finally, (7) it is difficult to ensure the administrative continuity needed to carry out long-term captive breeding programs (Clark *et al.* 1994).

Nevertheless, Captive Breeding and Reintroduction to the wild truly will be needed for a small percentage of endangered species recovery programs. Behavioral studies have much to contribute to present and future hopes for the successful use of this tool. Behavioral studies may be able to assist in addressing the first three limitations discussed above for specific situations. Behavioral studies have reduced barriers to reproduction in captivity and to survival upon reintroduction to the wild for some species (Kleiman 1989; Beck *et al.* 1994). There is also a real need to understand the process of domestication that inevitably occurs in captivity, and if it can be reversed to make a captive-bred animal into a wild one again, especially if the animal is to be reintroduced into an ecosystem that still has healthy populations of predators. Here studies of animal psychology and ethology may have an especially important role to play.

Ecosystem Restoration

Ecosystem Restoration can be considered the ecosystem analogue to recovering endangered species. It requires an understanding of the flow of energy and materials, and how organisms interact with each other and their abiotic environments in order to restore functions and biological diversity to degraded and damaged ecosystems (Cairns 1986; Jordan *et al.* 1988). Restoration can help decrease the rate of conversion of natural ecosystems (e.g., deforestation) by repairing land damaged by natural phenomena or human activities to a state that can again productively sustain economic development or biological diversity (Brown & Lugo 1994).

Ecosystem Restoration efforts often concentrate on reducing or removing the factors that limit plant establishment and succession by changing the frequency or intensity of abiotic stressors. For example, altering ecosystem hydrology by increasing flooding frequency can restore functions to wetland ecosystems. The addition or deletion of plant species is often used to create conditions that increase ecosystem retention of nutrients, or provide shade, soil moisture, or soil organic matter necessary for the establishment of desired communities.

While animal behavior has not been a cornerstone of Ecosystem Restoration, it can be incorporated into restoration approaches (e.g., Janzen 1988). Of particular interest is the use of animals to enhance vegetation recovery (e.g., seed dispersers and pollinators) and the role of species that retard vegetation recovery (e.g., seed predators). The regeneration of tropical forests from pastures or abandoned agricultural lands, for example, can be impeded not only by improper abiotic conditions but by a lack of propagules (Nepstad *et al.* 1991; Aide *et al.* 1995). Seedlings, sprouts, and seed banks of forest species are often eliminated after a few years of continuous grazing or cultivation (Uhl *et al.* 1990). Seed predators, such as ants and small rodents, can impede the establishment of trees and shrub by removing seeds that have reached abandoned pastures (Nepstad *et al.* 1991). Seed dispersers, such as birds, bats, and other mammals, play an important role in moving seeds away from the seed shadow of parent trees to new sites. Behavioral studies are investigating how movements of seed dispersers are affected by landscape elements like isolated dead trees, corridors, and living fences (Harris & Scheck 1991). Research is beginning to test ways to attract seed dispersers to visit restoration sites by adding desirable food sources or perches (Robinson & Handel 1992; McClanahan & Wolfe 1993).

Present and future spheres of influence of behavior in conservation biology

Consider for a moment the relative magnitude of the challenge facing modern conservation biologists. Threatened with local or global extinction are one-fourth of the 40 000 known invertebrates in the western half of Germany, one-third of the freshwater fish species in North America, about half of Australia's surviving mammals, 11% of the 9000+ species of the birds in the world, and more than two-thirds of the world's 150 species of primates (Ryan 1992; WCMC 1992). Obviously, a species-by-species approach to conservation problems of this magnitude will not be sufficient.

While there are potentially important ethological contributions to be made to nearly every tool being employed by modern conservation biologists, to date behavioral studies have made their most important contributions to those tools that recover threatened populations or species, rather than to tools that prevent the loss of biological diversity or that promote conservation compromises. Yet, by the very nature of scale and because of their multispecies approaches, most biological diversity will initially be conserved by the 'coarse filter' approach to conservation embodied in the tools that prevent the loss of biological diversity (Table 2.1), rather than the 'fine filter' approach which acts as a safety net to catch those species whose ranges fall between the protected and managed areas embodied in the coarse filter. This may be partly why behavior has presently taken a back seat to the domains of ecosystem, community, and population biology in the efforts to conserve biological diversity and the development of conservation biology theory. The exception to this trend has been the role of behavior in endangered species conservation problems, many of which are among the most critical and controversial conservation problems of our day! Behavioral studies have made important contributions to detecting and reversing factors that limit endangered populations and to successful captive breeding and reintroduction to the wild. This is the scale where behavior's present contributions to conservation dominate.

This is not to say that the implications of advances in behavioral sciences will not affect the business of conserving biological diversity, or that behavior cannot play a larger role in conservation biology in the future. From the above review, I see three very important areas that strategic research in the behavioral sciences could contribute directly to conservation biology. First, there is a need to integrate behavior into spatial and demographic models, whose applications in conservation are burgeoning. Individual-based models of population and landscape dynamics appear to be the wave of the future. They may affect how we practice Landscape and Reserve Design, Ecosystem Management, Population Viability Analysis, and sustainable harvesting programs. If these models are to achieve a high level of accuracy and precision, they will require a good understanding of the behavioral and ecological factors that influence animal dispersal, movements, and habitat choice. It is a golden opportunity to meld experimental and theoretical behavioral ecology and ethology into conservation biology, assuming these data-intensive models actually are an improvement over less demanding population- and community-level approaches. Just as important is the application of behavior to target species in models for PVA, Ecosystem Management, and sustainable harvesting programs. *To succeed, behavior must be directly translated into either demographic or spatial consequences.* This is one manner in which behavior can be translated across scales from individuals to populations to ecosystems, and become more accessible for making conservation policy decisions.

Second, the rate of establishment of national parks and reserves has been decreasing or is beginning to slow down in most countries around the world (WCMC 1992). I believe that the 'coarse filter' for conserving biological diversity will mostly be in place within the next two decades. What is not protected by 2040, when the total world population is projected to have doubled in size from 1990 levels to 11 billion people (UNFPA 1991; Tuckwell & Koziol 1992), is likely to be converted into urban, agricultural, or other managed ecosystems. Expect behavioral science to make greater contributions to conserving biological diversity as the rate of land preservation continues to slow. Species management issues are likely to grow within and around reserve areas, and in the matrix of agricultural and managed ecosystems. Species do not necessarily remain in protected areas. Wolves currently being reintroduced into Yellowstone National Park, for example, have been found on private lands, and bison have spread brucellosis to cattle on ranches adjacent to national parks (Aguirreo & Starkey 1994; Meagher & Meyer 1994). As top or keystone predators disappear from ecosystems, prey populations have grown out of control and in some cases threaten other species (Goodrich & Buskirk 1995). Behavioral ecology can play an important role in the management of species in human-dominated landscapes.

Third, successful long-term conservation of biological diversity will require multidisciplinary approaches based on the best possible biology, but developed with people in mind. In other words, we need a better understanding of economic, social, and political behavior of individuals and societies. Here, studies of behavior, especially human behavior, may have much to contribute (Heinen & Low 1992; Low & Heinen 1993).

Finally, behavioral scientists that want to contribute more directly to the effort to conserve what remains of the world's biological diversity need to consider two departures from the usual research program that we employ if we are to make our research relevant. First, incorporate human beings into the research questions that we ask. Humans are implicit in nearly every environmental problem that we face, and even the most remote field sites where we conduct our studies of animal behavior have been affected by modern or historical human factors. Whereas we usually strive in our basic research paradigms to purge all effects of people on our study systems, conservation requires incorporating anthropogenic and natural factors. Second, pick systems, animal models, or questions to study that have conservation or economic significance, and invest at least 20% of your effort in conservation issues. Often it is not hard to find interesting basic behavioral questions while studying threatened or endangered species to determine the factors that limit population growth (e.g., Beissinger 1986, 1987, 1995; Berger *et al.* 1993; Berger & Cunningham 1994, 1995). Even studies of common species or species with economic value can have immediate implications for conservation if they are conceived and executed as tests of conservation strategies or models (Beissinger & Bucher 1992). It does require making an extra commitment of time and energy to gather additional data, and to translate that work into products that are useful for conservation.

In conclusion, perhaps the greatest challenge for the biological sciences as we begin the 21st century is to understand the factors promoting the diversity of life and to find ways to conserve biological diversity before it disappears (Wilson 1992). Efforts from all areas of biology will be needed, both because the problem is so great and because there are contributions that can be made by every discipline. Behavioral ecologists and ethologists played an important early role in stimulating public interest in conserving biological diversity, long before conservation biology emerged, through studies of rare and endangered birds and mammals beginning in the early 1960s. Behaviorists can expand upon their present contributions to conservation biology and catalyze future conservation innovations by broadening their research to include the kinds of data needed to translate behavior into currencies relevant to conservation at large spatial scales and by conducting strategic studies that have direct relevance to conservation tools or policy.

Literature cited

Aguirreo, A. A., and E. E. Starkey. 1994. Wildlife diseases in US National Parks: historical and coevolutionary perspectives. *Conservation Biology* **8**: 654–661.

Aide, T. M., J. K. Zimmerman, L. Herrera, M. Rosario, and R. Serrano. 1995. Forest recovery in abandoned tropical pastures in Puerto Rico. *Forest Ecology and Management* **77**: 77–86.

Allendorf, F. W. 1993. Delay of adaptation to captive breeding by equalizing family size. *Conservation Biology* **7**: 416–419.

Balmford, A., N. Leader-Williams, and M. J. Green. 1995. Parks or arks: where to conserve large threatened mammals? *Biodiversity and Conservation* **4**: 595–607.

Bart, J. 1995. Acceptance criteria for using individual-based models to make management decisions. *Ecological Applications* **5**: 411–420.

Beck, B. B., L. G. Rapaport, M. S. Price, and A. Wilson. 1994. Reintroduction of captive-born animals. Pages 265–284 in P. J. S. Olney, G. M. Mace, and A. T. C. Feistner, editors. *Creative conservation: Interactive management of wild and captive animals*. Chapman and Hall, London.

Beissinger, S. R. 1986. Demography, environmental uncertainty, and the evolution of mate desertion in the Snail Kite. *Ecology* **68**: 1445–1459.

Beissinger, S. R. 1987. Mate desertion and reproductive effort in the Snail Kite. *Animal Behaviour* **35**: 1504–1519.

Beissinger, S. R. 1990. On the limits and directions of conservation biology. *BioScience* **40**: 456–457.

Beissinger, S. R. 1995. Modeling extinction in periodic environments: Everglades water levels and Snail Kite population viability. *Ecological Applications* **5**: 618–631.

Beissinger, S. R., and E. H. Bucher. 1992. Can parrots be conserved through sustainable harvesting? *BioScience* **42**: 164–173.

Belyaev, D. K. 1979. Destabilizing selection as a factor in domestication. *Journal of Heredity* **70**: 301–308.
Berger, J., and C. Cunningham. 1994. Active intervention in conservation: Africa's pachyderm problem. *Science* **263**: 1241–1242.
Berger, J., and C. Cunningham. 1995. Predation, sensitivity, and sex: why female black rhinoceroses outlive males. *Behavioral Ecology* **6**: 57–64.
Berger, J., C. Cunningham, A. A. Gawuseb, and M. Lindeque. 1993. 'Costs' and short-term survivorship of hornless black rhinos. *Conservation Biology* **7**: 920–924.
Blane, J. M., and R. Jaakson. 1994. The impact of ecotourism boats on the St. Lawrence beluga whale. *Environmental Conservation* **21**: 267–269.
Boo, E. 1990. *Ecotourism: The potentials and pitfalls*. World Wildlife Fund, Washington, DC.
Boyce, M. S. 1992. Population viability analysis. *Annual Review of Ecology and Systematics* **23**: 481–506.
Brown, S., and A. E. Lugo. 1994. Rehabilitation of tropical lands: a key to sustaining development. *Restoration Ecology* **2**: 97–111.
Burgman, M. A., H. R. Akçakaya, and S. S. Loew. 1988. The use of extinction models for species conservation. *Biological Conservation* **43**: 9–25.
Burgman, M. A., S. Ferson, and H. R. Akçakaya. 1993. *Risk assessment in conservation biology*. Chapman and Hall, London.
Butler, P. J. 1992. Parrots, pressures, people, and pride. Pages 25–46 in S. R. Beissinger and N. F. R. Beissinger, editors. *New World parrots in crisis: Solutions from conservation biology*. Smithsonian Institution Press, Washington, DC.
Cairns, J. Jr, 1986. Restoration, reclamation, and regeneration of degraded or destroyed ecosystems. Pages 465–484 in M. E. Soulé, editor. *Conservation biology: The science of scarcity and diversity*. Sinauer Associates, Sunderland, Massachusetts.
Cattan, P. E. 1989. Management of the Vicuña *Vicugna vicugna* in Chile: use of a matrix model to assess harvest rates. *Biological Conservation* **49**: 131–140.
Caughley, G. 1994. Directions in conservation biology. *Journal of Animal Ecology* **63**: 215–244.
Clark, T. W., R. P. Reading, and A. L. Clarke, editors. 1994. *Endangered species recovery*. Island Press, Washington, DC.
Conway, W. G. 1986. The practical difficulties and financial implications of endangered species breeding programmes. *International Zoo Yearbook* **24/25**: 210–219.
Derrickson, S. R., and N. F. R. Snyder. 1992. Potentials and limits of captive breeding in parrot conservation. Pages 133–163 in S. R. Beissinger and N. F. R. Snyder, editors. *New World parrots in crisis: Solutions from conservation biology*. Smithsonian Institution Press, Washington, DC.
Diamond, J. M. 1986. The design of a nature reserve system for Indonesian New Guinea. Pages 485–503 in M. E. Soulé, editor. *Conservation biology: The science of scarcity and diversity*. Sinauer, Sunderland, Massachusetts.
Dietz, J. M., L. A. Dietz, and E. Y. Nagagata. 1994. The effective use of flagship species for conservation of biodiversity: the example of lion tamarins in Brazil. Pages 32–49 in P. J. S. Olney, G. M. Mace, and A. T. C. Feistner, editors. *Creative conservation: Interactive management of wild and captive animals*. Chapman and Hall, London.

Dunning, D. B., Jr., D. J. Stewart, B. J. Danielson, B. R. Noon, T. L. Root, R. H. Lamberson, and E. E. Stevens. 1995. Spatially explicit population models: current forms and future uses. *Ecological Applications* **5**: 3–11.

FEMAT. 1993. *Forest ecosystem management: An ecological, economic, and social assessment*. USDA Forest Service, Washington, DC.

Fleming, I. A., and M. R. Gross. 1993. Breeding success of hatchery and wild coho salmon (*Oncorhynchus kisutch*) in competition. *Ecological Applications* **3**: 230–245.

Fleming, D. M., W. F. Wolff, and D. L. DeAngelis. 1994. Importance of landscape heterogeneity to Wood Storks in Florida Everglades. *Environmental Management* **18**: 743–757.

Foose, T. J., and J. D. Ballou. 1988. Management of small populations. *International Zoo Yearbook* **27**: 26–41.

Frazer, N. 1992. Sea turtle conservation and halfway technology. *Conservation Biology* **6**: 179–184.

Gibbs, J. P. 1993. Importance of small wetlands for the persistence of local populations of wetland-associated animals. *Wetlands* **13**: 25–31.

Gipps, J. H. W., editor. 1991. *Beyond captive breeding: Re-introducing endangered mammals to the wild*. Oxford University Press, Oxford.

Goodrich, J. M., and S. W. Buskirk. 1995. Control of abundant native vertebrates for conservation of endangered species. *Conservation Biology* **9**: 1357–1364.

Griffith, B., J. M. Scott, J. W. Carpenter, and C. Reed. 1989. Translocation as a species conservation tool: status and strategy. *Science* **245**: 477–486.

Groom, M. J., R. D. Podolsky, and C. A. Munn. 1991. Tourism as sustained use of wildlife: a case study of Madre de Dios, Southeastern Peru. Pages 393–412 in J. G. Robinson and K. H. Redford, editors. *Neotropical wildlife use and conservation*. University of Chicago Press, Chicago, Illinois.

Grumbine, R. E. 1994. What is ecosystem management? *Conservation Biology* **8**: 27–38.

Guindon, C. F. 1988. Protection of habitat critical to the Resplendent Quetzal (*Pharomachus mocinno*) on private land bordering the Monteverde Cloud Reserve. MS thesis, Ball State University, Muncie, Indiana.

Haas, C. A. 1995. Dispersal and use of corridors by birds in wooded patches on an agricultural landscape. *Conservation Biology* **9**: 845–854.

Harcourt, A. H. 1995. Population viability estimates: theory and practice for a wild gorilla population. *Conservation Biology* **9:** 134–142.

Harris, L. D., and J. Scheck. 1991. From implications to applications: the dispersal corridor principle applied to the conservation of biological diversity. Pages 189–220 in D. A. Saunders, and R. J. Hobbs, editors. *Nature Conservation 2: The role of corridors*. Surrey Beatty & Sons, Chipping North, NSW Australia.

Heinen, J. T., and R. S. Low. 1992. Human behavioural ecology and environmental conservation. *Environmental Conservation* **19**: 105–116.

Heppell, S. S., J. R. Walters, and L. B. Crowder. 1994. Evaluating management alternatives for Red-cockaded Woodpeckers: a management approach. *Journal of Wildlife Management* **58**: 479–487.

Hess, G. R. 1994. Conservation corridors and contagious disease: a cautionary note. *Conservation Biology* **8**: 256–262.

Jacobson, S. K., and A. F. Lopez. 1994. Biological impacts of ecotourism: tourists and nesting turtles in Tortuguero National Park, Costa Rica. *Wildlife Society Bulletin* **22**: 414–419.

Janzen, D. H. 1988. Tropical ecological and biocultural restoration. *Science* **239**: 243–244.

Jordan, W. R., III, R. L. Peters, II, and E. B. Allen. 1988. Ecological restoration as a strategy for conserving biological diversity. *Environmental Management* **12**: 55–72.

Kangas, P., M. Shave, and P. Shave. 1995. Economics of an ecotourism operation in Belize. *Environmental Management* **19**: 669–673.

Kleiman, D. G. 1989. Reintroduction of captive mammals for conservation. *BioScience* **39**: 152–161.

Knoder, E. 1959. Morphological indicators of heritable wildness in the turkey *(Meleagris gallopavo*) and their relation to survival. Pages 116–134 in *Proceedings of the First National Wild Turkey Symposium*. Southeast Section of the Wildlife Society, Memphis, Tennessee.

Lacy, R. C. 1991. Zoos and the surplus problem: an alternative solution. *Zoo Biology* **10**: 293–297.

Lande, R. 1988. Genetics and demography in biological conservation. *Science* **241**: 1455–1460.

Lindburg, D. G. 1991. Zoos and the 'surplus' problem. *Zoo Biology* **10**: 1–2.

Lindenmayer, D. B., and R. C. Lacy. 1995. A simulation study of the impacts of population subdivision on the mountain brushtail possum *Trichosurus caninus* Ogilby (Phalangeridae: Marsupialia) in south-eastern Australia. I. Demographic stability and population persistence. *Biological Conservation* **73**: 119–129.

Low, B. S., and J. T. Heinen. 1993. Populations, resources, and environment: implications of human behavioral ecology for conservation. *Population and Environment* **15**: 7–41.

Ludwig, D., R. Hilborn, and C. Walters. 1993. Uncertainty, resource exploitation and conservation: lessons from history. *Science* **260**: 17, 36.

Lyles, A. M., and R. M. May. 1987. Problems in leaving the ark. *Nature* **326**: 245–246.

Maille, P. and R. Mendelsohn. 1993. Valuing ecotourism in Madagascar. *Journal of Environmental Management* **38**: 213–218.

McClanahan, T. R., and R. W. Wolfe. 1993. Accelerating forest succession in a fragmented landscape: The role of birds and perches. *Conservation Biology* **7**: 279–288.

McDonald, D. B., and H. Caswell. 1993. Matrix methods for avian demography. *Current Ornithology* **10**: 139–185.

Meagher, M., and M. E. Meyer. 1994. On the origin of brucellosis in bison of Yellowstone National Park: a review. *Conservation Biology* **8**: 645–653.

Meffe, G. K. 1992. Techno-arrogance and halfway technologies: salmon hatcheries on the Pacific coast of North America. *Conservation Biology* **6**: 350–354.

Menges, E. S. 1992. Stochastic modeling of extinction in plant populations. Pages 253–275 in P. L. Fiedler and S. K. Jains, editors. *Conservation biology: The theory and practice of nature conservation, preservation and management*. Chapman and Hall, New York.

Miller, B., D. Bigins, C. Wemmer, R. Powell, L. Calvo, L. Hanebury, and T. Wharton. 1990. Development of survival skills in captive-raised Siberian *polecats (Mustela eversmanni*) II: predator avoidance. *Journal of Ethology* **8**: 95–104.

Miller, B., R. P. Reading, and S. Forrest. In press. *Prairie night: Recovery of the Black-footed ferret*. Smithsonian Institution Press, Washington, DC.

Moyle, P. B. 1969. Comparative behavior of young brook trout of domestic and wild origin. *Progressive Fish Culturist* **31**: 51–59.

Munn, C. A. 1992. Macaw biology and ecotourism, or 'when a bird in the bush is worth two in the hand.' Pages 47–72 in S. R. Beissinger and N. F. R. Snyder, editors. *New World parrots in crisis: Solutions from conservation biology*. Smithsonian Institution Press, Washington.

Myers, N. 1979. *The sinking ark: A new look at the problem of disappearing species*. Permagon Press, Oxford, England.

Nepstad, D. C., C. Uhl, and E. A. S. Serrao. 1991. Recuperation of degraded Amazonian landscape: forest recovery and agricultural restoration. *Ambio* **20**: 248–255.

Nicholls, A. O., and C. R. Margules. 1991. The design of studies to demonstrate the biological importance of corridors. Pages 49–61 in D. A. Saunders and R. J. Hobbs, editors. *Nature conservation 2: The role of corridors*. Surrey Beatty & Sons, Chipping North, NSW Australia.

Noss, R. F. 1990. Indicators of biodiversity: a hierarchical approach. *Conservation Biology* **4**: 355–364.

O'Brien, S. J., and J. F. Evermann. 1988. Interactive influences of infectious disease and genetic diversity in natural populations. *Trends in Ecology and Evolution* **3**: 254–259.

Powell, G. V. N., and R. Bork. 1995. Implications of intratropical migration on reserve design: a case study using *Pharomachrus mocinno*. *Conservation Biology* **9**: 354–362.

Rahbeck, C. 1993. Captive breeding – a useful tool in the preservation of biodiversity? *Biodiversity and Conservation* **2**: 426–437.

Reed, J. M., P. D. Doerr, and J. R. Walters. 1988. Minimum viable population size of the Red-cockaded Woodpecker. *Journal of Wildlife Management* **52**: 385–391.

Reid, W. V. C. 1989. Sustainable development: lessons from success. *Environment* **31**: 7–35.

Reid, W. V. 1992. How many species will there be? Pages 55–73 in T. C. Whitmore and J. A. Sayer, editors. *Tropical deforestation and species extinction*. Chapman and Hall, London.

Rinkevich, B. 1995. Restoration strategies for coral reefs damaged by recreational activities: the use of sexual and asexual recruits. *Restoration Ecology* **3**: 241–251.

Robinson, G. R., and S. N. Handel. 1992. Forest restoration on a closed landfill: rapid addition of new species by bird dispersal. *Conservation Biology* **7**: 271–278.

Robinson, J. G., and K. H. Redford, editors. 1991. *Neotropical wildlife use and conservation*. University of Chicago Press, Chicago, Illinois.

Ryan, J. C. 1992. Conserving biological diversity. Pages 9–26 in L. R. Brown, editor. *State of the World 1992*. W. W. Norton & Co., New York.

Shaffer, M. L. 1981. Minimum population sizes for species conservation. *BioScience* **31**: 131–134.

Shaffer, M. L. 1990. Population viability analysis. *Conservation Biology* **4**: 39–40.

Sheppard, C. 1995. Propagation of endangered birds in US institutions: how much space is there? *Zoo Biology* **14**: 197–210.

Simberloff, D., and J. Cox. 1987. Consequences and costs of conservation corridors. *Conservation Biology* **1**: 63–71.

Simberloff, D., J. A. Farr, J. Cox, and D. W. Mehlman. 1992. Movement

corridors: conservation bargains or poor investments? *Conservation Biology* **6**: 493–504.
Simon, D. 1989. Sustainable development: theoretical construct or attainable goal? *Environmental Conservation* **16**: 41–48.
Slocombe, D. S. 1993. Implementing ecosystem-based management. *BioScience* **43**: 612–622.
Smith, F. D. M., R. M. May, R. Pellew, T. H. Johnson, and K. R. Walter. 1993. How much do we know about the current extinction rate? *Trends in Ecology and Evolution* **8**: 375–378.
Snyder, N. F. R., and H. A. Snyder. 1989. Biology and conservation of the California Condor. *Current Ornithology* **6**: 175–267.
Snyder, N. F. R., S. R. Derrickson, S. R. Beissinger, J. W. Wiley, T. B. Smith, W. D. Toone, and B. Miller. 1996. Limitations of captive breeding in endangered species recovery. *Conservation Biology* **10**: 338–348.
Snyder, N. F. R., S. E. Koenig, H. A. Snyder, and T. B. Johnson. 1994. Thick-billed parrot releases in Arizona. *Condor* **96**: 845–862.
Soulé, M. E. 1985. What is conservation biology? *BioScience* **35**: 727–734.
Soulé, M. E., editor. 1986. *Conservation biology: The science of scarcity and diversity*. Sinauer Associates, Sunderland, Massachusetts.
Soulé, M. E., editor. 1987. *Viable populations for conservation*. Cambridge University Press, New York.
Soulé, M. E., and M. E. Gilpin. 1991. The theory of wildlife corridor capability. Pages 3–8 in D. A. Saunders, and R. J. Hobbs, editors. *Nature conservation 2: The role of corridors*. Surrey Beatty & Sons, Chipping North, NSW Australia.
Stiles, F. G. 1988. Altitudinal movements of birds on the Caribbean slope of Costa Rica. Pages 243–258 in F. Almeda, and C. M. Pringle, editors. *Tropical rainforests: Diversity and conservation*. California Academy of Sciences, San Francisco, California.
Swain, D. P., and B. E. Ridell. 1990. Variation in agnostic behaviour between newly-emerged juveniles from hatchery and wild populations of coho salmon, *Oncorhynchus kisutch*. *Canadian Journal of Fisheries and Aquatic Sciences* **47**: 566–571.
Thiollay, J.-M. 1995. The role of traditional agroforests in the conservation of rain forest bird diversity. *Conservation Biology* **9**: 335–353.
Thorne, E. T., and E. S. Williams. 1988. Disease and endangered species: the black-footed ferret as an example. *Conservation Biology* **2**: 66–74.
Tuckwell, H. C., and J. A. Koziol. 1992. World population. *Nature* **359**: 200.
Turner, M. G., R. H. Gardner, and R. V. O'Neill. 1995*a*. Ecological dynamics at broad scales. *BioScience* **1995**: S29–S35.
Turner, M. G., G. J. Arthaud, R. T. Engstrom, S. J. Hejl, J. Liu, S. Loeb, and K. McKelvey. 1995*b*. Usefulness of spatially explicit population models in land management. *Ecological Applications* **5**: 12–16.
Uhl, C., D. Nepstad, R. Buschbacher, K. Clark, B. Kauffman, and S. Subler. 1990. Studies of ecosystem response to natural and anthropogenic disturbances provide guidelines for designing sustainable land-use systems in Amazonia. Pages 24–42 in A. B. Anderson, editor. *Alternatives to deforestation: Steps toward sustainable use of the Amazon rain forest*. Columbia University Press, New York.
UNFPA (United Nations Population Fund). 1991. *The State of the World Population 1991*. UNFPA, New York.
Vannini, J. D. 1994. Nearctic migrants in coffee plantations and forest

fragments of south-western Guatemala. *Bird Conservation International* **4**: 209–232.

Vincent, J. R. 1992. The tropical timber trade and sustainable development. *Science* **256**: 1651–1655.

WCMC (World Conservation Monitoring Centre). 1992. *Global biodiversity: Status of the earth's living resources*. Chapman and Hall, London.

Wiley, J. W., N. F. R. Snyder, and R. S. Gnam. 1992. Reintroduction as a conservation strategy for parrots. Pages 165–200 in S. R. Beissinger and N. F. R. Snyder, editors. *New World parrots in crisis: Solutions from conservation biology*. Smithsonian Institution Press, Washington, DC.

Wilson, E. O. 1988. The current state of biological diversity. Pages 3–18 in E. O. Wilson, editor. *Biodiversity*. National Academy Press, Washington, DC.

Wilson, E. O. 1992. *The diversity of life*. Harvard University Press, Cambridge, Massachusetts.

Wolff, W. F. 1994. An individual-oriented model of a wading bird nesting colony. *Ecological Modelling* **72**: 75–114.

Chapter 3

Why hire a behaviorist into a conservation or management team?

PETER ARCESE, LUKAS F. KELLER, AND JOHN R. CARY

Of 17 textbooks on animal behavior, ethology and behavioral ecology in our departmental library, only two list 'conservation' in their indices and none lists 'management.' Prospecting students might conclude, therefore, that there are few reasons to seek out behavioral expertise when dealing with problems in these areas. Similarly, we have found that many young behaviorists are not being exposed to examples of behavior applied to conservation and management as often as they would like. This state of affairs is not, however, the result of behavioral knowledge being unimportant to the fields of wildlife conservation and management (hereafter referred to as 'conservation'). To the contrary, the work of Howard (1920), Elton (1927), Leopold (1933), Nicholson (1933), Nice (1937), and many other early ecologists show that studies of animal behavior have played a central role in discussions about the regulation of animal populations. Moreover, these discussions form the basis of modern conservation theory and practice.

Our purpose here, therefore, is not to provide further examples of why one must consider behavior in studies of production and loss in populations, or of community and ecosystem structure: the reasons for doing so are self-evident in the history and development of the field of ecology. Instead, we focus first on why behaviorists are often viewed as non-essential members of conservation teams. We then describe several lines of research that might allow behaviorists interested in pursuing a 'mixed strategy' of applied and empirical work to do so without sacrificing scientific rigor or progress. To those wishing to follow this path, we also suggest Fretwell's essay on a closely related topic (1972: p. viii–xix).

Conveying the value of behavioral knowledge

To counter the notion that behavioral research is a scientific luxury, and thus a non-essential part of conservation, a trickle of path-charting literature on the application of behavioral knowledge (e.g., Temple 1978; Kleiman 1980; Starfield *et al.* 1981; Wood-Gush 1983; Metcalfe & Monaghan 1987; Stanley-Price 1989) has recently risen to a flood (Smith & Peacock 1990; McDonald 1991; Walters 1991; Pearl 1992; Reed & Dobson 1993; Caro & Durant 1995; Dobson 1995; FitzGibbon & Lazarus 1995; Laurenson 1995; Curio 1996; this volume). Thus, it should now be clear that behavioral ideas often play a central role in conservation theory and practice.

We believe, however, that behaviorists themselves have played a smaller role than their abilities and interest might otherwise allow. As a result, many wildlife managers and conservationists, although fully-aware of the general importance of behavioral ideas, remain unconvinced that behaviorists themselves are essential members of conservation teams.

Understanding why this situation exists is a first step towards its amelioration. We therefore provide some background. First, behaviorists in significant numbers have only recently begun to emphasize studies that link the behavior of individual animals to the control of population processes (e.g., Wynne-Edwards 1962; Wilson 1975; Lomnicki 1988). As a result, classical behavioral research is often not presented in ways that are either interesting or relevant to biologists studying population, community, and ecosystem processes. This is particularly true for biologists conducting conservation-oriented work, where the focus is critically centered on ecological issues at these levels.

The paucity of early attempts to make the results of behavioral research relevant to conservationists has also had the effect of underselling good examples of where behavioral knowledge has been crucial to the success of high-profile conservation efforts. The use of sexual imprinting, cross-fostering, and behavioral training by surrogates to enhance and restore Peregrine falcon (*Falco peregrinis*), Trumpeter swan (*Cygnus biccinator*), and other endangered populations of birds offer such examples (reviews in Temple 1978). Many other examples are given in the articles cited above and elsewhere in this volume.

In addition, because behavioral knowledge often plays a key role in the design of programs aimed at manipulating populations or habitats to achieve conservation goals, non-behaviorally oriented biologists have long been incorporating behavior into management programs using their intuition, incidental behavioral training, and their own significant discoveries. A very few examples here include the discovery and incorporation of ideas about age-related social dominance, reproductive competition and infanticide to understand the population dynamics of large carnivores, including Grizzly bears (*Ursus horribilis*; Bunnell & Tait 1981; McCullough 1981), African lions (*Panthera leo*; Starfield *et al.* 1981) and Gray wolves (*Canis lupus*; Haber 1977; Walters *et al.* 1981).

Finally, a few published examples of the application of behavioral knowledge to conservation involve facts so simple that we believe that readers from other disciplines will be convinced more often of the non-essential nature of behaviorists than otherwise. Such examples include the incorporation of knowledge about the mating systems of harvested species to maintain favorable adult sex ratios, or the effects of competition for nest sites on the population sizes of hole-nesting species. It is widely appreciated that behavioral knowledge of this type has a role in conservation, but it is not at all clear that behaviorists are needed to either conceive of or successfully implement such knowledge.

Conservation and management as good science

Despite these sentiments, we feel strongly that behaviorists, particularly those who follow the evolutionary approach, have much to offer to conservation theory and practice. Moreover, blending the scientific method into conservation and management programs means that researchers are likely to be rewarded with results that will not only strengthen conservation, but will also advance understanding (e.g., Walters 1986; Hilborn & Ludwig 1993).

This is because the evolutionary approach to behavior focuses on how the actions of individual animals affect fitness and why these actions evolve (Alcock 1993). The first result of this approach will be the identification of key mechanisms that affect survival and reproduction. This is an essential step in understanding the control of production and loss in conserved populations.

Second, because evolutionary behaviorists are guided by the premise of evolution by natural selection, they are apt to ask questions and test hypotheses not considered by more traditional biologists (Gavin 1989, 1991). This means that evolutionary behaviorists often make predictions and observe events not anticipated by the mechanistic approach. We contend that this will sometimes lead to new insights that are uniquely important to conservation (see also Gavin 1991; Walters 1991).

We suggest, therefore, that a major step towards overcoming skepticism about hiring behaviorists into conservation teams could be taken by behaviorists themselves, by focusing their research on species or issues that are of concern for conservation. We now outline a few examples to encourage this transition. In doing so, we emphasize underdeveloped applications of behavioral theory. We hope that these will point out where a reader's own expertise could be applied to both strengthen conservation and advance general knowledge.

The application of behavioral knowledge in management and conservation

It is often argued that autecological studies are the first step in conservation planning, and that 'general rules' in ecology fail so often that they are applied at great risk to threatened species (reviews in Simberloff 1988; Pearl 1992). The number of species that require conservation measures to ensure their persistence, however, is rising much faster than the rate at which empirical studies are conducted or funded (Wilson 1988). We suggest, therefore, that rules are worthy goals as long as they more often lead to correct than incorrect predictions (National Academy of Sciences 1986). Ideally, behaviorists should be able to suggest rules of conservation practice that apply to types of species, such as those classed by mating system, grouping tactics or phenotypic plasticity, or perhaps even those that apply to most species.

An individual-based metapopulation model for a game species

Rules are clearly needed to build models of the movement of animals on landscapes. In particular, we need to know how individuals perceive and process information about habitats and conspecifics to choose settlement sites. This problem has taken on increased importance recently as researchers have tried to match theoretical predictions about animal dispersion in fragmented landscapes with observations about regional population dynamics (e.g., Hanski 1994; Donovan *et al.* 1995). We outline one approach that may have some generality.

In Wisconsin, hunting mortality in Ruffed grouse (*Bonasa umbellus*) is additive to natural mortality, and, because public land is hunted most heavily, grouse there have up to a 60% higher probability of death than those on private land (Small *et al.* 1991). As a result, simple population projection models suggested that Ruffed grouse should decline at a rate of 46% per year (Cary *et al.* 1992). This finding, however, was at odds with observations showing that the density of drumming male grouse in a large region under study was stable over the period 1980–87.

To rectify these results, Cary *et al.* (1992) speculated that landscapes inhabited by grouse might be considered as metapopulations, wherein public hunting areas constituted 'sinks' and private lands 'sources,' and that these sites were linked by dispersal and stabilized as a result (cf. Pulliam 1988). A critical question was what dispersal paradigm would maintain regional stability in population size given the known demographic and behavioral traits of grouse in the region?

The simplest assumption of animal dispersal is that individuals move randomly. This assumption is rarely supported in vertebrates (Johnson & Gaines 1990) but routinely made for modeling purposes (e.g., Gustafson & Gardner 1996). In the case above, modeling dispersal as occurring randomly between sites led to rapid population declines because dispersers too frequently left sites with positive potential rates of population increase (private) and entered those with negative ones (public).

An alternative assumption is that grouse use information about their present site and cues about adjacent ones (e.g., auditory, visual or otherwise) to make decisions about whether to settle or continue dispersal. Cary *et al.* (1992) suggested that site quality was positively related to the local density of conspecifics. Maximum local densities were defined by allowing resident grouse to possess exclusive territories, and dispersers were thus constrained to settle only in vacant sites. This assumption is reasonable because the presence of conspecifics will be positively related to mating opportunities, and it may indicate the presence of food or protective cover. At maximum density, however, settlement is impeded by the presence of territory owners.

By this approach, Cary *et al.* (1992) obtained a model that satisfied the question and conditions posed above. Moreover, because Cary *et al.* (1992) built their model to simulate a particular region with known attributes associated with real sites on the landscape and with data from individually identified grouse, they were able to use an individual-based model with an explicit description of the study area in a geographic information system (GIS) format. This approach is attractive because it is based on general principles but can be tailored to areas of concern for conservation.

Lomnicki (1988) argued that unequal resource partitioning via contest competition can affect the dispersion patterns of individuals on landscapes. Dynamic models of animals on landscapes based on ideas about unequal resource partitioning are beginning to appear (e.g., DeAngelis & Gross 1992). As for our example, these models use GIS technology to define landscapes appropriate to conservation questions at hand and then express these in units equivalent to the average home range size of the species in question. Rules for births, deaths, and dispersal are established for individuals based on their phenotype.

As demonstrated by Cary *et al.* (1992), and more recently by Reed & Dobson (1993) and Donovan *et al.* (1995), individual-based models may further benefit by incorporating conspecific attraction (Stamps 1988; Smith & Peacock 1990). This is because empirical studies of the distribution of individuals on landscapes often reveal a surplus of unoccupied but suitable habitat, and an aggregation pattern more extreme than expected by habitat heterogeneity. Traditional models of population dynamics do not predict these aspects of spatial heterogeneity in fragmented landscapes, but understanding the dynamics of species that inhabit such areas is of great interest to conservationists.

Simulation models cannot replace experiments and observation, but their utility in integrating the results of these approaches is clear (Walters 1986; Starfield & Bleloch 1991; Cary *et al.* 1992; DeAngelis & Gross 1992; Donovan *et al.* 1995; Hilborn *et al.* 1995; Starfield *et al.* 1995). To the extent that we are correct about the usefulness of individual-based models in conservation, behaviorists could play a key role in their development, application, and validation.

Habitat selection and illegal offtake in the Serengeti-Mara ecosystem

We now provide a very simple example of how habitat selection in ungulates may affect predictions about the impacts of illegal hunting in large African reserves. In this case, a simple idea suggests an equally simple rule about behavior and vulnerability.

The development of conservation policies for whole ecosystems, such as the Serengeti Ecosystem in Tanzania and Kenya, is an activity not typically undertaken by behaviorists. Many behaviorists, however, have worked in this system, especially while studying predators (e.g., Kruuk 1972; Schaller 1972; Bertram 1975; Packer & Pusey 1984; Caro 1994; Frank *et al.* 1995; Hanby *et al.* 1995; Hofer & East 1995) and their prey (e.g., Walther 1965; Estes 1969; Jarman & Jarman 1979; Arcese *et al.* 1995*a*; FitzGibbon & Lazarus 1995). As a result, behaviorists have collected a vast amount of data there, and in doing so, they have contributed significantly to our understanding of how the system operates and what is required for its future persistence.

Nevertheless, behaviorists and others have not traditionally communicated the results of their work in ways that local managers have found useful for predicting trends in the system of concern (Sinclair 1995). Making predictions, however, is essential if one is to use science to evaluate alternative management policies (Walters 1986; Hilborn & Ludwig 1993; Hilborn *et al.* 1995). Thus, without the ability to do so, managers in Serengeti have been frustrated by the apparent lack of applicability of research results, particularly in the case of behavioral research (Carol & Durant 1995; Sinclair 1995).

To remedy this problem, 40 researchers and managers assembled recently for a modeling workshop on the Serengeti (Sinclair & Arcese 1995*b*; Hilborn *et al.* 1995). Many of the former were behaviorists with several years of experience in the area. Other participants were ecologists, conservationists and managers, and the group's facilitator was a resource ecologist with experience in assembling models of poorly known systems.

Together, this group built a simulation model to evaluate several management scenarios of interest to local authorities. This model and its output are described elsewhere, and its results were probably the least important products of the overall exercise (Hilborn *et al.* 1995). In general, however, by the end of the workshop it was clear that behaviorists contributed much of the 'meat' of the model and, thus, were responsible for much of the success of the effort. This was the case because while conducting work on the evolution of behavior, most researchers had focused their efforts on estimating the consequences of behavior for survival and reproduction. Most behaviorists focus their studies on individual animals, so they were able to construct age and sex-specific schedules of reproduction and survival, and in most cases, to estimate population sizes and list habitat and other requirements of the species they studied.

Nevertheless, the results of the model also suggested that better behavioral information was required to obtain accurate predictions of the future of the ungulate community. For example, the results of most model scenarios predicted collapse in the migrant Wildebeest (*Connochaetes taurinus*) and Zebra (*Equus burchelli*) populations, because mortality from illegal hunting was expected to exceed the maximum sustained yield for these populations in the near future. The predicted effects of hunting on resident ungulates were less severe. This was remarkable because it is well known that snaring, the main way in which ungulates are illegally killed, is concentrated in bush habitats favored by human residents (Arcese *et al.* 1995*b*).

Clearly, ungulates in the Serengeti display pronounced behavioral preferences for particular habitats (Jarman & Jarman 1979), and resident species that primarily inhabit bush habitats are on average less abundant than migratory, plains-living species (Fryxell *et al.* 1988). Thus, it is reasonable to assume that habitat selection should affect mortality from snaring. Despite the plausibility of this idea, workshop participants assumed that snaring was random with respect to species because data on the interaction between snaring, habitat preference, and abundance was unavailable. The results of the model, however, pointed to the need for such data, so subsequent efforts to collect it were undertaken.

Does better behavioral information change model predictions about the vulnerability of plains versus bush-living ungulates to snaring? We addressed this question by modifying the model of Hilborn *et al.* (1995) to include results of Arcese *et al.* (1995*b*) on the relationship between snaring mortality and habitat selection, body size, and relative abundance. When snaring mortality was distributed as expected by selectivity indices (Table 24.1 in Arcese *et al.* 1995*b*), an opposite conclusion to that originally put forward was obtained: i.e., resident, bush-living species emerged as being under the greatest risk of decline from snaring (Fig. 3.1). We suggest that this is because plains-living species are being buffered from the negative affects of over-hunting by their reduced likelihood of encountering snares.

We caution against over-interpreting these results. More rigorous analyses are required to make strong predictions about the effects of snaring on individual species of ungulates in Serengeti. Nevertheless, our point here is that even simple behavioral modifications of complex models will sometimes play a key role in predicting the dynamics of even complex systems (see also DeAngelis & Gross 1992). Preliminary results also suggest that within species, behavioral information on sex-specific grouping patterns and habitat selection will also improve estimates of snaring impacts. Georgiadis (1988) reported that because male wildebeest are more likely to enter thickets than females with calves, males are also more likely to contact snares. Clearly, there are many avenues of research in this area that may be of interest to behaviorists and that are relevant to managers. These examples demonstrate that even simple rules about social behavior and vulnerability to snaring may have marked effects on population and community dynamics.

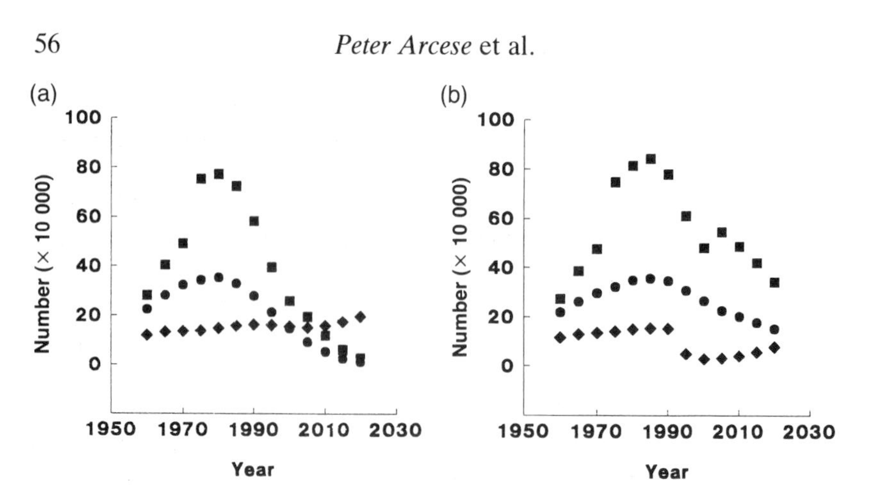

Fig. 3.1. Output from a simulation model produced at the Serengeti II workshop to explore alternate management scenarios (Hilborn *et al.* 1995). The number of migratory wildebeest (squares), zebra (diamonds), and resident ungulates (circles; collectively modeled as 'brown animals' for comparison with aerial census results) are modeled over the period 1960 to 2020. The model was constructed to simulate 30 years of census data from 1960 to 1990, and then run to 2020 to explore future trends based on current conditions. Details of the model are given in Hilborn *et al.* (1995). (a) Results based on the assumption that snaring mortality is equal in migratory (wildebeest and zebra) and resident ungulates (brown animals). (b) Snaring mortality is distributed as expected by the results of Arcese *et al.* (1995*b*) on the selectivity of snaring by species, body size, and habitat preference.

Reserves for species where infanticide is common

In many protected areas around the world, including Serengeti, managers regularly ask how large reserves should be to protect minimally viable populations (MVPs) of particular species. In the case of African lions, classical models focus on the size of the supporting prey base, the effects of interspecific competition, and the impacts of human interference as playing the main roles in demography (Schaller 1972) and affecting MVPs. It follows that one might calculate the number of prey to support a lion population of size 'x,' based mainly on the required rate of food consumption to maintain adult survival and reproduction, and juvenile survival and recruitment, such that recruitment balances natural mortality. Overall, prey density might be used as a predictor of habitat quality over a reserve, and the long-term expected size of the lion population would follow as a function of reserve size, prey density, and direct human impacts on lion numbers. Similar ideas are used to explore the population dynamics of Grizzly bears (Shaffer 1983; Knight & Eberhardt 1984, 1985).

Behavioral ideas about spacing and mating behavior may also affect MVPs, and by including such factors, one moves from a classical population model to one that includes the 'social fabric' of the species involved. For example, males and females of several species commit infanticide when this increases their chances of reproducing (Hausfater & Hrdy 1984). Infanticidal individuals manipulate the physiology and behavior of conspecifics to their own advantage, and they probably also reduce the population-level rate of recruitment. Infanticide occurs in lions (Schaller 1972; Packer & Pusey 1984) and other vertebrates (reviews in Hausfater & Hrdy 1984; Rohwer 1986), where polygynous males or polyandrous females compete for access to mates in territories, home ranges, and/or social groups.

Conservationists might view infanticide as a phenomenon of theoretical interest, but there may be circumstances where it is also a practical concern. The effect of infanticide on managed bear populations is widely discussed (e.g., Bunnell & Tait 1981; McCullough 1981; Taylor 1994; Wielgus & Bunnell 1994*a,b*). Evidence suggests it may also play a role in the population dynamics of lions (Caro & Durant 1995; Sinclair 1995).

For example, fee hunting for lions is an important activity in reserves bordering Serengeti National Park, and it regularly claims males in home ranges near park boundaries (Caro & Durant 1995). Males are also killed incidentally in snares (Arcese *et al.* 1995*b*). As a result, the rate of pride take-overs is accelerated by hunting, and the rate of recruitment may thus be reduced by infanticide. Hence, 'social fabric' models that include juvenile recruitment as a function of prey density and infanticide may predict larger MVPs than those that do not. These ideas are speculative but consistent with models of the impact of infanticide on lion demography (Starfield *et al.* 1981; Starfield & Bleloch 1991). Formal comparisons of social fabric and standard models will suggest tests of these ideas, and behaviorists could design and conduct this work.

Brood parasitism and host demography

Just as infanticidal individuals manipulate conspecifics to their own advantage, it is reasonable to expect parasites to manipulate the behavior of their hosts. Surprisingly, however, researchers have not pursued this general idea in studies of the impacts of cowbird (*Molothrus* spp.) parasitism on host demography (May & Robinson 1985; Pease & Grzybowski 1995; Robinson *et al.* 1996). We provide examples of how ideas about host–parasite evolution may enhance these efforts.

Brown-headed cowbirds (*M. ater*) are thought to reduce host productivity because adults remove host eggs and nestlings usurp host parental care (e.g., Friedmann 1963; Lowther 1993). Nest predation, however, also reduces productivity in cowbird hosts, and parasitism and predation are jointly linked to population declines in several species (Mayfield 1977; Grzybowski *et al.* 1986; Robbins *et al.* 1989; Askins *et al.* 1990; Hagan & Johnston 1992; Trail & Baptista 1993; Rothstein & Robinson 1994). This link is believed to be a coincidental result of the preferences of nest predators and cowbirds for fragmented habitats and ecological edges (Brittingham & Temple 1983; Terborgh 1989; Martin 1992; Robinson 1992; Sherry & Holmes 1992; Robinson *et al.* 1995).

An alternative hypothesis arises if one considers how cowbirds might facilitate parasitism by manipulating host behavior (Arcese *et al.* 1992, 1996). For example, cowbirds often find nests at unknown stages of incubation (references in Arcese *et al.* 1996). Parasites in this situation might ignore the nest, lay in it, or behave conditionally: i.e., by consuming or destroying the eggs if advanced beyond early incubation, or by laying in the nest if it is at a suitable stage for parasitism. Adaptationists will argue in favor of the last option. This is because depredation often initiates re-laying in hosts, thus increasing the future number of suitable nests available to parasites. Furthermore, because hosts often re-lay within a narrow time-frame (e.g., Friedmann 1963; Smith & Roff 1980; Trail & Baptista 1993), nest depredation may allow cowbirds to 'farm' their hosts by re-visiting territories after the re-laying interval. Nest depredation by cowbirds may thus enhance fitness in ways analogous to those favoring the evolution of infanticide (Rohwer 1986; Hausfater & Hrdy 1984).

With regard to conservation, this hypothesis offers a further explanation for the edge effect, and it suggests that the cumulative impact of cowbirds on hosts is greater than now assumed. Preliminary tests support this hypothesis (Arcese *et al.* 1996), but other factors also operate. For example, cowbirds employ different tactics to find nests (Norman & Robertson 1975) and habitat structure may affect which are employed most profitably (e.g., Gates & Gysel 1978; Gochfeld 1979; Freeman *et al.* 1990). Perching quietly and surveying for hosts building nests (Hann 1941) may thus be the most efficient tactic in forested habitats, where cowbirds go undetected in canopy cover. Studies of cowbird–host interactions clearly involve opportunities to link conservation and behavioral thinking.

Recent studies of a host–parasite system involving Great spotted cuckoos (*Clamator glandarius*) and magpies (*Pica pica*; Soler *et al.* 1995) offer further examples. Soler and co-workers report that hosts rapidly evolve counter-adaptations to parasitism and suggest that micro-evolution in behavioral traits may be observable over the duration of long-term studies, or by conducting comparative work in naive and experienced host populations (Soler 1990; Soler & Møller 1990; Soler *et al.* 1994). These studies suggest many avenues for future work.

Using behavior to predict a species' vulnerability to change

The ability to foresee conservation 'train wrecks,' such as that involving the Spotted owl (*Strix occidentalis*) in the Northwest US, could alleviate some of the social and political problems that result when species slide close to extinction prior to anyone noticing. This requires, however, that we be able to predict the degree to which species are put at risk by anthropogenic change so that we can prioritize our concerns most effectively. Prioritization will aid especially those conservationists that are forced to adopt species triage (Sinclair & Arcese 1995*a*). To these ends, rules for predicting a species' vulnerability to change could be enormously valuable conservation tools.

One possibility here is to use information about the degree to which species display flexibility in their behavior and ecology to predict their response to environmental change (see also Simberloff 1988; Walters 1991; Pearl 1992). Spotted owls offer a dramatic example of intolerance to change, probably because of their specialized use of habitat and selection of prey (e.g., Bart & Forsman 1990). Comparisons of the vagility, habitat specificity and prey selection of Spotted and Barred owls (*Strix varia*) show that variation in the ability to withstand change, or take advantage of it, exists in closely related species. For example, radio-tracking of juvenile Spotted owls has shown that dispersal is impeded by the presence of large clear-cut openings in forests, and that individual birds may starve at an edge rather than disperse to distant fragments (R. J. Small, personal communication). By comparison, Barred owls have colonized much of western North America in the last century by dispersing across the Great Plains.

The Mauritius kestrel (*Falco punctatus*) offers another example linking behavioral flexibility to species survival. Temple (1978, 1986) noted that the wild population of this species declined to only two pairs in 1974. This occurred because kestrels nested in tree-holes up to this point, and, following the introduction of Macaques (*Macaca fascicularis*), they suffered high rates of nest failure. In 1974, however, one pair nested on a cliff and successfully fledged young. Since then, recruits to the population have come from cliff-nests, and young from these nests also nest on cliffs. Flexibility in nest site selection has thus allowed a partial recovery of the species (Temple 1986).

These examples suggest that indices of species-specific differences in behavioral flexibility might help to prioritize conservation efforts in little-known species and communities. Recent papers by Wilson *et al.* (1994) on individual differences in risk-taking in animals, and by Greenberg (1995) on links between 'neophobia' (the aversive response to novel stimuli) and ecological plasticity, suggest avenues for further work (but see Wilson *et al.* 1995 for a cautionary note). For example, Wilson *et al.* (1994) linked ideas about phenotypic variation in risk-taking behavior to predation pressure, resource abundance, and population dynamics. Greenberg (1983) has shown that neophobia differs predictably among species and correlates with ecological plasticity, especially in terms of food and habitat selection. Indices of neophobia and risk-taking may therefore help to diagnose behavioral and ecological liabilities in little-known species.

Behavioral information may also help to predict vulnerability to stochastic extinction via the stabilizing effects of intraspecific aggression on population size (e.g., via enforced dispersal, territoriality, and dominance). For example, Clutton-Brock *et al.* (1991) showed that Soay sheep (*Ovis aries*) are scramble competitors displaying little intraspecific aggression and highly unstable population dynamics. In contrast, Red deer (*Cervus elaphus*) show intolerance via territorial and other spacing behaviors (Clutton-Brock *et al.* 1985) and exhibit stable populations in environments similar to those inhabited by Soay sheep. These are just two of many examples suggesting that social behavior affects population stability (e.g., Chitty 1967; Brown 1969; Lomnicki 1988; Sinclair 1989) and the likelihood of stochastic extinction (e.g., Gilpin & Soulé 1986; van Noordwijk 1994). In the absence of better information, therefore, species-specific differences in aggressive and spacing behavior may aid in predicting extinction risk.

Behavioral inputs to conservation genetics

Conservation genetics is the application of population genetic theory to the problem of conserving genetic diversity and minimizing inbreeding depression, and it is based on the genetics of individuals. Accordingly, the behaviors of individuals play fundamental roles. Nevertheless, although behavioral interactions underlie the propagation of genes in sexual species, only two classes of behavior seem to be of key importance to conservation genetics: mating and dispersal behavior. This is because dispersal affects an individual's potential set of mates, and mating behavior affects the actual mating pattern within this set.

Central to assessments of the rate of inbreeding and the loss of genetic variation is the concept of the effective population size, N_e (see also Parker & Waite, Chapter 10). N_e is the size of an 'ideal' population that behaves genetically in the same way as the real population, i.e., that exhibits the same rates of genetic drift and inbreeding as the population under consideration but that satisfies the assumptions for ideal populations in population genetic models (Crow & Kimura 1970). Low values of N_e indicate high rates of genetic drift and, therefore, decreased genetic variation and increased inbreeding.

Apart from the census population size, a number of behaviors affect N_e (e.g., Chepko-Sade & Shields 1987; Lande & Barrowclough 1987). These include the mating system; dominance hierarchies that affect mating and produce a larger variance in reproductive success than expected under random mating; members of the population that do not breed owing to immaturity, senescence, or reproductive suppression; and non-random mating. Many of these behaviors are regularly recorded by population ecologists, however, and their importance for conservation genetics therefore makes only a weak argument for including behaviorists in conservation teams. Examples of more sophisticated behaviors relevant to conservation genetics, however, do exist and we mention two of these below.

The first involves behaviors that function in inbreeding avoidance. These fall into two categories: those in which dispersal reduces contact between kin (Pusey 1987), and those in which animals that live with relatives actively recognize and avoid mating with kin (Blouin & Blouin 1988). Where inbreeding avoidance is present, the effects may be catastrophic for the conservation of very small populations. For example, Wayne *et al.* (1991) speculated that poor reproduction in Isle Royale Gray wolves resulted from incest avoidance. Following a population bottleneck, through which only 14 animals survived, all wolves were known to have descended from a single pair, but none of the survivors formed pair bonds. Since the winter of 1995, however, the population has begun to recover in the absence of immigration (R. Peterson, personal communication). Several questions relevant to behaviorists and conservation geneticists emerge: To what degree is inbreeding avoidance an integral part of most species' behavioral repertoires? Where inbreeding avoidance occurs, how rigidly it is maintained? Is inbreeding avoidance flexible?

A second example comes from attempts to incorporate behavior more explicitly into population genetic models. Van Noordwijk (1994) modeled interactions between inbreeding depression, environmental stochasticity, and extinction risk in small populations with different social systems, using Lomnicki's (1988) individual-based formulations of population dynamic theory. Van Noordwijk's work suggests that extinction probabilities vary markedly with social organization, and with egalitarian social systems experiencing higher extinction risk owing to the accumulation of weakly deleterious mutations. These models require a detailed knowledge of social organization and the distribution of resources among the individuals in populations, and they warrant empirical tests by behaviorists.

Human–animal interactions

A final topic that we can only touch on concerns the application of behavioral ideas to mediating the sometimes negative interactions between human and wild animal populations. The successful conservation of many animal populations requires the cooperation of the local people (e.g., Mbano *et al.* 1995; Sinclair & Arcese 1995*a*). Hence minimizing conflicts between humans and animals is imperative. Conflicts can arise because of human activities that pose a threat to the animals or because animals endanger humans or their possessions. In either case, this is an area where behaviorists might contribute tremendously to conservation.

For example, organizations involved in efforts to save the remaining tigers in Asia have had to deal with the occurrence of man-eaters among the protected tigers. The management of man-eaters is essential to the success of these conservation efforts (Sanyal 1987). Traditional management involved the removal of problem animals. More recently, behavioral approaches were taken with considerable success (Sanyal 1987). Adverse conditioning is now being used to condition tigers to associate pain with human beings. Clay models of humans charged to 230 volts are placed in critical areas to deliver electric shocks to the man-eaters attacking them. Since the introduction of electrified dummies, the occurrence of man-eaters has fallen by half.

Conclusions

The application of behavioral knowledge to manage animal populations in the wild is not new. Leopold (1933) used ideas about animal spacing and mating systems to predict the impacts of harvest. Chitty & Southern (1954) conducted detailed behavioral studies of rats and mice to aid in their control. Cade and Fyfe (1978) manipulated the behavior of peregrine falcons to facilitate their propagation and re-introduction. It is also clear that applied animal behavior is a cornerstone of captive animal husbandry (Kleiman 1980; Wood-Gush 1983). With the gelling of Conservation Biology as a discipline, however, behaviorists have also become more interested in applying what they know to the benefit of threatened species and systems.

Our examples of the application of behavioral knowledge to conservation offer a mix of well and little-known cases wherein behavioral knowledge provides the groundwork for thinking about how one might manage species or communities for conservation goals. We also wish to point out, however, that in non-behavioral circles many of these examples will not argue strongly in favor of hiring behaviorists into conservation or management teams. This is because, first, conservation biologists and managers regularly employ behavioral knowledge and techniques; just as behaviorists have recently done, for example, with new techniques in molecular genetics. Thus, most of the examples above deal with work regularly carried out by non-behaviorally oriented ecologists under the rubric of population ecology. Second, most conservation and management teams are strongly oriented towards application, while most behaviorists emphasize research. As a result, from the perspective of many conservation biologists and managers, there will be little that appears unique in the repertoire of most behaviorists.

We suggest that unique contributions to be made by behaviorists lie mainly in the types of questions asked and the approach taken. One reason for this is that evolutionary behaviorists are guided by the underlying premise of evolution by natural selection and thus are apt to ask questions and test hypotheses not normally considered by more traditional conservationists. This approach will sometimes lead to new insights that are uniquely important to conservation. Even when this does not occur, however, behaviorists that employ the evolutionary approach will focus on the identification of mechanisms that affect survival and reproduction. Once identified, these mechanisms will suggest how one might manipulate the rates of production and loss in conserved populations.

Demonstrating the value of the behavioral approach to the wider scientific community requires that behaviorists become more involved in applied problems in conservation and management. In doing so, the value of the behavioral approach will be clearest in those cases where the inclusion of behavioral theory, principles, or mechanisms can be shown to markedly alter predictions about the demography of populations or species of concern. In the absence of research results, simulation models offer a powerful means of testing for the importance of including additional behavioral ideas in applied models of population demography (cf. Walters 1986; Starfield & Bleloch 1991; Cary *et al.* 1992; Hilborn *et al.* 1995; Starfield *et al.* 1995). Behaviorists will appear most essential as participants in conservation teams when there is uncertainty about the behavioral responses of a species to management actions, and when the success or failure of a conservation project hinges on this uncertainty.

Literature cited

Alcock, J. 1993. *Animal behavior: An evolutionary approach*, 5th edition. Sinauer Associates, Inc., Sunderland, Mass.

Arcese, P., J. N. M. Smith, W. M. Hochachka, C. M. Rogers, and D. Ludwig. 1992. Stability, regulation and the determination of abundance in an insular song sparrow population. *Ecology* **73**: 805–822.

Arcese, P., G. Jongejan, and A. R. E. Sinclair. 1995*a*. Behavioural flexibility in a small African antelope: group size and composition in the oribi (*Ourebia ourebi*, Bovidae). *Ethology* **99**: 1–23.

Arcese, P., J. Hando, and K. Campbell. 1995*b*. Historical and present-day anti-poaching efforts in Serengeti. Pages 506–532 in A. R. E. Sinclair and P. Arcese, editors. *Serengeti II: Dynamics, management, and conservation of an ecosystem*. University of Chicago Press, Chicago.

Arcese, P., J. N. M. Smith, and M. I. Hatch. 1996. Nest predation by cowbirds and its consequences for passerine demography. *Proceedings of the National Academy of Sciences* **93**: 4608–4611.

Askins, R. A., J. F. Lynch, and R. Greenberg. 1990. Population declines in migratory birds in eastern North America. *Current Ornithology* **7**: 1–57.

Bart, J., and E. D. Forsman. 1990. Dependence of spotted owls on old-growth forests. *Biological Conservation* **62**: 95–100.

Bertram, B. C. R. 1975. The social system of lions. *American Scientist* **232**: 54–65.

Blouin, S. F., and M. Blouin. 1988. Inbreeding avoidance behaviors. *Trends in Ecology and Evolution* **3**: 230–233.

Brittingham, M. C., and S. A. Temple. 1983. Have cowbirds caused declines in passerine birds? *BioScience* **33**: 31–35.

Brown, J. L. 1969. Territorial behavior and population regulation in birds. *Wilson Bulletin* **81**: 293–329.

Bunnell, F. L., and D. E. N. Tait. 1981. Population dynamics of bears – Implications. Pages 75–98 in C. W. Fowler and T. D. Smith, editors. *Dynamics of large mammal populations*. John Wiley and Sons, New York.

Cade, T. J., and W. Fyfe. 1978. What makes peregrines breed in captivity? Pages 251–262 in S. A. Temple, editor. *Endangered birds: Management techniques for preserving threatened species*. University of Wisconsin Press, Madison, WI and Croom Helm Limited, London.

Caro, T. M., 1994. *Cheetahs of the Serengeti plains: Group living in an asocial species*. University of Chicago Press, Chicago.

Caro, T. M., and S. M. Durant 1995. The importance of behavioral ecology for conservation biology: examples from Serengeti carnivores. Pages 451–472 in A. R. E. Sinclair and P. Arcese, editors. *Serengeti II: Dynamics, management, and conservation of an ecosystem*. University of Chicago Press, Chicago.

Cary, J. R., R. J. Small, and D. H. Rusch. 1992. Dispersal of ruffed grouse: A large-scale individual-based model. Pages 727–737 in D. R. McCullough and R. H. Barrett, editors. *Wildlife 2001: Populations*. Elsevier Applied Science, London and New York.

Chepko-Sade, B. D., and W. M. Shields, with J. Berger, Z. T. Halpin, W. T. Jones, L. L. Rogers, J. P. Rood, and A. T. Smith. 1987. The effects of dispersal and social structure on effective population size. Pages 287–321 in B. D. Chepko-Sade and Z. T. Halpin, editors. *Mammalian dispersal patterns: The effect of social structure on population genetics*. University of Chicago Press, Chicago.

Chitty, D., and H. N. Southern. 1954. *The control of rats and mice*. Vols. 1–3. Oxford University Press, Oxford.

Chitty, D. 1967. The natural selection of self-regulatory behavior in animal populations. *Proceedings of the Ecological Society of Australia* **2**: 51–78.

Clutton-Brock, T. H., M. Major, and F. E. Guinness. 1985. Population regulation in male and female red deer. *Journal of Animal Ecology* **56**: 56–67.

Clutton-Brock, T. H., O. F. Price, S. D. Albon, and P. A. Jewell. 1991. Persistent instability and population regulation in Soay sheep. *Journal of Animal Ecology* **60**: 593–608.

Crow, J. F., and M. Kimura. 1970. *An introduction to population genetics theory*. Burgess Publishing Company / Alpha Editions, Minneapolis.

Curio, E. 1996. Conservation needs ethology. *Trends in Ecology and Evolution* **11**: 260–263.

DeAngelis, D. L., and L. J. Gross. 1992. *Individual-based population models and approaches in ecology: Populations, communities and ecosystems*. Chapman and Hall, London, England.

Dobson, A. P. 1995. The ecology and epidemiology of rinderpest virus in Serengeti and Ngorongoro Conservation Area. Pages 485–504 in A. R. E. Sinclair and P. Arcese, editors. *Serengeti II: Dynamics, management, and conservation of an ecosystem*. University of Chicago Press, Chicago.

Donovan, T. M., R. H. Lamberson, A. Kimber, F. R. Thompson III, and J. Faaborg. 1995. Modelling the effects of habitat fragmentation on source and sink demography of Neotropical migrant birds. *Conservation Biology* **9**: 1396–1407.

Elton, C. S. 1927. *Animal ecology*. Macmillan, London.

Estes, R. D. 1969. Territorial behavior of the wildebeest (*Connochaetes taurinus* Burchell, 1823). *Zeitschrift für Tierpsychologie* **26**: 284–370.

FitzGibbon, C. D., and J. Lazarus. 1995. Antipredator behavior of Serengeti ungulates: individual differences and population consequences. Pages 274–297 in A. R. E. Sinclair and P. Arcese, editors. *Serengeti II: Dynamics, management, and conservation of an ecosystem*. University of Chicago Press, Chicago.

Frank, L. G., K. E. Holekamp, and L. Smale. 1995. Dominance, demography, and reproductive success of female spotted hyenas. Pages 364–384 in A. R. E. Sinclair and P. Arcese, editors. *Serengeti II: Dynamics, management, and conservation of an ecosystem*. University of Chicago Press, Chicago.

Freeman, S., D. F. Gori, and S. Rohwer. 1990. Red-winged blackbirds and brown-headed cowbirds: some aspects of a host parasite relationship. *Condor* **92**: 336–340.

Fretwell, S. D. 1972. Populations in a seasonal environment. *Monographs in Population Biology* **5**: 1–217.

Friedmann, H. 1963. Host relations of the parasitic cowbirds. *United States National Museum Bulletin* **233**: 1–276.

Fryxell, J. M., J. Greever, and A. R. E. Sinclair. 1988. Why are migratory ungulates so abundant? *American Naturalist* **131**: 478–498.

Gates, J. E., and L. W. Gysel. 1978. Avian nest dispersion and fledging success in field-forest ecotones. *Ecology* **59**: 871–883.

Gavin, T. A. 1989. What's wrong with the questions we ask in wildlife research? *Wildlife Society Bulletin* **17**: 345–350.

Gavin, T. A. 1991. Why ask 'why': the importance of evolutionary biology in wildlife science. *Journal of Wildlife Management* **55**: 760–766.

Georgiadis, N. 1988. *Efficiency of snaring the Serengeti migratory wildebeest*. Unpublished manuscript deposited with the Serengeti Ecological Monitoring Programme, Serengeti National Park.

Gilpin, M. E., and M. E. Soulé. 1986. Minimum viable populations: processes of species extinction. Pages 19–34 in M. E. Soulé, editor. *Conservation biology: The science of scarcity and diversity*. Sinauer Associates, Inc., Sunderland, MA.

Gochfeld, M. 1979. Brood parasite and host co-evolution: interactions between shiny cowbirds and two species of meadowlarks. *American Naturalist* **113**: 855–870.

Greenberg, R. 1983. The role of neophobia in determining the degree of foraging specialization in some migrant warblers. *American Naturalist* **122**: 444–453.

Greenberg, R. 1995. Novelty responses: the bridge between psychology, behavioral ecology and community ecology? *Trends in Ecology and Evolution* **10**: 165–166.

Grzybowski, J. A., J. T. Marshall, and R. B. Clapp. 1986. Interactions between black-capped vireos and brown-headed cowbirds. *American Birds* **40**: 1151–1161.

Gustafson, E. J., and R. H. Gardner. 1996. The effect of landscape heterogeneity on the probability of patch colonization. *Ecology* **77**: 94–107.

Haber, G. C. 1977. *Socio-ecological dynamics of wolves and prey in a subarctic ecosystem*. PhD Dissertation, University of British Columbia, Vancouver, Canada. 786 pages.

Hagan, J. M. III, and D. W. Johnston, editors. 1992. *Ecology and conservation of neotropical migrant land birds*. Smithsonian Institution Press, Washington, DC.

Hanby, J. P., J. D. Bygott, and C. Packer. 1995. Ecology, demography, and behavior of lions in two contrasting habitats: Ngorongoro Crater and the Serengeti plains. Pages 315–331 in A. R. E. Sinclair and P. Arcese, editors. *Serengeti II: Dynamics, management, and conservation of an ecosystem*. University of Chicago Press, Chicago.

Hann, H. W. 1941. The cowbird at the nest. *Wilson Bulletin* **53**: 211–221.
Hanski, I. 1994. A practical model of metapopulation dynamics. *Journal of Animal Ecology* **63**: 151–162.
Hausfater, G., and S. B. Hrdy, editors. 1984. *Infanticide: Comparative and evolutionary perspectives.* Aldine Publishers, New York.
Hilborn, R., and D. Ludwig. 1993. The limits of applied ecological research. *Ecological Applications* **3**: 550–552.
Hilborn, R., N. Georgiadis, Joy J. Lazarus, J. M. Fryxwell, M. D. Broten, *et al.* 1995. A model to evaluate alternative management policies for the Serengeti-Mara ecosystem. Pages 617–637 in A. R. E. Sinclair and P. Arcese, editors. *Serengeti II: Dynamics, management, and conservation of an ecosystem.* University of Chicago Press, Chicago.
Hofer, H., and M. East. 1995. Population dynamics, population size, and the commuting system of the Serengeti spotted hyena. Pages 332–363 in A. R. E. Sinclair and P. Arcese, editors. *Serengeti II: Dynamics, management, and conservation of an ecosystem.* University of Chicago Press, Chicago.
Howard, H. E. 1920. *Territory in bird life.* John Murray, London.
Jarman, P. J., and M. V. Jarman. 1979. The dynamics of ungulate social organization. Pages 185–220 in A. R. E. Sinclair and M. Norton-Griffiths, editors. *Serengeti: Dynamics of an ecosystem.* The University of Chicago Press, Chicago.
Johnson, M. L., and M. S. Gaines. 1990. Evolution of dispersal: theoretical models and empirical tests using birds and mammals. *Annual Review of Ecology and Systematics* **21**: 449–480.
Kleiman, D. G. 1980. The sociobiology of captive propagation. Pages 243–262 in M. E. Soulé and B. A. Wilcox, editors. *Conservation biology: An evolutionary and ecological perspective.* Sinauer Associates, Inc., Sunderland, Mass.
Knight, R. R., and L. L. Eberhardt. 1984. Projected future abundance of the Yellowstone grizzly bear. *Journal of Wildlife Management* **48**: 1434–1438.
Knight, R. R., and L. L. Eberhardt. 1985. Population dynamics of Yellowstone grizzly bears. *Ecology* **66**: 323–334.
Kruuk, H. 1972. *The Spotted Hyena: A study of predation and social behavior.* University of Chicago Press, Chicago.
Lande, R., and G. F. Barrowclough. 1987. Effective population size, genetic variation, and their use in population management. Pages 87–123 in M. E. Soulé, editor. *Viable populations for conservation.* Cambridge University Press, Cambridge.
Laurenson, K. M. 1995. Implications of high offspring mortality for cheetah population dynamics. Pages 385–398 in A. R. E. Sinclair and P. Arcese, editors. *Serengeti II: Dynamics, management, and conservation of an ecosystem.* University of Chicago Press, Chicago.
Leopold, A. 1933. *Game management.* Charles Scribner's Sons, New York.
Lomnicki, A. 1988. *Population ecology of individuals.* Princeton University Press, Princeton.
Lowther, P. 1993. The brown-headed cowbird. In A. Poole and F. Gill, editors. *The birds of North America*, Vol. 47. The Academy of Natural Sciences, Philadelphia: The American Ornithologist's Union, Washington, DC.
Martin, T. E. 1992. Breeding productivity considerations: What are the appropriate habitat features for management? Pages 455–473 in J. M. Hagan

III and D. W. Johnston, editors. *Ecology and conservation of neotropical migrant land birds*. Smithsonian Institution Press, Washington, DC.
May, R. M., and S. K. Robinson. 1985. Population dynamics of avian brood parasitism. *American Naturalist* **126**: 475–495.
Mayfield, H. 1977. Brown-headed cowbird, agent of extermination? *American Birds* **31**: 107–113.
Mbano, B. N. N., R. C. Malpas, M. K. S. Maige, P. A. K. Symonds, and D. M. Thompson. 1995. Pages 605–616 in A. R. E. Sinclair and P. Arcese, editors. *Serengeti II: Dynamics, management, and conservation of an ecosystem*. University of Chicago Press, Chicago.
McCullough, D. R. 1981. Population dynamics of the Yellowstone grizzly bear. Pages 173–196 in C. W. Fowler and T. D. Smith, editors. *Dynamics of large mammal populations*. John Wiley and Sons, New York.
McDonald, D. W. 1991. Behavioral studies as a tool for vertebrate conservation. Pages 129–133 in *Wildlife conservation: Present trends and perspectives for the 21st century*. Proceedings of the International Symposium on Wildlife Conservation in Tsukuba and Yokohama, Japan, August 21–25, 1990.
Metcalfe, N. B., and P. Monaghan. 1987. Behavioral ecology: theory into practice. *Advances in the Study of Behavior* **17**: 85–120.
National Academy of Sciences/National Research Council. 1986. *Ecological knowledge and environmental problem-solving: Concepts and case studies*. National Academy Press, Washington, DC.
Nice, M. M. 1937. Studies in the life history of the song sparrow. Vol. 1. A population study of the song sparrow. *Transactions of the Linnaean Society of New York* **5**: 1–247.
Nicholson, A. J. 1933. The balance of animal populations. *Journal of Animal Ecology* **2**: 132–178.
Norman, R. F., and R. J. Robertson. 1975. Nest-searching behavior in the brown-headed cowbird. *Auk* **92**: 610–611.
Packer, C., and A. E. Pusey. 1984. Infanticide in carnivores. Pages 31–42 in G. Hausfater and S. B. Hrdy, editors. *Infanticide: Comparative and evolutionary perspectives*. Aldine Publishers, New York.
Pearl, M. 1992. Conservation of Asian primates: aspects of genetics and behavioral ecology that predict vulnerability. Pages 297–320 in P. L. Fiedler and S. K. Jain, editors. *Conservation biology: The theory and practice of nature conservation, preservation and management*. Chapman and Hall, New York.
Pease, C. M., and J. A. Grzybowski. 1995. Assessing the consequences of brood parasitism and nest predation on seasonal fecundity in passerine birds. *Auk* **112**: 343–363.
Pulliam, H. R. 1988. Sources, sinks and population regulation. *American Naturalist* **123**: 652–661.
Pusey, A. E. 1987. Sex-biased dispersal and inbreeding avoidance in birds and mammals. *Trends in Ecology and Evolution* **2**: 295–299.
Reed, M. P., and A. P. Dobson. 1993. Behavioral constraints and conservation biology: conspecific attraction and recruitment. *Trends in Ecology and Evolution* **8**: 132–135.
Robbins, C. S., J. R. Sauer, R. S. Greenberg, and S. Droege. 1989. Declines in neotropical migrant passerines. *Proceedings of the National Academy of Sciences* **86**: 7658–7662.

Robinson, S. K. 1992. Population dynamics of breeding neotropical migrants in a fragmented Illinois landscape. Pages 408–418 in J. M. Hagan III and D. W. Johnston, editors. *Ecology and conservation of neotropical migrant land birds*. Smithsonian Institution Press, Washington, DC.

Robinson, S. K., F. R. Thompson III, T. M. Donovan, D. R. Whitehead, and J. Faaborg. 1995. Regional forest fragmentation and the nesting success of migratory birds. *Science* **267**: 1987–1990.

Robsinson, S. K., S. I. Rothstein, M. C. Brittingham, L. J. Petit, and J. A. Grzybowski. 1996. Ecology and behavior of cowbirds and their impact on host populations. Pages 428–460 in T. Martin and D. M. Finch, editors. *Ecology and conservation of neotropical migrant birds*. Cambridge University Press, Cambridge.

Rohwer, S. A. 1986. Selection for adoption versus infanticide by replacement 'mates' in birds. *Current Ornithology* **3**: 353–396.

Rothstein, S. I., and S. K. Robinson. 1994. Conservation and coevolution: implications of brood parasitism by cowbirds. *Trends in Evolution and Ecology* **9**: 162–164.

Sanyal, P. 1987. Managing the man-eaters in the Sundarbans Tiger Reserve of India – case study. Pages 427–434 in R. L. Tilson and U. S. Seal, editors. *Tigers of the world: The biology, biopolitics, management, and conservation of an endangered species*. Noyes Publications, New Jersey.

Schaller, G. B. 1972. *The Serengeti lion: A study of predator–prey relations*. The University of Chicago Press, Chicago.

Shaffer, M. L. 1983. Determining minimum viable population sizes for the grizzly bear. *International Conference on Bear Research and Management* **5**: 133–139.

Sherry, T. W., and R. T. Holmes. 1992. Population fluctuations in a long-distance neotropical migrant: demographic evidence for the importance of breeding-season events in the American redstart. Pages 431–442 in J. M. Hagan III and D. W. Johnston, editors. *Ecology and conservation of neotropical migrant land birds*. Smithsonian Institution Press, Washington, DC.

Simberloff, D. 1988. The contribution of population and community biology to conservation science. *Annual Review of Ecology and Systematics* **19**: 473–512.

Sinclair, A. R. E. 1989. Population regulation in animals. Pages 197–241 in J. M. Cherrett, editor. *Ecological concepts, the contribution of ecology to an understanding of the natural world*. 29th Symposium of the British Ecological Society. Blackwell Scientific, Oxford.

Sinclair, A. R. E. 1995. Serengeti past and present. Pages 3–30 in A. R. E. Sinclair and P. Arcese, editors. *Serengeti II: Dynamics, management, and conservation of an ecosystem*. University of Chicago Press, Chicago.

Sinclair, A. R. E., and P. Arcese. 1995*a*. Serengeti in the context of worldwide conservation efforts. Pages 31–46 in A. R. E. Sinclair and P. Arcese, editors. *Serengeti II: Dynamics, management, and conservation of an ecosystem*. University of Chicago Press, Chicago.

Sinclair, A. R. E., and P. Arcese, editors. 1995*b*. *Serengeti II: Dynamics, management, and conservation of an ecosystem*. University of Chicago Press, Chicago.

Small, R. J., J. C. Holzwart, and D. H. Rusch. 1991. Predation and hunting mortality of ruffed grouse in central Wisconsin. *Journal of Wildlife Management* **55**: 513–521.

Smith, A. T., and M. M. Peacock. 1990. Conspecific attraction and the determination of metapopulation colonization rates. *Conservation Biology* **4**: 320–323.
Smith, J. N. M., and D. A. Roff. 1980. Temporal spacing of broods, brood size and parental care in song sparrows (*Melospiza melodia*). *Canadian Journal of Zoology* **58**: 1007–1015.
Soler, M. 1990. Relationship between the great spotted cuckoo, *Clamitor glandarius*, and its corvid hosts in a recently colonized area. *Ornis Scandinavica* **21**: 212–223.
Soler, M., and A. P. Møller. 1990. Duration of sympatry and coevolution between the great spotted cuckoo and its magpie host. *Nature* **343**: 748–750.
Soler, M., J. J. Soler, J. G. Martinez, and A. P. Møller. 1994. Microevolutionary change in host response to a brood parasite. *Behavioral Ecology and Sociobiology* **35**: 295–301.
Soler, M., J. J. Soler, J. G. Martinez, and A. P. Møller. 1995. Magpie host manipulation by great spotted cuckoos: evidence for an avian Mafia? *Evolution* **49**: 770–775.
Stamps, J. A. 1988. Conspecific attraction and aggregation in territorial species. *American Naturalist* **131**: 329–347.
Stanley-Price, M. R. 1989. *Animal reintroductions: The Arabian Oryx in Oman*. Cambridge University Press. Cambridge.
Starfield, A. M., P. R. Furnis, and G. L. Smuts. 1981. A model of lion population dynamics as a function of social behavior. Pages 121- 134 in C. W. Fowler and T. D. Smith, editors. *Dynamics of large mammal populations*. John Wiley and Sons, New York.
Starfield, A. M., and A. L. Bleloch. 1991. Building models for conservation and wildlife management. Second edition. Burgess Publishing, Edina, Minnesota.
Starfield, A. M., J. D. Roth, and K. Ralls. 1995. 'Mobbing' in Hawaiian monk seals (*Monachus schauinslani*): the value of simulation modelling in the absence of apparently crucial data. *Conservation Biology* **9**: 166–174.
Taylor, M., editor. 1994. Density-dependent population regulation of black, brown, and polar bears. *Ninth International Conference on Bear Research and Management Monograph Series, No. 3.*
Temple, S. A., editor. 1978. *Endangered birds: management techniques for preserving threatened species*. University of Wisconsin Press and Croom Helm Limited.
Temple, S. A. 1986. The recovery of the endangered Mauritius Kestrel. *Auk* **103**: 632–633.
Terborgh, J. 1989. *Where have all the birds gone*? Princeton University Press, Princeton, NJ.
Trail, P. W., and L. F. Baptista. 1993. Demography of a declining population of white-crowned sparrows. *Conservation Biology* **7**: 309–315.
Van Noordwijk, A. J. 1994. The interaction of inbreeding depression and environmental stochasticity in the risk of extinction of small populations. Pages 131–146 in V. Loeschcke, J. Tomiuk and S. K. Jain, editors. *Conservation genetics*. Birkhäuser Verlag, Basel, Switzerland.
Walters, C. J. 1986. *Adaptive management of renewable resources*. Macmillan Publishing Co., New York.
Walters, J. R. 1991. Application of ecological principles to the management of endangered species: the case of the red-cockaded woodpecker. *Annual Review of Ecology and Systematics* **22**: 505–523.

Walters, C. J., M. Stocker, and G. C. Haber. 1981. Simulation and optimization models for a wolf–ungulate system. Pages 317–338 in C. W. Fowler and T. D. Smith, editors. *Dynamics of large mammal populations*. John Wiley and Sons, New York.

Walther, F. 1965. Verhaltensstudien an der Grantgazelle (*Gazella granti* Brooke, 1872) in Ngorongoro Krater. *Zeitschrift für Tierpsychologie* **22**: 167–208.

Wayne, R. K., D. A. Gilbert, N. Lehman, K. Hansen, A. Eisenhawer, D. Girman, R. O. Peterson, L. D. Mech, P. J. P. Gogan, U. S. Seal, and R. J. Krumenaker. 1991. Conservation genetics of the endangered Isle Royale gray wolf. *Conservation Biology* **5**: 41–51.

Wielgus, R. B., and F. L. Bunnell. 1994*a*. Dynamics of a small, hunted brown bear *Ursus arctos* population in southwestern Alberta, Canada. *Biological Conservation* **67**: 161–166,

Wielgus, R. B., and F. L. Bunnell. 1994*b*. Sexual segregation and female grizzly bear avoidance of males. *Journal of Wildlife Management* **53**: 405–413.

Wilson, D. S., A. B. Clark, K. Coleman, and T. Dearstyne. 1994. Boldness and shyness in animals. *Trends in Ecology and Evolution* **9**: 442–446.

Wilson, D. S., A. B. Clark, K. Coleman, and T. Dearstyne. 1995. Reply to R. Greenberg. *Trends in Ecology and Evolution* **10**: 166.

Wilson, E. O. 1975. *Sociobiology: The new synthesis*. Belknap Press, Cambridge, MA.

Wilson, E. O. 1988. The current state of biological diversity. Pages 3–20 in E. O. Wilson, editor. *Biodiversity*. National Academy Press, Washington DC.

Wood-Gush, D. G. M. 1983. *Elements of ethology: A textbook for agricultural and veterinary students*. Chapman and Hall, London.

Wynne-Edwards, V. C. 1962. *Animal dispersion in relation to social behavior*. Oliver and Boyd, Edinburgh.

Chapter 4

Conservation, behavior, and 99% of the world's biodiversity: Is our ignorance really bliss?

HUGH DINGLE, SCOTT P. CARROLL, AND
JENELLA E. LOYE

If all vertebrates went extinct, there would for all practical purposes be no statistically significant effect on biodiversity. We have only the roughest notion of how many invertebrate species there are (Burk 1994), but even the highly inexact estimates we have confirm this point. Burk (1994) summarizes several attempts to determine the approximate number of insect species. Depending on the assumptions and sources of data for calculations, these estimates range from 2.3 million to possibly as many as 80 million species. We will assume the lower figure. To this can be added the more than 60 000 arachnids, 30 000 crustaceans (with probably four times that number undescribed), 12 000 nematodes (with who knows how many undescribed), and an unknown number of species from other Phyla (Brusca & Brusca 1990). A reasonable estimate of the total number of vertebrates would be about 50 000 species (Hickman *et al.* 1993). If we take 3 million as an estimate of the total invertebrates, vertebrates represent only one-sixtieth of the planet's animal species. We doubt, however, that any invertebrate zoologist or entomologist believes the world's total invertebrates number only 3 million.

In spite of the enormous disparity in numbers of species it is also the case that vertebrates receive a vastly disproportionate amount of attention from both animal behaviorists and conservation biologists. We surveyed the last three volumes of *Conservation Biology* for papers on invertebrates and behavior with the results shown in Table 4.1. Half the papers were devoted exclusively to vertebrates or nearly so and only 5% to invertebrates (the remainder were addressed to plants or habitats in general). No papers discussed the behavior of invertebrates, and very few, for that matter, the behavior of vertebrates. Taxon bias is equally evident among animal behaviorists. At the 1995 meeting of the Animal Behavior Society there were 348 papers or posters; 58 were on invertebrates. This

Table 4.1 *Distribution of subject matter for papers in* Conservation Biology *1993–95*

Year	All papers	Vertebrate	Invertebrate	Behavior
1993	129	59	9	0
1994	150	75	5	0
1995	134	73	7	0
Total	413	207	21	0

represented a slight improvement on the 1994 meeting where the ratio was 47 out of 432. Meetings devoted to invertebrate conservation are sparsely attended, especially by vertebrate biologists (Kellert 1993). The chapters in this volume are consistent with these biases. Entomologists in particular, with their focus on pest species, are not without blame in the neglect of invertebrates in conservation and behavior (van Hook 1994), but it is also the case that the education of both biologists and the public focuses primarily on charismatic fauna. We think some of that education needs changing with respect to conservation and behavior, and here we attempt to show why.

Accordingly, a special part of our task is to present evidence for and advocate the biological importance of invertebrate behavioral ecology. We use a case history approach to illustrate four main points concerning the importance of invertebrates and their behavior in conservation biology. The paucity of empirical studies on the relation between invertebrate behavior and conservation makes most generalization premature. Based in part on the case histories examined, however, we propose some tentative generalizations in the last section. We concentrate on arthropods because they are most numerous, we know them best, and they are better known with respect to conservation problems than other invertebrates. (We fully agree with colleagues who pointed out that other invertebrates provided equally good illustrative examples.) Our first point is that the behavior of invertebrates plays an important, if not a dominant, role in governing ecosystem function, and that it is important to understand that role in ecosystem conservation strategies. We argue that the invertebrates in question are likely more important to ecosystems than the charismatic vertebrates often used as a ploy for habitat preservation (van Hook 1994), and their role needs to be more completely understood. Our second point is that invertebrate behaviors may impact and even place at risk certain endangered vertebrates. Our example is the host seeking and host choice behavior of parasites. Third we argue that some behaviors are so important, so little understood, and so at risk that they deserve conservation efforts even though the species in which they occur are not in danger of demise. Fourth, we suggest that invertebrates have traits, such as rapid life cycles and small size, that make them especially amenable to experimental analysis and therefore useful for testing general theory in behavior and conservation biology, as they have been employed in evolutionary biology and ecology. We conclude by emphasizing the need for more attention to invertebrate behavioral ecology in general, discussing the types of behavioral studies of particular relevance to conservation strategies, and suggesting how complex invertebrate life cycles that require more than one habitat to complete could be a boon to reserve design.

Invertebrate behavior and ecosystem function

The role of ant behavior

Not only are invertebrates diverse, they also contribute enormously in terms of biomass. Perhaps the clearest examples of biomass contribution come from ants and termites. These social insects are the dominant organisms in many habitats, and in tropical regions the colonies of some species can reach into the hundreds of thousands or millions of individuals. Their above ground structures or 'mounds' are often but a small portion of colonies that are many times the size revealed at the surface; most of the colony structure is underground. Some colonies, such as those of leafcutter ants (*Atta* spp.) are large enough to enclose a quite reasonably sized house (see example on p. 602 in Hölldobbler & Wilson 1990). When one considers the number of colonies present in an average patch of rainforest, it is not hard to see why invertebrates constitute most of the animal biomass. In the Brazilian Amazon, for example, the dry weight of ants alone is about four times that of all the land vertebrates present in the forest, without even considering any of the other invertebrates that occur there (Hölldobbler & Wilson 1994).

The construction of ant and termite colonies and the maintenance of organization and function within them requires a complex and sophisticated set of social behaviors that have a significant bearing on the structure and function of ecosystems (Hölldobbler & Wilson 1990). The colony construction and foraging behavior of ants is responsible for much of soil and litter turnover. In the course of their annual foraging a colony of leafcutter ants will cut, bring back to the colony, and process for use in the fungus 'garden', several tons of leaf and flower material. In addition a square meter of tropical forest floor has an ant colony of one of several species in the hollows of many fallen twigs (Byrne 1994). These mostly tiny creatures scavenge and forage over every centimeter of the ground surface, in the process serving as a major factor in litter decomposition. The foraging phalanxes of army ants (*Eciton*) in the New World Tropics and of driver ants (*Dorylis*) in Africa are major predators of both invertebrates and small vertebrates. In European forests a wood ant (*Formica polyctena*) colony can take 100 000 prey items a day (Hölldobler & Wilson 1994). In temperate forests ants are major distributors of seeds for germination, and fleshy eliasomes attached to the seeds and other ant–plant mutualisms have evolved as a result (Howe & Wesley 1988). Ant colonies host developing larvae of certain endangered British lycaenid butterflies. The ant–butterfly relationship is dependent on a specialized ant behavior by which the butterfly larvae are carried into the ant nests (Thomas 1991; New *et al.* 1995). It is ironic that although there are grant funds to study the attractive butterflies, money to study the behavior of the ants on which they are utterly dependent is virtually nonexistent (Jeremy Thomas, personal communication).

It is apparent from these examples that at both large and small scales ant foraging and scavenging behavior can be a dominant contributor to the processes that ecologists have long regarded as keys to herbivore interactions, decomposition, and mutualism. Small wonder, then, that in the absence of invertebrates the vertebrates would go extinct and the world would rot (Wilson 1992). Terborgh (1988), responding to Wilson (1987), has argued that 'big things' drive ecosystems. Indeed they do, but the big things are likely to be ants. An individual army ant is small, but a colony is a larger predator than a jaguar.

Most conservation efforts ultimately are directed at the preservation of biodiversity or the preservation of an endangered species, usually a vertebrate. The above summary of ant behavior in ecosystems implies that a knowledge of invertebrate behavior may be key to several conservation decisions. Will a given rainforest reserve include enough ants, individuals or species, to carry out adequately processes such as litter turnover? Is there some missing invertebrate behavior that contributes to the decline of an ecosystem or, for that matter, the decline of a more charismatic species such as a lycaenid butterfly? In an extension of Caughley's (1994) argument, we argue that behavior is a crucial part of the study of natural history that is the essential first step in applying the scientific method to species or ecosystem preservation. Data on behavior should thus be integrated into a framework for understanding demography and/or species interactions that can guide us in conservation efforts.

Pollination biology: Behavior of both plants and insects is important to conservation

Many conservation efforts are directed at the preservation of ecosystems that contain unique features of biological interest. Rainforests are an example because of the biodiversity they support, but there are also many examples of small segments of nature that have probably never been particularly numerous. One such example is the vernal pools that occur in the Central Valley of California (Thorp & Leong 1995). These occur in small depressions in a generally flat topography that fill up with the winter rains and linger well into spring when the water disappears almost exclusively by evaporation. As the pools shrink in size, concentric rings of brilliant wildflowers, of species rare or absent away from the pools, come into bloom as the soil is progressively exposed and then dries. The pools themselves are home to endangered fairy shrimp (Order Anostraca) and Tiger salamanders (*Ambystoma tigrinum*) as well as other interesting faunal elements such as tadpole shrimp (*Triops*: Notostraca).

With its purple and white flowers, one of the most attractive of the small plants that encircle pools in the spring is valley Downingia (*Downingia concolor*: Campanulaceae). Pollination in Downingia is accomplished by the andrenid bee, *Panurginus atriceps*, and the complex mutualistic interaction between the behavior of the bee and the behavior of the flower means each is fully dependent on the other (Robin Thorp, 1990 and personal communication). When a bee lands on a flower, it begins the pollination process by probing into the corolla and tripping trigger hairs on the flower's 'sexual' column. The column then opens and the stigma/style acts as a piston pushing pollen out via the anthers. This pollen is deposited on the head and back. The bee is then rewarded as it collects the pollen, using combs to groom it into the pollen baskets on the hind legs. The behavior of the flower does not end here, however, because it also saves some pollen for deposition on 'safe sites' on the body of the bee, inaccessible to the grooming pollen combs. This pollen serves the plant's end because it is transferred to other flowers where it is deposited and fertilizes them.

These complex mutualistic behaviors of bee and flower make the survival of each dependent on the other. Simply conserving the vernal pools, however, does not ensure this survival because of two other aspects of the behavior of the bees: they require drier upland nesting sites to construct their nest holes, and they will colonize new flower patches over only short distances (Thorp & Leong 1995). These aspects of the bee's behavior mean that the vernal pool ecosystem in effect also includes adjacent upland areas, and that fragmentation of clusters of pools by agriculture will reduce the likelihood that isolated vernal pool plant communities will be visited or colonized by their required pollinators. Other vernal pool wildflowers, and their andrenid bee pollinators, have similar sorts of requirements, but each bee species has its own nesting microhabitat.

Once again the information on vernal pool bees and flowers indicates that a knowledge of natural history is a required first step in a successful conservation program (Caughley 1994). An understanding of the full interactive behavioral repertoire of bees and flowers is a necessity to preserve the rings of flowers that are so much a part of the vernal pool system. Until the behavioral ecology was known, there was no reason to suspect that nearby uplands were an integral component of vernal pools. The exchange between upland and vernal pool also reinforces another point well known to ecologists and conservation biologists, namely that ecosystems rarely stand alone, but depend on inputs from adjacent ecosystems. What is sometimes forgotten or ignored is that the behaviors of invertebrates (and vertebrates) are likely important components of inputs and exchanges.

Other bee pollinators further illustrate the problems of habitat fragmentation. A case in point is the Euglossine bee fauna of forest fragments in Amazonas, Brazil (Powell & Powell 1987). Visits were recorded of 15 species of these 'orchid bees' to chemical baits placed in forest stands of 100, 10, and 1 hectare in size. The visitation rates were compared with those at baits at the same site the previous year before any of the trees were removed. Before fragmentation, visits occurred at a mean rate of 56.3 ± 21.5 bees per hour. Rates after fragmentation (Table 4.2) demonstrate that a marked reduction in visits occurred. The trend as a function of forest area is clear. A closer examination of the foraging behavior of the bees indicates that they do not cross open areas between habitat fragments, preferring instead to move through forest understory. Euglossines also differ from some other bees in that they are 'trapliners,' pollinating along more or less linear paths through the forest, often for several kilometers. They are thus likely to be particularly susceptible to habitat fragmentation both because they will not move between

Table 4.2 *Influence of habitat fragmentation on pollination visits by Amazonian Euglossine bees*

Habitat size	Visits (bees/hour)
Before fragmentation	56.3±21.5
100 hectares	29.2±12.4
10 hectares	13.6±2.5
1 hectare	7.0±1.4

fragments and because they may require long stretches of forest understory. Other bees, such as highly territorial megachilids (Johnson & Hubbell 1974) may be less at risk in small fragments, as long as an adequate supply of the flowers they pollinate is also included. A general point that emerges from the above discussions of vernal pool andrenid and tropical forest euglossine bees is that to conserve a particular plant community, it is necessary to understand the foraging behavior, nesting habits, and probably other aspects of the behavioral ecology of the pollinators that service it.

Invertebrate behavior and a classic vertebrate conservation problem

Here we consider the behavior of a blood-feeding ectoparasite and how habitat changes may have increased its impact on endangered vertebrate hosts, the Puerto Rican parrot (*Amazona vittata*) and the Puerto Rican population of the Sharp-shinned hawk (*Accipiter striatus*). The effect of habitat disturbance on vertebrate populations is a chief focus of conservation biology (e.g., Gates & Geysel 1978; Soulé 1986; Yahner 1988 and references therein; Caughley 1994). That parasites may limit host populations is now widely appreciated (e.g., papers in Loye & Zuk 1991), but the potential for habitat disturbance to alter parasite behavior, and thus change the interaction between vertebrate hosts and their invertebrate parasites, has received little attention (Loye & Carroll 1995).

A provocative system for examining the effects on endangered vertebrate species of parasite host-searching behavior in disturbed habitats is that of botfly (*Philornis* sp.) infestation and its spread from Pearly-eyed thrashers (*Magarops fuscatus*) (Snyder *et al.* 1987) to the endangered Puerto Rican parrot and Sharp-shinned hawk (Delannoy & Cruz 1991). Thrashers invaded the parrot's remnant habitat, the Luquillo forest reserve in Puerto Rico, beginning in about 1950 (Snyder *et al.* 1987), when extensive road building greatly increased edge habitats and perhaps provided corridors for thrasher colonization. Pearly-eyed thrashers are abundant in disturbed habitats throughout Puerto Rico, and now occur in all remaining forests. This species typically supports large numbers of blood-feeding botfly maggots. Adult female flies fly among bird nests depositing eggs, and the larvae develop subcutaneously, principally on the nestlings, which may die from large infestations (Arendt 1985).

With the fragmentation of forest habitats, more thrashers have gained proximity to the diminishing nesting areas of both the parrots and hawks. At the same time the botflies have expanded their choice of hosts to include these birds. These endangered species have demonstrated less resilience than the thrashers when challenged with botfly parasites: hawk chicks infested with more than two botfly larvae experienced 100% mortality in some years, an infection rate that is quite common (Delannoy & Cruz 1991), and botfly infestation has become a novel source of mortality for the parrot as well (Snyder *et al.* 1987). Thus a parasite endemic in a common host may become epidemic in rare ones (Loye & Carroll 1995), and the sponsorship of a large population of botflies by the thrashers may make control efforts more difficult.

The change in forest structure as a result of road building may have increased the detectability of hosts for botflies both by providing flight corridors and increasing the proportion of habitat edge to interior (Loye & Carroll 1995). Habitat structure is known to influence the orientation behavior of flying insects, and this includes orientation toward edges (e.g., Couturier & Robert 1962). Yet the design of reserves seldom considers the orientation behavior of host-seeking parasites, or their willingness to switch hosts to endangered species.

Habitat fragmentation may also encourage the introduction of new species of disease vectors that are better adapted to the new environment and display more aggressive host seeking and feeding behavior. Mosquito vectors of avian malaria are typically adapted to intact forests with tree-holes and epiphytes for larval development. When disturbance increases standing water, as in logged gaps and along roadways (e.g., barrow pits built in the Lacondon forests of southern Mexico), new and more successful species of malaria vectors may invade (Rejmankova *et al.* 1991), and these can increase the infection rate among bird populations with serious consequences for endangered species (cf. Warner 1968; van Riper *et al.* 1986).

Such unforeseen consequences argue for the value of increasing our awareness of the behavior of the invertebrates that parasitize vertebrate species of conservation concern. A great many conservation biologists are specialists in vertebrate biology, but they may be less sensitive to issues involving invertebrate parasitism. Filling the information gap concerning the behavior of parasites in changing environments is an important challenge for behavioral ecologists interested in conservation (see Dobson & May 1986).

Behaviors as endangered phenomena

Many invertebrates that are not themselves at risk of extinction include populations that display characteristic or unique behaviors that are at risk. For both scientific and esthetic reasons, these populations may deserve special attention from conservation biologists. The behaviors at risk (and other characteristics or traits) have been given the designation 'endangered phenomenon,' defined as "a spectacular aspect of the life history of a . . . species . . . that is threatened with impoverishment or demise; the species per se need not be in peril, rather, the phenomenon it exhibits is at stake" (Brower & Malcolm 1991). Similar criteria have also been used to define a 'red data book status category' of 'threatened phenomena,' and invertebrates provide many examples of especially interesting populations that are endangered (Wells *et al.* 1983). Indeed the notion that phenomena, as well as species and ecosystems, can be both endangered and worth preserving seems to be a significant contribution to conservation biology deriving from studies of invertebrate behavior. We would argue that the concept deserves more attention from conservationists than it has so far received.

To illustrate our contention we present two examples. The first is the highly migratory population of the Monarch butterfly, *Danaus plexippus*, occurring in eastern North America; the second, discussed below, concerns song patterns in a genus of tree crickets, *Laupala*, of the Hawaiian Islands. The extensive range of the monarch butterfly in the New World and, following introductions, in Hawaii, other Pacific islands, and Australia means that the species itself is at no risk of extinction. The migratory populations of the butterfly on both the east and west coasts of North America are, however, at considerable risk, and fully qualify as endangered phenomena (Brower & Malcolm 1991; Malcolm 1993 and other papers in Malcolm & Zalucki 1993; Dingle 1996). We concentrate here on the much larger eastern population, because it so dramatically makes the point about interesting behaviors worth preserving. Not only is the migration aesthetically pleasing, but the enormous numbers of individuals involved offer opportunities present in few other organisms to study the details of migratory behaviors such as orientation (Schmidt-Koenig 1985) or the use of thermals for soaring (Gibo & Pallett 1979; Gibo 1986) plus the detailed physiology of and requirements for successful overwintering (Masters *et al.* 1988). Almost the entire population is concentrated during the winter into a few specific locations in mountains in the Transvolcanic Range in Central Mexico because of the requirements for suitable overwintering sites (Urquhart 1976; Malcolm 1993). Only ten permanent Mexican sites are known, with a smattering of temporary sites (de la Maza & Calvert 1993), and the five largest, harboring up to several million individuals each, occur within an area of only 800 square kilometers in the high altitude fog belt dominated by forests of Oyamel fir (*Abies religiosa*). Microclimate is the key to survival of the butterflies in the wintering sites (Masters *et al.* 1988), and even small disturbances to the forest can tip the balance against the butterflies. The fact that the local economy is highly dependent on these forests puts the monarchs at considerable risk and poses severe problems for preservation efforts. The small number of suitable overwintering sites makes the threat posed by anthropogenic activity acute because the probability of extinction is inversely proportional to the number of refuges (Quinn & Hastings 1987).

Both the Mexican government and the private conservation association Monarca, AC (which unfortunately no longer exists) have devoted much effort to preserving the monarch overwintering sites by working for practical solutions grounded in the local economy and culture (Ogarrio 1993 and other papers in Malcolm & Zalucki 1993). In 1984 a trust was established by Monarca, AC in cooperation with the federal government that resulted in the purchase of 700 hectares in the monarch overwintering area. Joint action has also produced a presidential decree defining the extent of five overwintering areas and their protection. Unfortunately a complex set of laws and controls by several government entities and a local community property system make it difficult to purchase areas for protection, and work against successful conservation efforts. The local people also depend on many forest products such as cordwood and roof shingles for their livelihood and on clearing forest for agriculture (Snook 1993). There has been considerable success in promoting the butterfly concentrations as tourist attractions, but the needs and activities of tourists also put pressure on the forests (Snook 1993). The monarch project serves as an excellent model for how conservation should be carried out, because of the effort to work with both local people and governments and to exploit the phenomenon as an ecotourist attraction. That the monarch populations are still at risk indicates how difficult conservation can be in the face of political and population pressures. At the same time the project shows what can be done when a local people understand the value of the conservation effort.

Tree crickets of the genus *Laupala* are one of several groups of Hawaiian crickets that have undergone extensive adaptive radiation following colonization of this isolated Pacific archipelago (Otte 1989). Their adaptive radiation mimics many other elements of the native Hawaiian flora and fauna, but these are of particular interest because in certain parts of their range it may be possible to observe speciation in progress, an observation that would be much more difficult to achieve with vertebrates. Crickets have been particularly useful for studying the role of song patterns in sexual selection, even to the details of genetic variation (Hedrick 1994), and the *Laupala* group offers opportunities to probe a step further to examine the role of song variation in behavioral isolation among populations. For example, in a 3 km transect along a ridge line on Kauai, Otte (1989) noted song differences among individuals that suggested the presence of three putative 'species.' One sings relatively slowly, occurs mostly at the lower end of the transect, and is more common in summer than winter. A second sings more rapidly, occurs only in summer, and ranges over the entire length of the transect. Finally, there is a winter species of intermediate song type, completely syntopic with the summer species. These song types strongly suggest behavioral mechanisms maintaining the integrity of different species. At several places along the transect the summer and winter species overlap temporally to some degree, and at these spots there is evidence from the song patterns that hybridization has occurred, suggesting speciation may not be complete. These crickets are highly sedentary so that isolation among the ridges and valleys of the Hawaiian islands can produce considerable local adaptation and, evidently, speciation. These song patterns provide rich material for studying the role of behavior in speciation and adaptive radiation as active processes, as well as finished products, in the evolutionary laboratory that is the Hawaiian islands (Kaneshiro 1988). As with much of the Hawaiian biota, *Laupala* species are very local in distribution and are under threat from development and commercialization, but their considerable scientific value as models for the process of speciation make them and their songs endangered phenomena well worth preserving.

Given the enormous diversity of invertebrates, we can expect great variety in the endangered behavioral phenomena worthy of the attention of conservationists. Many of these phenomena provide nearly unique opportunities for studying interesting and important biological problems, of which speciation and migration are but two. Compared with vertebrates, many invertebrates are resource specialists, and may therefore be more likely to be involved in coevolved symbioses (e.g., the behavioral interactions between rare lycaenid butterflies and the ants in whose nests their larvae dwell (New *et al.* 1995)). Other behavioral phenomena worthy of conservation consideration include such examples as the unique and often bizarre predatory behavior of certain endangered freshwater bivalves, and the mass lunar-cycle matings of endangered Japanese populations of horseshoe crabs (*Limulus* sp.). The concentration by conservation biologists and animal behaviorists on the very small proportion of the world's biota represented by large vertebrates means that unique opportunities for insight and understanding are being missed and possibly lost.

Invertebrates as model organisms: Anthropogenic change as a cause of behavioral evolution

Compared with vertebrates, many invertebrate taxa are ideal for testing models in conservation ecology and genetics by virtue of smaller body size, greater manipulability and briefer generation time. The value of invertebrates as ecological models is thoroughly established, and behavior may have a basic influence on population dynamics and community interactions (e.g., Huffaker 1958; Werner 1992). Invertebrates, however, have been little used for studies of the demographic importance or evolutionary dynamics of behavior in a conservation context. Instead, experimental work with invertebrates as conservation models is mainly limited to studies in *Drosophila* of the influence of small effective population size and breeding structure on the maintenance of gross genetic variation believed critical to long-term adaptive potential (e.g., Borlase *et al.* 1993; summarized by Ralls & Meadows 1993).

Many other aspects of conservation, ecology and genetics may be explored with regard to model invertebrate populations. For example, experimental populations in artificial environments could be used to test how patterns of metapopulation fragmentation (e.g., the number of fragments and the size and composition of each) interact with ranging and migratory behavior to influence extinction probabilities. Likewise, invertebrates have great potential for examining the conservation implications of contemporary evolution within focal populations or species. Natural selection is increasingly understood to be an ongoing process (e.g., Grant 1986; Gibbs & Grant 1987; Carroll & Boyd 1992); rapid phenotypic evolution is expected, and indeed observed, to be most significant in cases of intense, hard selection, as in the evolution of pesticide resistance (reviewed by Rosenheim & Tabashnik 1991), and in response to many other anthropogenic changes in the environment. In clear contrast to the concern that depauperate genetic variation will prevent threatened species from adapting to long-term environmental change, rapid evolution of threatened species, in response to ongoing environmental change, may render certain habitat restoration efforts inappropriate for newly adapted populations.

Habitat changes can challenge organisms at all levels, in contexts ranging from immunological resistance to habitat choice behavior. Behavioral evolution may often be a first step in populations that successfully cope with the challenge of finding new ways to persist in altered habitats. Genetic divergence is implicated in foraging behavior differences between closely related species of sparrows and wood warblers (Greenberg 1985, 1987, 1992), suggesting that trophic divergence is correlated with speciation. Likewise, genetically-based differences in host preference have been described for a number of plant-associated insect species (reviewed in Via 1990).

Compared with behavioral differences among species, patterns among populations within species provide an even better opportunity to measure the rate and direction of evolution. Here we will describe rapidly evolving differences in host preference between populations of an insect, the Soapberry bug (*Jadera haematoloma*; Rhopalidae) that has colonized an intentionally introduced host plant species in the past few decades. While populations of this insect are not currently threatened, this species is a model for understanding the potential for and significance of rapid behavioral evolution in nature. Isolated populations of this species which still occur on different native host plant species in the southwestern versus southeastern US show intrinsic differences in adult host preference (Carroll & Boyd 1992). In evolutionary terms, host choice behavior is a particularly important trait because of its potential for influencing gene flow and, thus, local adaptation.

In addition to the differences in host preference between southwestern and southeastern populations, our preliminary data indicate that hatchling host preference behavior is rapidly evolving in Florida, where one population still occurs on the native host plant, the Balloon vine (*Cardio-*

spermum corindum), and the other occurs on a host plant that was principally introduced beginning in the 1950s, the goldenrain tree (*Koelreuteria elegans*). Soapberry bugs are endemic to the Americas and feed on the seeds of sapindaceous plants (Carroll & Loye 1987). Compared with bugs on the native host, populations on the new host exhibit adaptive, genetically based changes in mouthpart length, which relate to fruit size (Carroll & Boyd 1992), and in life history characters, which relate to host phenology and nutritional quality (Carroll, Klassen & Dingle, in press).

We conducted our host preference test with naive hatchling bugs in order to control the effects of experience on choice. They were studied in sibling groups and were the second laboratory generation from populations collected from the native host on Plantation Key near Tavernier, Florida, and from the introduced host 350 km to the north in Lake Wales, Florida. For each full sib group we measured host preference as determined by the number of individuals feeding on the seeds of each species during the first three days after hatching. For each observation we scored preference based on the host with more bugs feeding on it.

Hatchlings from the population on the introduced host (the "derived" population) clearly showed a higher feeding frequency on the introduced tree than did hatchlings from the population on the native host (Fig. 4.1). Over the course of the past 30–40 years (equivalent to about 100 generations), this host preference has evolved one-to-two phenotypic standard deviations toward favoring the introduced host. Host preference may both reduce gene flow between host races and determine the selective environment, so its rapid evolution has the potential to profoundly influence evolution in the entire suite of behavioral, morphological and life history characteristics associated with resource utilization in this species.

The main value of this study is that it suggests that persistent directional selection may cause rapid evolution in important behavioral traits in 100 generations or less. This finding is thus relevant to both invertebrate and vertebrate taxa. It is a crucial result because endangered populations must be managed as they *are*, rather than as they *were* before their habitats were altered. Moreover, to the extent that behavioral evolution may alter selection on other traits related to fitness, behavioral adaptations to environmental change may engender the rapid evolution of novel populations poorly suited to their ancestral, undisturbed habitats. In general, declining populations or species are most likely not to be in the habitat originally most favorable to them, but rather in the habitats least

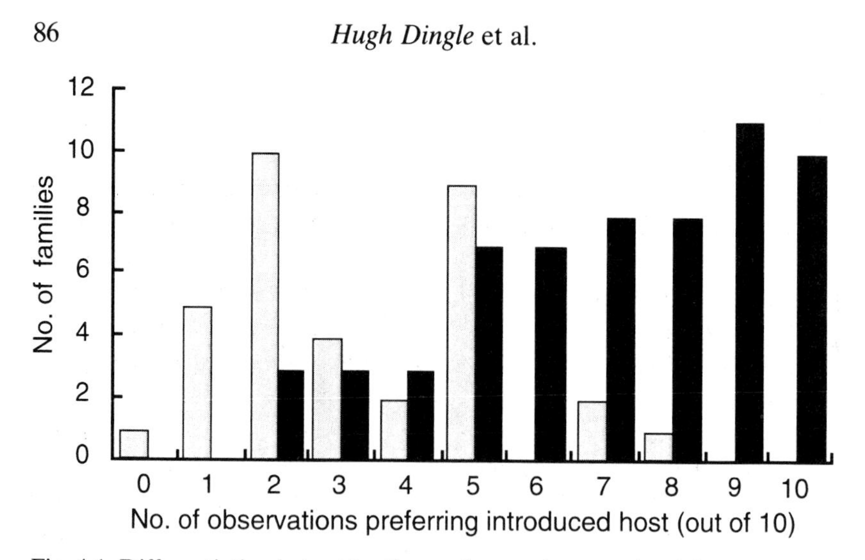

Fig. 4.1. Differentiation in host feeding preference between hatchling soapberry bugs from a population associated with a native host (ancestral-type population, light bars) and hatchlings from a population associated with a recently introduced host (derived population, dark bars).

favorable to their agents of decline (Caughley 1994). In these latter habitats they will probably have undergone evolution. At the very least, this possibility suggests that management plans should rely on data on behavior in contemporary, rather than historical, populations.

Conclusions: Toward cautious generalization

The examples we have presented indicate some of the ways in which invertebrate behavior is central to questions and concerns of great conservation interest. The behavior of invertebrate taxa is fundamental to the functioning of ecosystems, the reproductive success of endangered plants, and the conservation of endangered vertebrates. Understanding invertebrate behavior is basic to questions in both applied and theoretical biology, including the design of reserves (as in the migratory monarch butterfly), the detection of species and the process of speciation (Hawaiian *Laupala* cricket song), and evolution in altered habitats (soapberry bugs on introduced host plants). Even in cases where communities are faced with abiotic changes (e.g., changes in temperature and humidity associated with habitat fragmentation, global climate change), factors undermining the viability of local populations may be based as much in altered biotic interactions as in abiotic changes affecting, for example, physiological performance. As our examples demonstrate, such changing biotic interactions, including predation, mutualism, and parasitism, may all be analyzed from the perspective of behavior.

What generalizations may be made from the cases and issues that we have considered? Variation predominates in nature, and so generalization is challenging in conservation biology for the same reasons that it is difficult in ecology and evolution. Moreover, not only must conservation biologists make predictions in the face of natural variation, but they must also institute long-term management plans in the context of ongoing human–environment interactions. Clearly, any long-term management scheme will require first, an appreciation for ecological roles of all major taxa, and second, frequent updates from the field. Regarding the first point, conservation biology may benefit from a re-emphasis on the study of natural history rather than theory (Caughley 1994; but see Arcese *et al.*, Chapter 3). Nowhere is this need more apparent than in the near omission, in practice, of consideration for invertebrate ecology. Regarding the second point, behavioral data are among the most important information to collect, and, as we have shown, behavioral information from invertebrates cannot be neglected.

A large portion of meaningful natural history is behavioral, because behavior may be construed as an organism's set of solutions to the problems (and opportunities) of survival and reproduction. In this way behavioral data directly complement data collected about the demographic status of populations. Behavioral observations may provide readily detectable clues about individual and population health and status (Smith & Logan, Chapter 12; Wingfield *et al.*, Chapter 5). They may also provide scientists with the opportunity to make key observations about the natural history of endangered taxa that would otherwise be missed. Certain behaviors will be especially valuable to conservation biologists, including movement behavior in relation to the distribution of resources, habitat choice behavior, breeding behavior, and behaviors involved in specialized symbioses. The importance of these behaviors will in many cases be similar for both invertebrate and vertebrate taxa. Movement behavior such as ranging and migration is likely to be related to understanding the full suite of resources that individuals require to complete the life cycle, and it is likely to provide information about gene flow and the spatial design requirements of reserves. Data on sensory behavior may lead to a greater understanding of habitat choice, and details of breeding behavior may be critical to an organism's ability to reproduce in altered environments. Likewise, specialized symbioses may be particularly susceptible to disturbance, and serve as barometers of environmental quality.

How may the study of invertebrate behavior contribute in special ways to the development of conservation biology? Currently, 'umbrella species,' those species whose habitat requirements, that when met, may serve to protect entire biotic communities, are regularly defined without any reference to invertebrate animals at all. Yet the conservation of invertebrate species may often do the greatest good for the greatest species aggregate. Invertebrates are often characterized by complex life cycles, in which the life styles and behaviors of the juvenile stages are replaced by entirely different adult natural histories. Moreover, the relatively minimal parental care exhibited by most invertebrates means that juvenile stages are no more buffered from environmental perturbations than are the adults. Few vertebrates could promote the conservation of habitats more diverse than those required by the migratory monarch butterfly populations, by coral reef invertebrates with pelagic larvae, or insects that rely on pristine aquatic habitats for development and intact terrestrial habitats for adult function. The habitat choice strategies of such invertebrates provide powerful models for reserve designers seeking to link multiple ecosystems into conservation networks.

Another important practical attribute of invertebrates is their tractability as experimental subjects. Progress in genetic and demographic theory in conservation biology may be tested with model invertebrate species, the behavior patterns and ecologies of which are often closely analogous to those of the less tractable vertebrate species that hypotheses normally address. Moreover, invertebrate taxa, by virture of brief generation times, permit more direct analyses of behavioral evolution, and invite us to explore concepts of evolutionary planning that go well beyond the notion of simply protecting genetic diversity. In certain circumstances, guided behavioral evolution of threatened taxa may be a useful conservation strategy, and experiments with invertebrates are a logical first step in developing such an approach.

Clearly, then, it is important to include invertebrate behavior in conservation schemes. Our reluctance to rely on generalization necessarily promotes a more labor-intensive approach to conservation science, but we are not recommending the impossible. Some conservation strategies will of necessity always be scientifically crude, but crude studies must nonetheless include invertebrates and their behavior both specifically (focal projects) and in general. Beyond the conservation status of most invertebrates being completely unknown, it would be additionally unwise to assume that the things we do not yet know are not important. Behaviorists, invertebrate biologists and ecologists need to build collaborations with the express purpose of identifying problems and taxa to develop as objects and models of conservation. Within the unstudied invertebrata are many important lessons yet to be discovered, and our relative ignorance of them is a special liability. Akin to prospecting for pharmaceuticals in the world's disappearing flora, the behavioral abilities and limitations of invertebrates need to be examined for the basic and applied information they hold.

Literature cited

Arendt, W. J. 1985. *Philornis* ectoparasitism of pearly-eyed thrashers. II. Effects on adults and reproduction. *Auk* **102**: 281–292.

Borlase, S. C., D. A. Loebel, R. Frankham, R. K. Nurthen, *et al.* 1993. Modeling problems in conservation genetics using captive *Drosophila* populations – consequences of equalization of family sizes. *Conservation Biology* **7**: 122–131.

Brower, L. P., and S. B. Malcolm. 1991. Animal migrations: endangered phenomena. *American Zoologist* **31**: 265–276.

Brusca, R. C., and G. J. Brusca. 1990. *Invertebrates*. Sinauer Associates, Sunderland, Massachusetts.

Burk, T. 1994. Diversity status of terrestrial insects. Pages 326–343 in S. K. Majundar, F. J. Brenner, J. E. Lovich, J. F. Schalles, and E. W. Miller, editors. *Biological diversity: Problems and challenges*. The Pennsylvania Academy of Sciences, Philadelphia.

Byrne, M. M. 1994. Ecology of twig-dwelling ants in a wet lowland tropical forest. *Biotropica* **26**: 61–72.

Carroll, S. P., and C. Boyd. 1992. Host race radiation in the soapberry bug: natural history with the history. *Evolution* **46**: 1052–1069.

Carroll, S. P., and J. E. Loye. 1987. Specialization of *Jadera* species (Hemiptera:Rhopalidae) on seeds of the Sapindaceae and coevolution of defense and attack. *Annals of the Entomological Society of America* **80**: 373–378.

Carroll, S. P., S. Klassen, and H. Dingle. Rapidly evolving adaptations to host ecology and nutrition in the soapberry bug. *Evolutionary Ecology* (in press).

Caughley, G. 1994. Directions in conservation biology. *Journal of Animal Ecology* **63**: 215–244.

Couturier, A., and P. Robert. 1962. Observations sur le comportement du Hanneton commun (*Melolontha melolontha* L.) (Coléoptere Scarabidae). *Revue de Zoologie Agricol et Appliquée* **3**: 99–108.

De la Maza, E. J., and W. H. Calvert. 1993. Investigations of possible monarch butterfly overwintering sites in central and southern Mexico. Pages 295–298 in S. B. Malcolm and M. P. Zalucki, editors. *Biology and conservation of the monarch butterfly*. Los Angeles County Natural History Museum, Science Series, No. 38.

Delannoy, C. A., and A. Cruz. 1991. *Philornis* parasitism of nestlings. Pages 93–103 in J. E. Loye and M. Zuk, editors. *Bird–parasite interactions: Ecology, behavior and evolution*. Oxford University Press, Oxford.

Dingle, H. 1996. *Migration: The biology of life on the move*. Oxford University Press, New York.

Dobson, A.P., and R.M. May. 1986. Disease and conservation. Pages 345–365 in M. E. Soulé, editor. *Conservation biology*. Sinauer Associates, Sunderland, Massachusetts.

Gates, J. E., and J. W. Geysel. 1978. Avian nest dispersion and fledging success in field-forest ecotones. Ecology **59**: 871–883.

Gibbs, H. L., and P. R. Grant. 1987. Oscillating selection on Darwin's finches. *Nature* **327**: 511–513.

Gibo, D. L. 1986. Flight strategies of migratory monarch butterflies (*Danaus plexippus* L.) in southern Ontario. Pages 172–184 in W. Danthanarayana, editor. *Insect flight: Dispersal and migration*. Springer, Berlin.

Gibo, D. L., and M. J. Pallett. 1979. Soaring flight of Monarch butterflies, *Danaus plexippus* L. (Lepidoptera:Danainae) during the late summer migration in southern Ontario. *Canadian Journal of Zoology* **57**: 1393–1401.

Grant, P. R. 1986. *Ecology and evolution of Darwin's finches*. Princeton University Press, Princeton.

Greenberg, R. 1985. A comparison of foliage discrimination learning in a specialist and a generalist species of migrant wood warbler (Aves: Parulidae). *Canadian Journal of Zoology* **63**: 773–776.

Greenberg, R. 1987. Development of dead leaf foraging in a tropical migrant warbler. *Ecology* **68**: 130–141.

Greenberg, R. 1992. Differences in neophobia between naive song and swamp sparrows. *Ethology* **91**: 17–24.

Hedrick, A. V. 1994. The heritability of mate-attractive traits: a case study on field crickets. Pages 228–250 in C. R. B. Boake, editor. *Quantitative genetic studies of behavioral evolution*. University of Chicago Press, Chicago.

Hickman, C. P, Jr., L. S. Roberts, and A. Larson. 1993. *Integrated principles of zoology*. Ninth Edition. Wm. C. Brown Publishers, Dubuque, Iowa, USA.

Hölldobbler, B., and E. O. Wilson. 1990. *The ants*. Belknap Press, Harvard, Cambridge Massachusetts.

Hölldobbler, B., and E. O. Wilson. 1994. *Journey to the ants*. Belknap Press, Harvard, Cambridge, Massachusetts.

Howe, H. F., and L. C. Wesley. 1988. *Ecological relationships of plants and animals*. Oxford University Press, New York.

Huffaker, C. B. 1958. Experimental studies on predation: dispersion factors and predator-prey oscillations. *Hilgardia* **27**: 344–383.

Johnson, L. K., and S. P. Hubbell. 1974. Aggression and competition among stingless bees: field studies. *Ecology* **55**: 120–127.

Kaneshiro, K. Y. 1988. Speciation in the Hawaiian *Drosophila*. *Bioscience* **38**: 258–263.

Kellert, S. R. 1993. Values and perceptions of invertebrates. *Conservation Biology* **7**: 845–855.

Loye, J. E., and S. P. Carroll. 1995. Birds, bugs and blood: avian conservation and parasitism. *Trends in Ecology and Evolution* **10**: 231–235.

Loye, J. E., and M. Zuk, editors. 1991. *Bird–parasite interactions: Ecology, behavior and evolution*. Oxford University Press, Oxford.

Malcolm, S. B. 1993. Conservation of monarch butterfly migration in North America: an endangered phenomenon. Pages 357–361 in S. B. Malcolm and M. P. Zalucki, editors. *Biology and conservation of the monarch butterfly*. Los Angeles County Natural History Museum, Science Series, Number 38.

Malcolm, S. B., and M. P. Zalucki, editors. 1993. *Biology and conservation of the monarch butterfly*. Los Angeles County Natural History Museum, Science Series, Number 38.

Masters, A. R., S. B. Malcolm, and L. P. Brower. 1988. Monarch butterfly (*Danaus plexippus*) thermoregulatory behavior and adaptations to overwintering in Mexico. *Ecology* **69**: 458–467.

New, T. R., R. M. Pyle, J. A. Thomas, C. D. Thomas, and P.C. Hammond. 1995. Butterfly conservation management. *Annual Review of Entomology* **40**: 57–83.

Ogarrio, R. 1993. Conservation actions taken by Monarca, AC, to protect the overwintering status of the monarch butterfly in Mexico. Pages 377–379 in S. B. Malcolm and M. P. Zalucki, editors. *Biology and conservation of the monarch butterfly*. Los Angeles County Natural History Museum, Science Series, Number 38.

Otte, D. 1989. Speciation in Hawaiian crickets. Pages 482–526 in D. Otte and J. A. Endler, editors. *Speciation and its consequences*. Sinauer Associates, Sunderland, Massachusetts.

Powell, A. H., and G. V. N. Powell. 1987. Population dynamics of male Euglossine bees in Amazonian forest fragments. *Biotropica* **19**: 176–179.

Quinn, J. F., and A. Hastings. 1987. Extinction in subdivided habitats. *Conservation Biology* **1**: 198–208.

Ralls, K., and R. Meadows. 1993. Breeding like flies. *Nature* **361**: 689–690.

Rejmankova, E., H. M. Savage, M. Rejmanek, M. Arredondo-Jimenez, and D. R. Roberts. 1991. Multivariate analysis of relationships between habitats, environmental factors and occurrence of Anopheline mosquito larvae *Anopheles albimanus* and *A. pseudopunctipennis* in southern Chiapas, Mexico. *Journal of Applied Ecology* **28**: 827–841.

Rosenheim, J. A., and B. E. Tabashnik. 1991. Influence of generation time on rate of response to selection. *American Naturalist* **137**: 527–541.

Schmidt-Koenig, K. 1985. Migration strategies of monarch butterflies. Pages 786–798 in M. A. Rankin, editor. *Migration: Mechanisms and adaptive significance*. Contributions to Marine Science 27 (Supplement).

Snook, L. C. 1993. Conservation of the monarch butterfly reserves in Mexico: focus on the forest. Pages 363–375 in S. B. Malcolm and M. P. Zalicki, editors. *Biology and conservation of the monarch butterfly*. Los Angeles County Natural History Museum, Science Series, Number 38.

Soulé, M. E. 1986. *Conservation biology*. Sinauer Associates, Sunderland, Massachusetts.

Snyder, N. F. R., J. W. Wiley, and C. B. Kepler. 1987. *The parrots of Luquillo: natural history and conservation of the Puerto Rican parrot*. Western Foundation of Vertebrate Zoology, Los Angeles, California.

Terborgh, J. 1988. The big things that run the world – a sequel to E.O. Wilson. *Conservation Biology* **2**: 402–403.

Thomas, J. A. 1991. Rare species conservation: case studies of European butterflies. Symposium, *British Ecological Society* **31**: 149–197.

Thorp, R. W. 1990. Vernal pool flowers and host specific bees. Pages 109–122 in D. H. Ikeda and R. A. Schlising, editors. *Vernal pool plants – their habitat and ecology*. Studies from the Herbarium Number 8, California State University, Chico.

Thorp, R. W., and J. M. Leong. 1995. Native bee pollinators of vernal pool plants. *Fremontia* **23**: 3–8.

Urquhart, F. A. 1976. Found at last: the monarch's winter home. *National Geographic* **150**: 160–173.

Van Hook, T. 1994. The conservation challenge in agriculture and the role of entomologists. *Florida Entomologist.* **77**: 42–73.

Van Riper, C., III., S. G. van Riper, M. L. Goff, and M. Laird. 1986. Epizootiology and ecological significance of malaria in Hawaiian birds. *Ecological Monographs* **56**: 327–344.

Via, S. 1990. Ecological genetics and host adaptation in herbivorous insects: the experimental study of evolution in natural and agricultural systems. *Annual Review of Ecology and Systematics* **35**: 421–446.

Warner, R. E. 1968. The role of introduced diseases in the extinction of the endemic Hawaiian avifauna. *Condor* **70**: 101–120.

Wells, S. M., R. M. Pyle, and N. M. Collins. 1983. *The IUCN invertebrate red data book*. International Union for the Conservation of Nature and Natural Resources. Gland, Switzerland.

Werner, E. E. 1992. Individual behavior and higher-order species interactions. *American Naturalist* **140 S**: S5-S32.

Wilson, E. O. 1987. The little things that run the world (the importance and conservation of invertebrates). *Conservation Biology* **1**: 344–346.

Wilson, E. O. 1992. *The diversity of life*. Harvard University Press, Cambridge, Massachusetts.

Yahner, R. H. 1988. Changes in wildlife communities near edges. *Conservation Biology* **2**: 333–339.

Part II:

Conservation and the four levels of behavioral study

As we discussed in Chapter 1, behavioral research is divided into four complementary levels of study about the origin and causation of behavior patterns. In this section we have chosen four chapters to demonstrate how conservation problems can fit into the ethological framework with the result of increasing our understanding of factors affecting population viability. The first two chapters of this section provide examples of research areas devoted to the study of proximate questions – the study of factors that contribute to the immediate causation of behavior and its ontogenetic development. Chapter 5 (Wingfield, Hunt, Breuner, Dunlap, Fowler, Freed, and Lepson) shows how techniques developed by endocrinologists to study physiological and behavioral responses to naturally occurring changes in the environment may now find useful application in monitoring populations under real or potential stress to anthropogenically-induced changes. One advantage of stress analyses for conservation is that they can provide early warning signs of populations or ecological areas under stress or most susceptible to stress before conditions become dangerous. Chapter 6 (McLean) considers the potential and necessity of ontogenetic studies for conservation biology. He describes how altered environments may result in inappropriate (i.e., non-adaptive) behavior of wild animals and how developmental principles from psychology and ethology can be integrated to bring about more appropriate behavior of animals in fragmented, translocated, or reintroduced populations.

The last two chapters of this section present fields of study that consider ultimate questions about behavior at the level of populations: perpetuation (including the adaptive significance) of behavior patterns, and phylogeny, or historical factors contributing to the evolution of behavior patterns. In Chapter 7, Stoleson and Beissinger explore the possible reasons why parrots display hatching asynchrony, in which siblings hatch in staggered sequence and the last-hatched young usually do not survive. In some species hatching asynchrony appears to confer an advantage in the form of 'insurance' offspring, whereas in other species hatching asynchrony appears to have evolved in response to physiological constraints. The authors discuss the conservation implications of the different hypotheses explaining hatching asynchrony.

In the last chapter of this section (Chapter 8, Buchholz and Clemmons), we explore the current limits of the traditional systematic methods and approaches used in conservation biology for understanding the scope of biological diversity and its protection. Behavior patterns provide clues to the phylogenetic history of animals but are neglected because of largely outmoded professional biases. Integrating behavior into biodiversity analyses would result in the protection of unique and evolutionarily critical behavior patterns, facilitate the identification of cryptic species, and protect a greater variety of diversity of value in our everyday lives.

Chapter 5

Environmental stress, field endocrinology, and conservation biology

JOHN C. WINGFIELD, KATHLEEN HUNT, CREAGH BREUNER, KENT DUNLAP, GENE S. FOWLER, LEONARD FREED, AND JAAN LEPSON

Virtually all vertebrates use environmental signals such as day length, temperature, and availability of food to time life history events, such as reproduction and moult (e.g., Wingfield & Kenagy 1991). These cues provide predictable information about future events, such as onset of a breeding season. Reproductive maturation is thus initiated several weeks in advance so that breeding can begin as soon as environmental conditions allow. It is also critical, however, that individuals respond to unpredictable events in the environment. Severe storms can disrupt breeding, resulting in abandonment of the nest and young, or precipitate movements away from the home range in search of alternative food resources and shelter. These facultative physiological and behavioral responses to unpredictable environmental events (*modifying factors,* Wingfield 1983) occur in response to not only a wide spectrum of natural events described above, but also to human disturbance, pollution, and introduced parasites. Human-induced changes such as these are 'unpredictable events' for free-living populations of animals and elicit the same response as natural modifying factors – a facultative physiological and behavioral pattern. The relevance to conservation biology here is that unpredictable disruptions of human origin, as well as natural perturbations, will interrupt breeding. Additionally, highly mobile vertebrates such as birds and large mammals, may leave the restricted confines of reserves. These potential events are of major concern for conservation biology in general, and for refuge managers in particular.

Many modifying factors can be classified as stressors. They trigger a cascade of hormone secretions typical of stress in all vertebrates studied to date (e.g., Wingfield 1994). By measuring circulating levels of glucocorticosteroids (one of the major types of hormones involved in the orchestration of physiological and behavioral responses to stress) in free-living individuals, we can determine whether an individual is indeed stressed, or monitor populations in relation to unpredictable and disruptive environmental change. Patterns of these hormone levels circulating in blood can provide reliable information indicative of potential and actual environmental stress, either natural or human-induced. Samples for monitoring these hormones are easy to obtain, do not harm the subject animal, and provide incisive information in a few days. Behavioral observations may require weeks or even longer to provide the same information. Another advantage of using field endocrinology techniques to monitor stress in free-living populations is that the data can provide predictive information, i.e., they allow us to anticipate likely problems *before* the population becomes stressed, and allow us to identify sub-populations that may be more vulnerable to disturbance or prolonged inclement weather than others. This type of information will be particularly useful to refuge managers.

Before going on to discuss how these techniques can be applied in the field, it is useful to reflect on the interface of what has been traditionally regarded as a 'laboratory science,' endocrinology, and conservation biology. Over the past two decades, techniques have been developed by which samples (such as blood, tissue, feces, urine, etc.) can be collected from free-living individuals without debilitating them. These samples can then be transported to the laboratory for hormone analysis. One strength of 'field endocrinology' techniques is that individuals continue to live normally, and so can be followed through their life cycle. Further samples for hormone analysis can then be collected when desirable (see Wingfield & Farner 1976).

Application of endocrinology to conservation biology: A general perspective

The role of endocrinology in the conservation of vertebrates, especially in captive breeding programs, is not new. Nevertheless, the application of endocrine techniques to conservation biology is still very much in its formative stages. It is pertinent here to discuss how endocrinology can contribute to conservation biology in general, and then focus on one aspect of these applications – monitoring stress. There are at least five ways in which hormone physiology can contribute to conservation issues:

1. Hormone therapy can be used in veterinary medicine to treat endocrine disorders and other diseases that may affect reproductive

function (Iatropoulos 1994). This may be particularly important for endangered species, both in captivity and free-living.

2. Hormones can be used in otherwise healthy animals to enhance reproduction (e.g., Wakabayashi *et al.* 1992). More recently, endocrine treatment has been applied to decrease reproductive output in captive populations that have become overcrowded, or even to reduce populations of exotic introduced species and pests that may compete with endangered species (e.g., Kirkpatrick & Turner 1991; Seal 1991).
3. Field endocrinology protocols can be used to identify specific characteristics of the breeding cycle within or among populations. Temporal patterns of hormone secretion in free-living individuals, for example, can identify potential environmental requirements for successful breeding and whether a threatened population is having problems during the nesting season. Additionally, profiles of circulating reproductive hormones in free-living populations can provide a template for monitoring reproductive performance in captive breeding programs of the same or closely related species (e.g., Ishii *et al.* 1994).
4. Endocrine techniques can be used to determine how pollutants may affect free-living populations. Several components of petroleum and related products, for example, have been shown to have estrogenic properties. These 'environmental estrogens' may have potent effects on reproductive maturation and function (e.g., Hose & Guillette 1995).
5. Field endocrinology protocols can be used to identify populations vulnerable to human disturbance, development, or even natural stressors such as prolonged severe weather (drought, El Niño events, etc.). Such protocols are also able to identify populations already under stress and, thus, likely to decline rapidly, as well as monitor animals that have been reintroduced to natural conditions (e.g., Dunlap & Schall 1995; Fowler *et al.* 1995).

All of these ways in which endocrinology can be applied to conservation issues deserve attention, but here we will focus on only one: the application of field endocrinology protocols to assess whether free-living populations may be vulnerable to stress, or are actually exposed to debilitating stress (i.e., number 5 above).

Before presenting examples of how measuring hormonal responses to stress can be applied to conservation biology, we will next discuss the role of glucocorticosteroids in the generalized vertebrate stress response, how appropriate samples can be collected in the field, and how these hormones can be measured. Then we will outline several examples of how these techniques can be used to identify free-living populations under stress and monitor them for vulnerability to potential future stresses.

Glucocorticosteroids and the vertebrate response to stress

In the vast majority of vertebrates studied to date, a broad spectrum of unpredictable and potentially stressful stimuli elicit the following responses: (1) release of catecholamines from the adrenal medulla within seconds, followed by (2) activation of the hypothalamo–adenohypophysial–adrenal cortex axis a few minutes later (culminating in the synthesis and secretion of glucocorticosteroids such as corticosterone from cortical cells of the adrenal gland), and accompanied by (3) secretion of cytokines from cells of the immune system (e.g., Axelrod & Reisine 1984; Munck *et al.* 1984; Greenberg & Wingfield 1987; Sapolsky 1987). Rapid secretion of catecholamines influences the vascular system, particularly cardiac output, in preparation for massive exertion in immediate social and/or physical stresses (e.g., predator avoidance) and promotes mobilization of glucose (Axelrod & Reisine 1984; Sapolsky 1987). Simultaneously, cytokines activate the immune system to reduce susceptibility to infection. Glucocorticosteroids then trigger gluconeogenesis, the formation of new glucose usually at the expense of protein and fat (Chester-Jones *et al.* 1972; Wingfield 1994; Table 5.1). Chronic effects of stress-induced high levels of glucocorticosteroids, usually over many days or even weeks, may reverse some earlier responses and produce several potentially detrimental effects. For example, suppression of the immune system, neuronal cell death, and disruption of second cell messenger systems occurs (Table 5.1). There are many other hormonal responses known, usually characteristic of the type of perturbation. During osmotic stress, osmoregulatory hormones are secreted, whereas in cases of metabolic stress, hormones associated with energy balance are released. Despite this complex mixture of hormone release tailored to the specific type of stress, the rapid increase of catecholamines followed by the elevation of glucocorticosteroid secretion appears common to all hormone responses to stress, regardless of what that stress may be. Catecholamine levels rise in blood extremely rapidly (a few seconds) after the onset of a sudden stress, so the adrenocortical response, as measured by release of glucocorticosteroids, is a more practical marker for stress in general and will be the focus of the discussions below.

Table 5.1 *Effects of corticosterone during a "stress" response (from references cited in text)*

Rapid (i.e., short term)	Chronic (i.e., long term)
Suppress reproductive behavior	Inhibit reproductive system
Suppress territorial behavior	Suppress immune system
Increase gluconeogenesis	Promote severe protein loss
Increase foraging behavior	Disrupt second messenger systems
Promote daytime escape (irruptive) behavior	Neuronal cell death
Promote night restfulness by lowering standard metabolic rate	Suppress growth and metamorphosis

Originally, it was proposed that stress hormones adapt the organism to the stress and restore homeostasis. The 'exhaustion' phase and its attendant deleterious effects (Table 5.1) set in if homeostasis is not restored and the stress becomes chronic (Selye 1971). More recently Axelrod & Reisine (1984) stated that "stress hormones serve to adapt the body to stressors ranging from mildly psychological to intensely physical." Sapolsky (1987) pointed out that stress hormones orchestrate an emergency reaction that suppresses unnecessary processes and promotes survival until the stress passes. Reproduction and territorial behavior, for example, are temporarily suspended following the onset of stress (Wingfield 1984, 1988; Table 5.1), but energy mobilization is elevated. When the stress has subsided, glucocorticosteroid levels decline, then reproduction and territorial behavior can resume (Wingfield 1984, 1988). The adrenocortical response to stress is important in the short term (hours to a few days) but can become highly deleterious if it is repeatedly activated (see Table 5.1).

A rather different view has been suggested by Munck *et al.* (1984). They state that the function of stress hormones may be to protect against the normal defense reactions activated by stress (i.e., the immune system). These defenses may be inappropriate reactions to stressors and could be a source of stress itself. Glucocorticosteroids thus subdue immune responses activated by the onset of stress that may themselves threaten homeostasis. This view is not necessarily mutually exclusive from the others cited above, but we feel that the hypotheses of Sapolsky (1987) are most relevant to an organism in its natural environment.

In an ecological context it is difficult to see any adaptive value of the long-term (chronic) effects of elevated glucocorticosteroids listed in

Table 5.1. In its natural habitat an organism would be at a considerable disadvantage in any one of these chronic stressed states (e.g., massive protein loss from skeletal muscle, or prolonged suppression of the immune system, Table 5.1). The short-term actions of glucocorticosteroids appear more relevant to an organism in the field. The stress response is an emergency reaction that promotes survival while temporarily suspending other activities (Sapolsky 1987; Wingfield 1994). In other words, the stress-induced increase in glucocorticosteroids triggers a facultative behavioral pattern with attendant physiological changes that allow an individual to cope with the unpredictable and potentially stressful event. In this way the individual can avoid the effects of chronic high levels of glucocorticosteroids outlined in Table 5.1. On occasion chronic stressors do occur under natural conditions; indeed the literature is full of accounts of natural stressors such as extreme weather resulting in starvation and mortality (e.g., for birds, Gessamen & Worthen 1982). Facultative behavioral and physiological patterns, however, do appear effective in re-directing behavior and physiology toward survival. One obvious benefit is increased life-time fitness.

We prefer to use the term 'modifying factors' to refer to unpredictable events in the environment rather than 'stressors' because only under severe or chronic conditions are they truly stressful. In many cases the facultative behavioral and physiological patterns triggered by initial rises in glucocorticosteroids result in a physical state that is not stressed at all. Only if the modifying factor becomes chronic, thereby constantly triggering increases in glucocorticosteroid secretion, does the individual finally progress to a stressed and debilitated state.

Short-term effects of glucocorticosteroids and facultative behavioral and physiological patterns in birds

The chronic effects of prolonged high circulating levels of glucocorticosteroids are of great importance to clinical, agricultural, and aquacultural research. The short-term effects have been less well studied, possibly because their importance as a stress-avoiding mechanism has not been fully appreciated. Here we review evidence for the effects of glucocorticosteroids in the short term. The emphasis will be on birds, although there is no reason *per se* why these effects should not be found in other vertebrates as well.

Glucocorticosteroids and reproductive/territorial behavior

Elevated circulating levels of corticosterone suppress reproductive and associated behavior in birds. Subcutaneous implants of corticosterone into breeding male and female Pied flycatchers (*Ficedula hypoleuca*), in the field decrease the frequency with which they feed young resulting in lower reproductive success (Silverin 1986). Additional implants that increase circulating corticosterone even further result in all birds abandoning their nests and territories. Similarly in the Song sparrow (*Melospiza melodia)* experimental treatment with corticosterone reduces the responsiveness of free-living males to simulated territorial intrusion within 12–18 hours indicating that territorial behavior has been suppressed (Wingfield & Silverin 1986). Note that although plasma levels of testosterone are decreased in males given corticosterone implants compared with controls, they are still within the normal limits that would support territorial aggression. Plasma levels of luteinizing hormone (LH) are unaffected. Additional evidence comes from field endocrine studies of territorial Side-blotched lizards (*Uta stansburiana*). Male lizards given implants of corticosterone are less territorial than controls. Furthermore, simultaneous implants of testosterone fail to restore territorial status in corticosterone treated animals (DeNardo & Sinervo 1994). Clearly, high plasma levels of corticosterone, similar to those induced by modifying factors, suppress territorial behavior and, perhaps, reproductive behavior as well, even when testosterone levels are within the range that would normally activate these behaviors. This response may be adaptive in the long term if survival is favored by temporarily re-directing behavior away from reproduction and territoriality. Meanwhile, the reproductive system is maintained in a near functional state so that breeding can resume immediately after the perturbation has passed and corticosterone levels have subsided (Wingfield 1984, 1988). If, as indicated in Table 5.1, the modifying factor became chronic, i.e., stressful, then continued high levels of corticosterone would result in suppression of the reproductive system that could not be reversed easily within a single breeding season. In this case long-term (chronic) effects would essentially result in complete reproductive failure for that season, whereas the short-term effects and activation of facultative behavioral and physiological patterns would enable another attempt at breeding (Wingfield 1988).

Glucocorticosteroids and mobilization of glucose

Gluconeogenesis, especially from the breakdown of protein in muscle, is a major action of elevated circulating corticosterone levels (e.g., Holmes & Phillips 1976). For example, implants of corticosterone in the House sparrow (*Passer domesticus*) significantly decreases the mass of the large pectoralis muscles (a potential source of endogenous protein) indicative of possible mobilization for gluconeogenesis (Honey 1990). Further analysis reveals that the loss of protein appears to be primarily from soluble components of the pectoralis muscles as the contractile component is not affected. Whether this suggests that the massive pectoralis muscles of birds have a protein store independent of contractile components is unclear (Honey 1990). Corticosterone implants increase fat score dramatically in male Song sparrows without a change in body mass, suggesting a 'shunt' of energy derived from protein into a more readily accessible form – fat (Wingfield & Silverin 1986). Treatment of Dark-eyed juncos (*Junco hyemalis*) with corticosterone in winter also results in wasting of the pectoralis muscles and increases the deposition of fat (Gray *et al.* 1990). Note, however, that muscle lipoprotein lipase activity was not affected, suggesting that the ability of muscle to take up free fatty acids and glycerol is not impaired. It is possible that muscle protein is mobilized in response to corticosterone treatment but without increases in energy expenditure that would be experienced during an actual response to a modifying factor. The 'new energy' is then re-deposited as fat. Loss of body mass owing to protein mobilization is thus compensated for by increases of fat depot.

Glucocorticosteroids and foraging behavior

Glucocorticosteroids may be involved in the regulation of food intake (e.g., Leibowitz *et al.* 1984) and could be an important component of the facultative behavioral and physiological patterns triggered by modifying factors. In male White-crowned sparrows (*Zonotrichia leucophrys*), metyrapone, a blocker of corticosterone synthesis in adrenocortical cells, reduces foraging behavior: a combination of searching, scratching, pecking and food intake. Replacement therapy with corticosterone implants increases foraging (Wingfield *et al.* 1990). Later studies by Astheimer *et al.* (1992) show a trend for increases in foraging, but this is not significant. In the song sparrow, experimental manipulation of corticosterone levels has a greater effect on foraging but has no influence on food searching behavior when food is removed for 24 hours to simulate the effects of a passing storm. When food is returned, however, corticosterone treatment significantly increases foraging behavior. These data raise the interesting possibility that elevated circulating levels of corticosterone induced by modifying factors may not only trigger facultative behavioral and physiological patterns, but also facilitate recovery when environmental conditions ameliorate (Astheimer *et al.* 1992). This point deserves more investigation. Studies of other species, such as Domestic fowl (*Gallus domesticus*) support a role for corticosterone in the control of food intake, but effects may vary with the season (Astheimer *et al.* 1992). Clearly, additional experiments are needed to clarify the role of hormones on foraging behavior.

Glucocorticosteroids and activity

In birds, and perhaps other vertebrates, modifying factors such as extreme weather result in the abandonment of a breeding territory or winter home range. These movements are called 'irruptive migrations,' and they serve to remove the bird from the source of the modifying factor to refugia that provide shelter or adequate food (Elkins 1983; Wingfield 1984, 1988). This movement may be local or over hundreds of kilometers. In other cases individuals may become inactive and/or torpid, thus reducing energy expenditure, in an attempt to 'ride-out' a period of inclement weather (Elkins 1983). Combinations of these two cases are also possible in which a bird exposed to extreme weather first becomes inactive, but if inclement conditions continue, then leaves. Note that if an individual attempts to leave, it should do so while it still has sufficient reserves of fat to fuel the movement.

Corticosterone implants significantly decrease perch-hopping activity in male White-crowned sparrows (a measure of possible irruptive migratory activity) consistent with a strategy of reduced activity and 'riding out' the storm. Alternatively, if food is removed from the cages to simulate the effects of severe weather on food resources (Wingfield 1988), corticosterone-treated birds show greatly enhanced perch-hopping, including escape-like behavior in which individuals are attempting to get out of the cage (Astheimer *et al.* 1992). These data suggest that elevated circulating levels of corticosterone, similar to those induced by direct modifying factors, activate behavioral strategies consistent with the observations of free-living birds exposed to extreme weather. Additional factors obviously must be involved in the transition from inactivity to greatly enhanced perch-hopping and escape behavior. Interestingly, increased irruptive migration occurs during the day, whereas normal migration occurs mostly at night in White-crowned sparrows (e.g., Wingfield *et al.* 1990). Thus, irruptive migration may have different ecological and hormonal bases from regular seasonal migration.

Glucocorticosteroids and metabolic rate

Evidence to date indicates that rises in plasma levels of corticosterone may increase perch-hopping activity, at least when combined with reduced availability of food. The question then arises as to whether the metabolic rate is also affected. Implants of corticosterone in White-crowned sparrows, Pine siskins (*Carduelis pinus*) and other species have no effect on basal metabolic rate compared with birds sampled before implants were given. All birds were sampled at night when activity levels were very low. Control birds show periodic surges in oxygen consumption perhaps as a result of sleep–wake rhythms. In contrast, corticosterone-treated birds show virtually no surges in oxygen consumption during the night (Buttemer *et al.* 1991). Corticosterone treatment appears to suppress extended metabolic rate when asleep and promote 'nocturnal restfulness,' with a projected 20% savings of energy over the 16 hour night. This conserved energy can then be utilized the following day to recommence irruptive migration or continue to ride out the storm.

Facultative behavioral and physiological patterns and the 'emergency state'

The experimental evidence cited above is consistent with the role of elevated corticosterone secretion, induced by direct modifying factors, in the activation of facultative behavioral and physiological patterns. High plasma levels of corticosterone trigger an emergency life history state designed to maximize fitness during *and* after temporary disruption of activities normally expressed at that stage in an individual's life cycle. There are three major components to this emergency reaction in birds (modified from Wingfield 1994):

1. Corticosterone may suppress 'unnecessary' physiological and behavioral functions such as reproductive behavior and territoriality. This suppression may be a direct action rather than through the inhibition of reproductive hormones.

2. Corticosterone may activate temporary, facultative behavioral and physiological patterns that promote survival. For example, foraging behavior may be elevated, and activity either decreased ('ride it out' strategy), or increased (irruptive migration) depending upon severity and duration of a direct modifying factor such as extreme weather. At night corticosterone promotes restfulness with up to 20% savings in energy.
3. Corticosterone promotes gluconeogenesis, including the mobilization of protein as a substrate. It is possible that protein is stored in flight muscles, as the initial loss of pectoralis muscle mass appears to be from soluble fractions rather than contractile elements.

Facultative behavioral and physiological patterns triggered by short-term elevations of circulating corticosterone serve to promote survival in the best physical condition and avoid the deleterious long-term effects characteristic of a stressed state. Note that there is no reason why these components of short-term behavioral and physiological responses to modifying factors should not be applicable to other vertebrates.

How to measure adrenocortical responsiveness to modifying factors

Clearly, changes in circulating levels of glucocorticosteroids indicate responsiveness to modifying factors, some of which could potentially be stressful. Investigators interested in conservation biology, however, may not be familiar with field and laboratory techniques that have been designed specifically to monitor the adrenocortical responses of individuals under free-living conditions. In this section we outline some techniques that have proved practical, relatively inexpensive, and highly reproducible. It should be borne in mind that these techniques require a period of learning and adjustment, but we feel that the results obtained provide unique and highly useful information that could be vitally important for monitoring free-living populations of threatened and endangered species. These techniques have the potential to provide predictive information concerning (1) whether the population under investigation is stable and healthy, (2) more vulnerable to modifying factors, or (3) currently in trouble and likely to be chronically stressed. Such information could be of considerable importance for management decisions.

The stress series protocol

Responses to modifying information can be assessed by measuring changes in plasma levels of glucocorticosteroids secreted by the adrenal gland (Wingfield 1994). This is also known as the 'adrenocortical response to stress' and is common to virtually all gnathostome vertebrate classes. A simple comparison of baseline plasma levels of glucocorticosteroids in free-living versus captive populations could indicate quickly whether non-breeding captive birds had higher plasma levels of corticosterone than free-living birds sampled at the same time. If so, then captive conditions may be 'stressful.' One well-known effect of prolonged (chronic) high levels of glucocorticosteroids is complete inhibition of reproductive function. Certainly, if captive populations are stressed, then reproduction will be compromised – information that is vital to captive breeding programs. Recent work suggests that fecal and urinary levels of steroids also can be used to indicate responsiveness to modifying factors, especially in cases when reproduction is impaired (e.g., Creel *et al.* 1992, 1996; Wasser 1995). Fecal and urine samples usually can be collected without disturbing the subjects, so the technique may prove especially useful for monitoring critically endangered species.

The hormonal response to modifying factors is common to a wide spectrum of stimuli, including capture and handling. Capture and restraint results in a rapid increase of glucocorticosteroids, usually within 5–10 minutes, and reaches a maximum within 30–60 minutes (Wingfield 1994). Serial bleeding adds to the cumulative effects of handling stress. As an example, a very small (20–30 μl) sample of blood from the wing vein of a bird may be collected as soon as possible after capture (e.g., Wingfield & Farner 1976). The bird is then held in a cloth bag for 60 minutes and further samples collected at 5, 10, 30, and 60 minutes after capture for measurement of changing glucocorticosteroid levels. Under field conditions, blood samples can be stored on ice until return to the laboratory where it must be centrifuged, at about 2000 rpm for 5 minutes, and the plasma then harvested (Wingfield & Farner 1976). Plasma can be stored at −20 °C for several months until assayed for glucocorticosteroid content. If field sites are remote, then blood can be centrifuged in the field using an automobile battery as a power source, or a hand operated centrifuge if power is not available. Plasma can be stored on dry ice in a cooler (2–3 kg will last about three days in a regular cooler) or in a liquid nitrogen container (some are designed for field use and can keep plasma samples frozen for several weeks, depending upon size of the container).

Examples of the pattern of glucocorticosteroid secretion during the capture stress protocol are given in Fig. 5.1. Note that plotted baseline levels, such as at the time of capture, versus time of day indicate whether natural changes in glucocorticosteroid secretion could confound our results. This is as close to a true control as we can obtain for this procedure in the field. Extensive data thus far show that diurnal changes in baseline levels of glucocorticosteroids are not responsible for the marked elevations following capture and restraint (Wingfield 1994; Wingfield *et al.* 1994*a*,*b*). The collection of capture stress series in free-living birds and captive populations can, therefore, tell us a great deal about sensitivity to modifying factors and current conditions. Changes in circulating levels of glucocorticosteroids are sensitive to human disturbance as well as other types of disturbances, such as natural environmental conditions, disease, or exposure to pollution. Data in the field indicate that sensitivity to the capture stress protocol may change with season and reproductive state as well as with the body condition of individuals (e.g., Wingfield *et al.* 1994*a*,*b*).

In general we feel that a rapid and steep increase in plasma levels of glucocorticosteroids during the capture stress protocol indicates greater sensitivity of the adrenocortical axis to stress than does a shallow and slow increase. We have found, for example, that although baseline (i.e., within 1 minute of capture) glucocorticosteroid levels may be similar in captive and free-living birds, captives tend to have a greatly enhanced elevation of glucocorticosteroids following restraint. The collection of stress series from males and females in breeding and non-breeding seasons would also be an essential start in determining sensitivity to modifying factors at two important stages in the life cycle.

Further analysis in relation to body mass, condition, fat score, and other variables reveals additional information at the individual level (Wingfield 1994). Within a sampled cohort of individuals, there may be little or great variation in the pattern of glucocorticosteroid levels during the capture stress protocol (Fig. 5.2). Such data are accumulated over time and eventually provide critical evaluation at the individual level, especially as lower sensitivity to modifying factors (i.e., a flat pattern of glucocorticosteroid release following capture and restraint) is often correlated with good body condition, whereas a steep pattern of glucocorticosteroid release (Fig. 5.2) may be related to poor body condition (e.g., Smith *et al.* 1994; Wingfield *et al.* 1994*a*,*b*; Dunlap & Wingfield 1995).

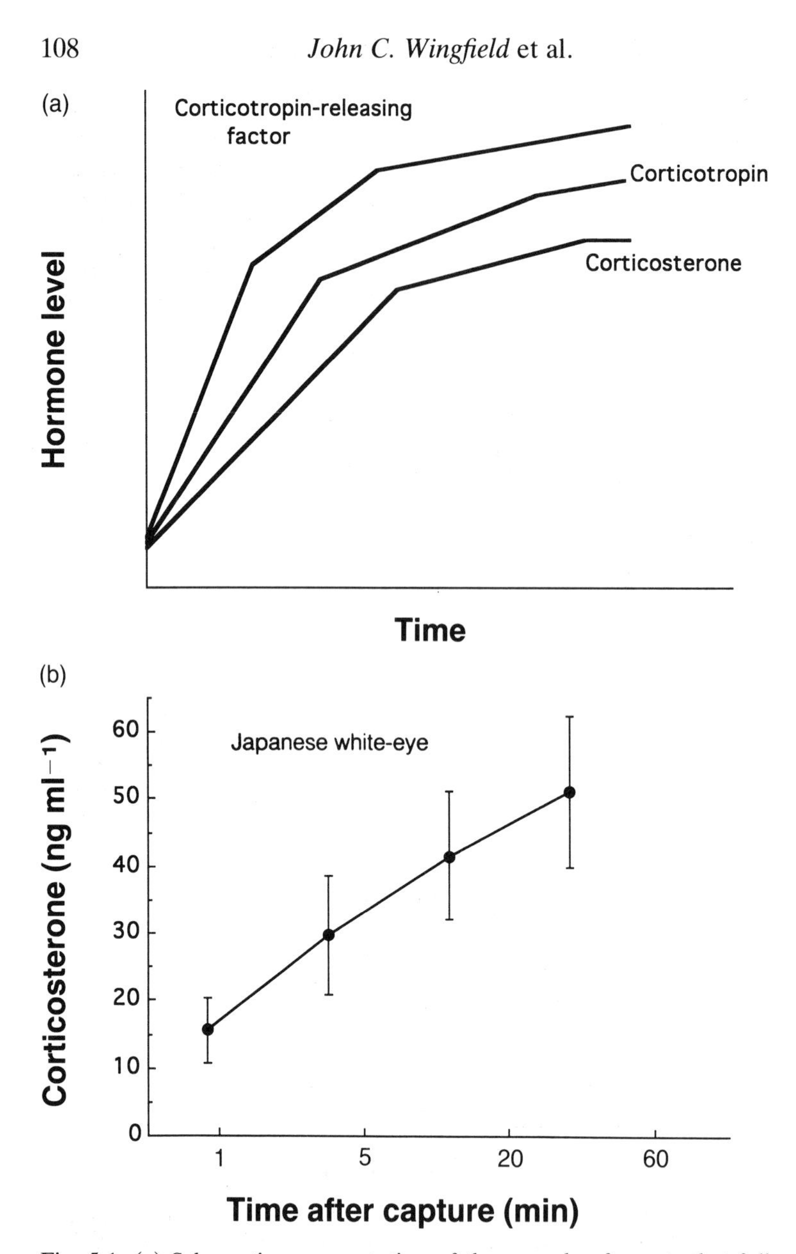

Fig. 5.1. (a) Schematic representation of the cascade of events that follows an increase in corticotropin-releasing factor secretion triggered by a modifying factor such as habitat modification. Corticotropin levels rise a few seconds later followed by release of corticosterone from the adrenal gland (after Wingfield 1994). (b) The profile of corticosterone levels in the plasma of the Japanese white-eye (*Zosterops japonica*) during capture, handling, and restraint for 1 hour. Points are the means and vertical bars standard errors. Note how the level of corticosterone increases as suggested in (a).

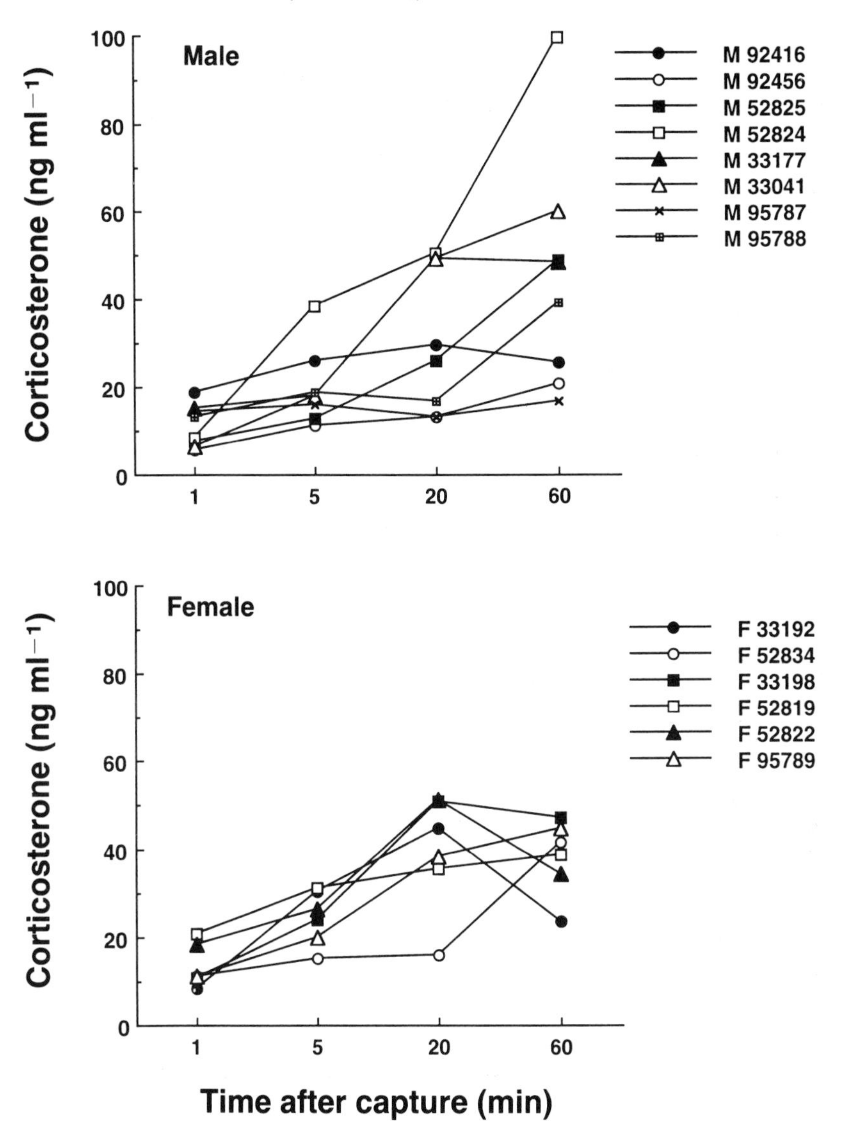

Fig. 5.2. Individual variation in the corticosterone profiles during capture, handling, and restraint in male and female Common Amakihi (*Hemignathus virens*). Numbers in the figure key represent individual band numbers.

Measurement of glucocorticosteroid levels in plasma

An important initial step is to determine which glucocorticosteroid is found in the species of interest. Cortisol is the major glucocorticosteroid of humans and some other mammals and all teleost fish, whereas corticosterone is found in other mammals and all non-mammalian tetrapods (e.g., Gorbman *et al.* 1983). Cortisol and corticosterone differ by the addition of a single hydroxyl group, but their biological actions are identical. The antisera generated against corticosterone, however, usually do not cross-react with cortisol (and vice versa). Thus it is critical to determine which hormone is the major glucocorticosteroid in your subject species. Consulting Gorbman *et al.* (1983) as a standard textbook is one way to do this. More detailed information can be found in Idler (1972).

The two most sensitive methods to measure circulating levels of steroid hormones are by radioimmunoassay (RIA) and enzyme-linked immunosorbant assay (ELISA). Both can use small samples (down to 5 μl in volume), but in our experience RIA is still a preferable method in terms of inter-assay variability. Certain items of equipment required, however, are expensive, and the assay itself requires some training. It is strongly advised that an investigator unfamiliar with the technique makes contact with an endocrinology laboratory and inquires about possible collaborations. Most larger research universities will have at least one laboratory of this type. Others may have to inquire further afield to make appropriate contact. The effort is well worth the initial outlay because the technique provides large quantities of information once it is established. Detailed descriptions of the RIA procedure for steroid hormones, and particularly glucocorticosteroids, can be found in Wingfield & Farner (1975) and Wingfield *et al.* (1994*a*,*b*).

Examples in free-living populations

Thus far, we have described the changes of plasma glucocorticosteroid levels observed during stress, the effects of these steroids on facultative behavioral and physiologic patterns, and the methods by which investigators can collect appropriate samples and measure hormone levels. We will next give several examples of how the patterns of glucocorticosteroid secretions following stress can be applied to monitor free-living populations. Where data exist, we point out their direct relevance to conservation biology.

Adrenocortical responses to stress in Hawaiian honeycreepers

From January to March 1993 we collected blood samples and data on reproductive function from endemic forest birds of the Island of Hawaii (Hawaiian honeycreepers, Drepanidinae, Fringillidae). This subfamily has many critically endangered species, and we hope these studies will provide useful baseline information on the regulation of their breeding cycles. The following taxa are the focus of investigations: Apapane (*Himatione sanguinea*); Common Amakihi (*Hemignathus virens*); and Iiwi (*Vestiaria coccinea*), in the Hakalau Forest National Wildlife Refuge, Hawaii. Our questions are: do these insular island birds have an adrenocortical response to stress that is similar to related continental species (i.e., testing the generality of the capture stress protocol), and is it indeed possible to collect these types of samples without seriously debilitating the free-living birds?

We use the stress series protocol as a measure of responsiveness to stress in general in pre-breeding populations of Drepanidinids. In the Common Amakihi there is a significant rise in corticosterone following capture and handling (Fig. 5.3a, $F = 16.161$, $p < 0.0001$, two-way ANOVA for repeated measures). There is no difference in the response between males and females ($F = 3.514$, $p = 0.0936$). Similarly in Iiwi, the increase of corticosterone levels following capture and handling (Fig. 5.3b, $F = 11.286$, $p < 0.0001$) is similar in males and females ($F = 2.477$, $p = 0.1466$) but rather less dramatic than in the other species. Very few female Apapanes were captured, and so both sexes are combined in Fig. 5.3c). Again, there is a highly significant increase in corticosterone following capture and handling ($F = 10.503$, $p < 0.0001$).

Relationships of the adrenocortical response to stress with body condition

Although capture and other stresses can result in marked increases in adrenocortical hormones such as corticosterone, there is also considerable variation in the degree to which individuals respond to the same stress. Hawaiian honeycreepers appear to show similar variation. In Common Amakihis (Fig. 5.2), both males and females exhibit large variations in not only the rate of increase in corticosterone following capture, but also the maximum corticosterone level generated by 60 minutes of capture and handling stress. Note that some (e.g., bird numbers M 92416 and M 92456, Fig. 5.2) have very little or no adrenocortical response to stress, whereas others (e.g., M 52824 and F 33198)

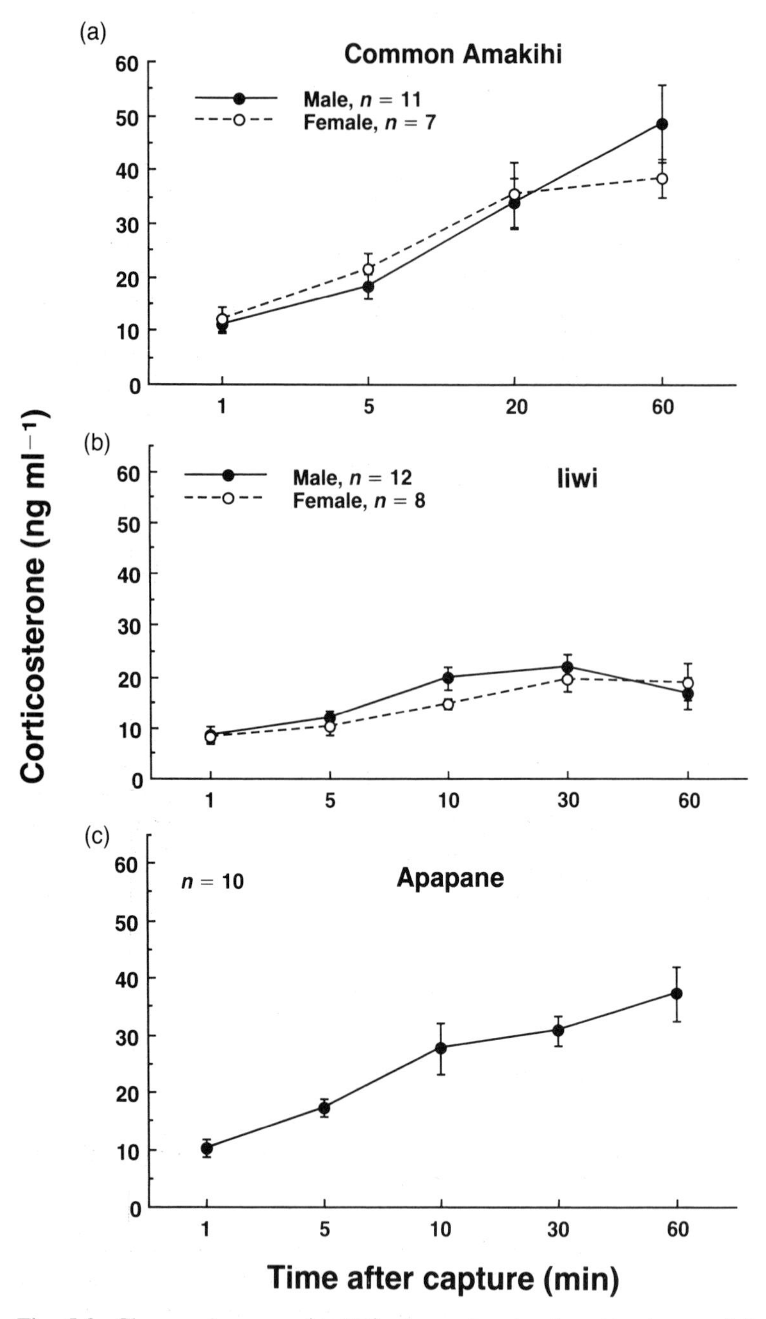

Fig. 5.3. Changes in mean (± SE) plasma levels of corticosterone following capture, handling, and restraint in free-living Drepanidinids at Hakalau National Wildlife Refuge, Hawaii.

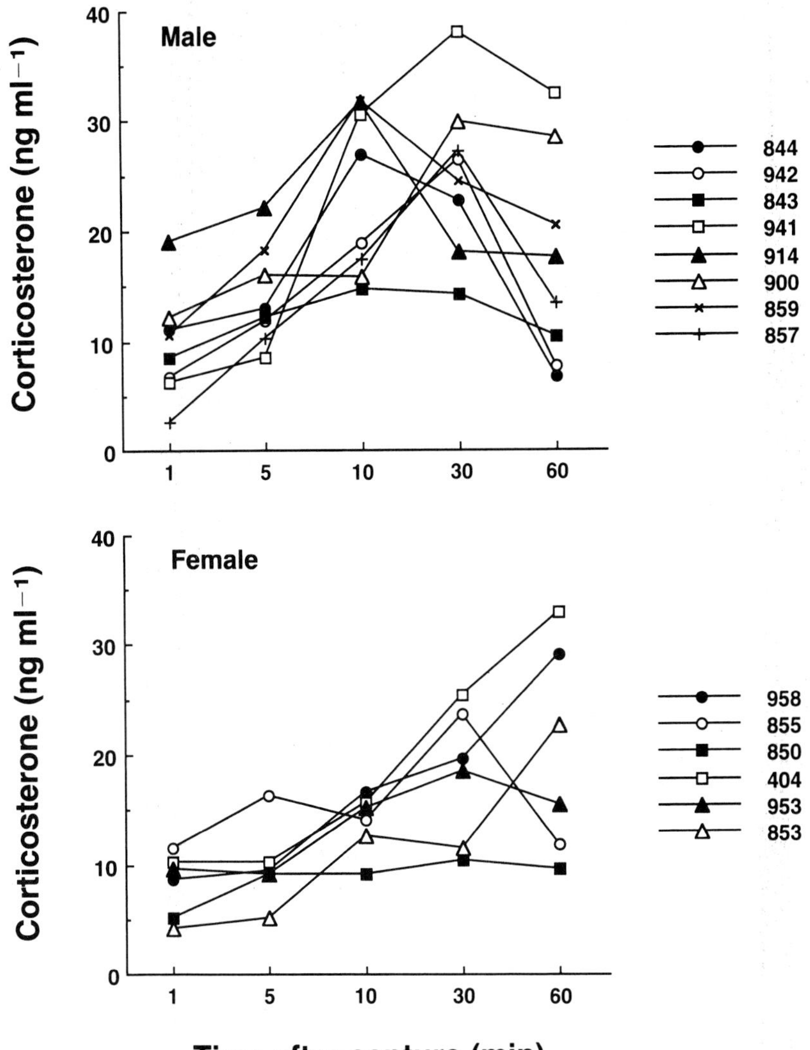

Fig. 5.4. Individual variation in the corticosterone profiles during capture, handling, and restraint in male and female Iiwi (*Vestiaria coccinea*). Numbers in the figure key represent individual band numbers.

undergo a 5- to 20-fold increase in circulating corticosterone levels following similar treatment. Similarly, in the Iiwi (Fig. 5.4), both sexes show wide variation in rate of corticosterone increase and the maximum corticosterone level generated following stress (e.g., compare numbers 843 and 850 with 941 and 404).

Wingfield (1994) suggests that this variation may represent different degrees of resistance to stress possibly correlated with factors such as body condition, sex, and reproductive state. Accordingly, we have correlated different aspects of the adrenocortical response to stress during capture and handling with body condition (i.e., body mass and fat score) in Hawaiian honeycreepers. In these species, the adrenocortical response to stress can be separated into distinct components, including initial level of corticosterone (i.e., at capture), rate of corticosterone increase over the 60 minute sampling period, percent increase in corticosterone, and maximum level generated during the stress series. These data can then be correlated with sex, body mass, fat score and reproductive state (see Wingfield 1994 for review of other species). As all birds in this study were pre-breeding, we cannot assess the effects of breeding status on responsiveness to stress, but we can investigate the effects of body condition.

The body masses of male and female Common Amakihi are similar (Fig. 5.5, $t = 1.233$, $p = 0.2364$, unpaired, DF = 15), whereas females have a higher fat score ($t = 2.483$, $p = 0.0253$, unpaired, DF = 15). In the Iiwi, males are heavier than females (Fig. 5.5, $t = 4.638$, $p = 0.0002$, unpaired, DF = 18), but females have a higher fat score ($t = 3.205$, $p = 0.0049$, unpaired, DF = 18). A comparison of fat score across species indicates that there is significant variation ($F = 16.896$, $p = 0.0001$, total DF = 46). Female Iiwi appear to have higher fat scores than all other species and sexes ($p < 0.05$, Fisher's PLSD test). Male Common Amakihi have significantly less fat than all other species and sexes ($p < 0.05$, Fisher's PLSD test). Correlations of body condition and stress responses are considered within species below.

Common Amakihi

There is a significant correlation of body mass and fat score in males, but this is not significant in females. When data for both sexes are combined, however, a significant relationship reappears (Table 5.2). The only correlations of body mass or fat score with components of the adrenocortical response to stress that are significant are the negative relationships of body mass to initial corticosterone level in males and females combined (Table 5.2). Birds with the least body mass appear to have the highest baseline levels of corticosterone, but the subsequent response to stress is not affected. Variation in the adrenocortical response to stress in this species appears to be independent of body condition.

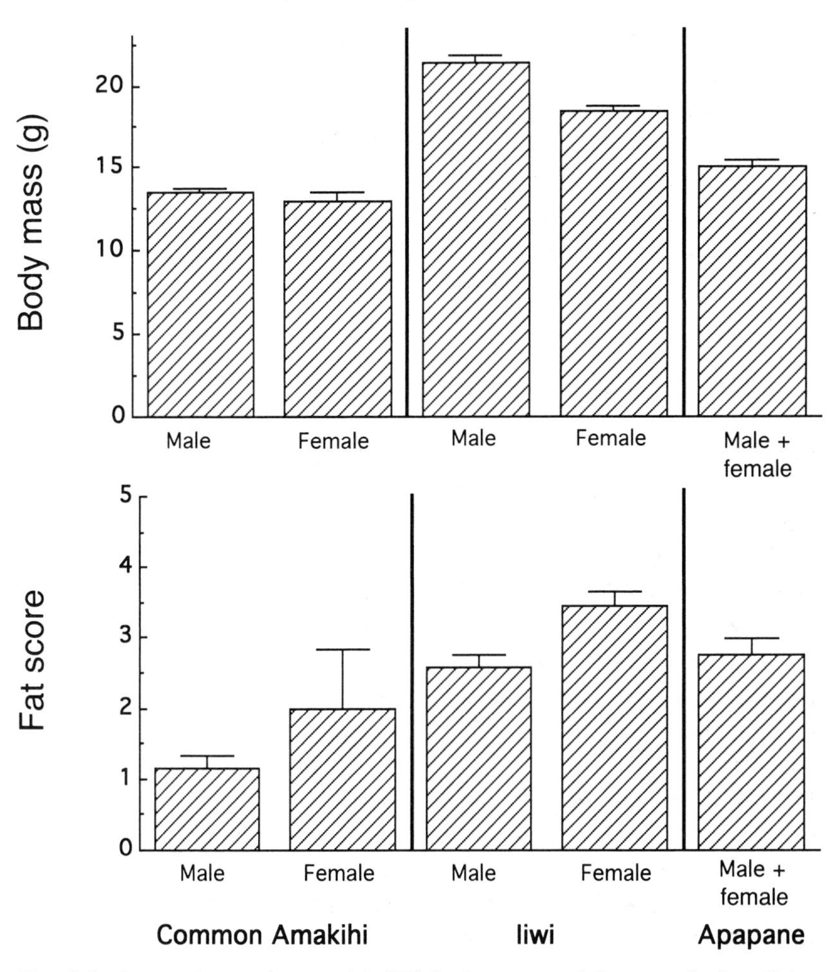

Fig. 5.5. Comparisons of mean (± SE) body mass and fat score in free-living Drepanidinids at Hakalau National Wildlife Refuge, Hawaii. Fat score is based on an arbitrary scale in which 0 = no visible fat beneath the skin of the furculum (pectoral girdle or 'wishbone') and abdomen, and 5 = gross bulging fat bodies in the same area (see Wingfield & Farner 1978 for details).

Iiwi

In the Iiwi, there also is a positive correlation of body mass and fat score, but only in females (Table 5.2). In males there is a negative relationship of the rate of corticosterone increase following capture stress with body mass. Females show no tendency to modulate the adrenocortical response to stress in relation to body condition. If data from both males and females are combined, then a negative relationship of fat score and rate of

Table 5.2 *Relationships of the adrenocortical response to stress and body condition in Hawaiian Drepanidinids*

Species	Correlation	R^a	P value	N
Common Amakihi				
Male	Body mass vs. fat score	0.803	0.001*	11
	Body mass vs. initial B[b] level	−0.655	0.039*	11
Female	Body mass vs. fat score	0.794	0.074	6
Male + female	Body mass vs. fat score	0.591	0.018*	17
	Body mass vs. initial B level	−0.492	0.049*	17
Iiwi				
Male	Body mass vs. fat score	−0.143	0.635	12
	Body mass vs. rate of B increase	−0.664	0.036*	11
Female	Body mass vs. fat score	0.844	0.026*	8
Male + female	Fat score vs. rate of B increase	−0.459	0.058	18
Apapane				
Male + female	Body mass vs. fat score	0.416	0.212	10
	Fat score vs. maximum B level	0.565	0.090	10

[a]Spearman Rank Correlation Coefficient.
*Denotes significance at less than 0.05 level.
[b]Corticosterone. Sample size (*N*) varies because not all birds were laparotomized and/or bled.

corticosterone is suggested but is not significant (Table 5.2). Thus, in this species, body condition may influence the rate of adrenocortical responsiveness to stress, at least in males. Those birds with higher body mass show a slower increase in corticosterone secretion following application of a stressor stimulus.

Apapane

As very few females were subjected to the capture stress protocol, we combined the male and female data for multiple correlations with body condition. In this species there is no correlation of body mass and fat score (Table 5.2), unlike the case in many other mainland species studied (Wingfield 1994). Furthermore, there are no significant correlations among components of the adrenocortical response to stress and body condition, with the possible exception of fat score and maximum corticosterone levels generated during the stress series paradigm (Table 5.2). This relationship is not quite significant. Interestingly, the Apapane is the first species studied in which there is a possible positive relationship between a component of the stress response and body condition. In all other species there are negative relationships (e.g., Wingfield 1994). This point deserves further study in this species.

The adrenocortical response to stress, as applied by capture and handling, in endemic Hawaiian honeycreepers appears to be similar to that seen in continental species and other vertebrates in general (see Wingfield 1988, 1994). In at least the Common Amakihi and Iiwi, there is no apparent sexual difference in the response, unlike many species from the North American mainland (e.g., Wingfield 1994*a*,*b*; Wingfield *et al.* 1995). This may be because virtually none of the birds was in breeding condition, when most sex differences in the response occurs in other species (Wingfield *et al.* 1995). Nevertheless, it is clear that Hawaiian honeycreepers are sensitive to the effects of stress, which could have extensive deleterious effects on reproductive success, resistance to disease, and endogenous reserves of protein and fat for survival during severe storms or other extreme modifying conditions. Chronic high levels of corticosteroids may also trigger facultative behavioral patterns that may force birds to leave the reserve and wander over a large area (Astheimer *et al.* 1992; Wingfield 1994). These possible effects will have important ramifications for management and monitoring populations in reserves.

Correlates of the stress response and body condition can be particularly revealing. In the Drepanidinids studied, sex differences in body mass are independent of fat score (Fig. 5.5). Although there appears to be great individual variation in responsiveness to stress, this variation is only partly correlated with body condition (Table 5.2, unlike in many northern temperate species, see Wingfield 1994). Particularly fascinating is the possibility that Apapane show a positive correlation of body condition and the response to stress. As far as we are aware, this is a unique finding. Whether this is a phenomenon related to nomadism (in which flocks of birds wander looking for favorable food resources) remains to be seen. Note that high levels of corticosterone are known to promote irruptive behavior (see Astheimer *et al.* 1992). On the other hand, we must be cautious because with so many correlations, it is possible that some will be significant at the $p < 0.05$ level by chance. Further studies on variation in the stress response in relation to body condition, degree of infection, and reproductive state will shed further light on this potentially important phenomenon. If it turns out that variables such as body condition are reliable indicators of susceptibility to stress, then another potential tool can be applied to captive breeding programs and the management of free-living populations. It is important in passing to mention that although these techniques are invasive, they do not appear to debilitate free-living birds. In our study of Hawaiian honeycreepers, the percentages of unsampled birds that were recaptured are 8.8% for Common Amakihi, 3.7% for Apapane, and 8.0% for Iiwi. The percentages of sampled birds that we recaptured during the capture stress study are 18.9% for common Amakihi, 2.7% for Apapane, and 13.2% for Iiwi. These numbers are all either above or similar to recapture percentages for non-sampled birds. This study indicates clearly that collection of these kinds of data is entirely possible in endemic Hawaiian honeycreepers. It is critical that these techniques now be applied to threatened and endangered populations promptly. The information to be gleaned will shed light on many potential problems that may not be apparent from simple behavioral observations.

Baboons in an East African savannah

Olive baboons (*Papio anubis*) live in large groups (20–200 individuals) that have complex social organizations. Highest ranking males dominate these social groups. Obtaining high rank is essential if a male is to gain maximum reproductive success. Reaching that high rank, however, and then maintaining it can be extremely demanding – indeed stressful (Sapolsky 1987). There are frequent aggressive interactions to determine and maintain rank. The degree to which agonistic interactions occur varies with season, such as in a stable season when aggression is at a minimum and in an unstable season when many interactions occur, including fights and even injuries (Sapolsky 1987).

Male baboons with varying rank are captured from free-living troops by darting with an anesthetic (Sapolsky 1983). Once anesthetized, a series of blood samples is collected from the animals to assess responsiveness to capture and restraint (i.e., a form of the capture stress protocol described above). During the stable season, high ranking males have lower baseline levels of circulating cortisol (the major glucocorticosteroid in many primates) than low ranking males. In response to anesthesia and restraint, high ranking males show a greater rise in cortisol than low ranking males. During the unstable season, however, these differences disappear (Sapolsky 1983). Baseline plasma levels of testosterone also increase during the unstable season, suggesting that increased aggression increases both sex steroid levels and glucocorticosteroids. Curiously, high ranking males also have an impaired adrenocortical responsiveness to stress during the unstable season. Chronically high baseline levels of cortisol circulating in the blood may decrease sensitivity of the testis to gonadotropins, which regulate testicular function and thus fertility, as well as increase susceptibility to disease by suppressing the immune system (Sapolsky 1987). Either of these effects could result in marked impairment of reproductive function and fertility. These findings were followed by a number of elegant field experiments to determine mechanisms (Sapolsky 1987) that do not concern us here, but this study indicates the utility of monitoring glucocorticosteroids with a capture stress protocol to assess the effects of extreme social pressure, or lack of, on reproductive function and pathology. Loss of habitat, shrinking reserves, or any environmental change that results in concentration of individuals in less and less space, therefore, could have profound detrimental effects on survival and reproduction. Baseline circulating levels of glucocorticosteroids and responsiveness to stress are highly informative ways of monitoring wild populations for the development of potential problems.

Fence lizards in western North America

The Western fence lizard (*Sceloporus occidentalis*) is an abundant lizard in western North America with a vast range extending from the Mojave Desert in Southern California to the beaches of Puget Sound in Washington State. Baseline levels of circulating corticosterone, the major glucocorticosteroid in reptiles, and responsiveness to the capture stress protocol were measured in six populations from around this species' range and during two different seasons. Baseline levels of corticosterone do not vary with population or season. As in several avian species, however, the amplitude of the increase in corticosterone levels during the capture stress protocol is correlated with physiological condition, with length adjusted for body mass. These data indicate that the capture stress protocol is useful for monitoring the health of reptiles in the wild. When physiological condition is factored out of the analysis, strong population and season effects remain. Adrenal responsiveness is higher in populations at the periphery of the species' range and is higher during the late summer than in spring (Dunlap & Wingfield 1995; Fig. 5.6). Adrenocortical responsiveness, thus, is modulated not only by changes in the external environment that affect physiology, but also by intrinsic changes associated with the annual cycle (e.g., reproductive state,

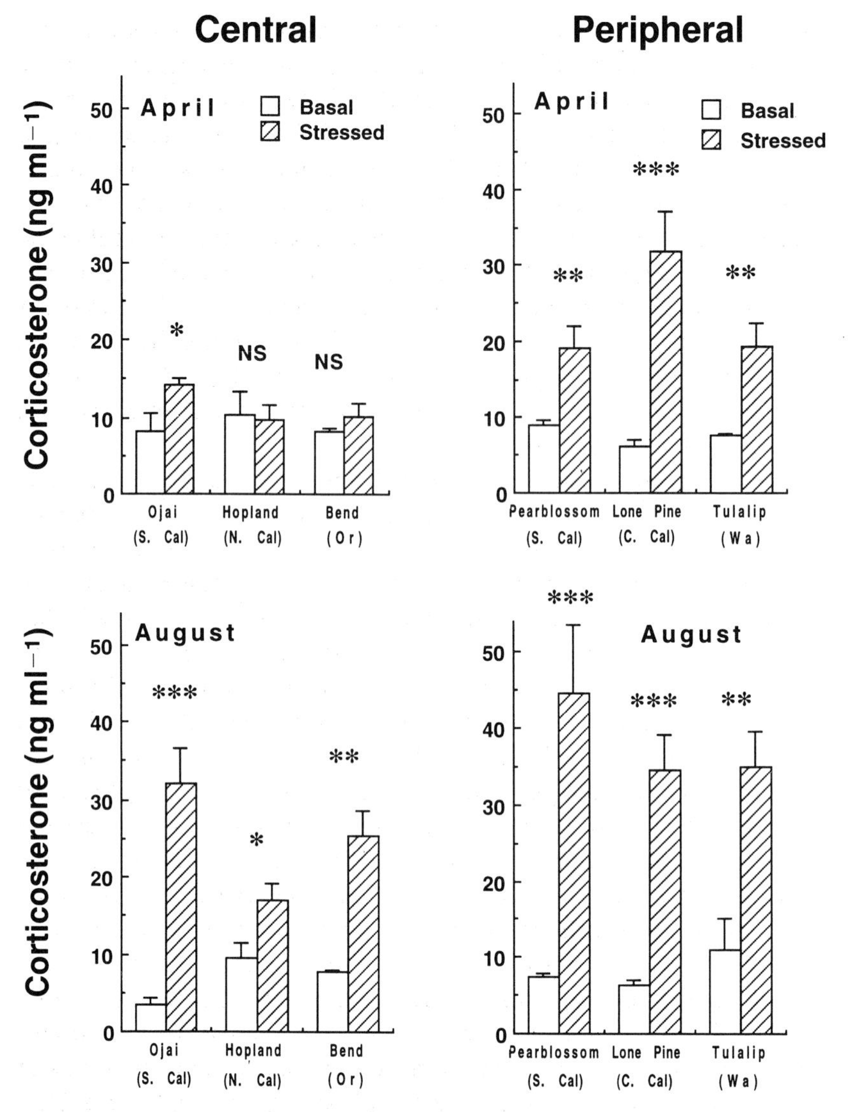

Fig. 5.6. Plasma levels of corticosterone (mean ± SE) at capture and after 60 minutes capture stress in Western fence lizards (*Sceloporus occidentalis*) sampled in April and August in the center and at the periphery of their normal range. From Dunlap & Wingfield (1995).

emergence from hibernation, etc.), and by population differentiation (e.g., possible genetic differences).

Although Western fence lizards are not a threatened species, the population differences in adrenocortical responsiveness to stress have considerable importance for conservation biology. If individuals of a population at the edge of the species' range, or normal habitat, are more susceptible to environmental stress, then habitat fragmentation could make that population more vulnerable to environmental stress, even if population size has not changed. Under 'average' or favorable conditions, populations in fragmented habitats may survive well, but the data above suggest that any detrimental environmental change resulting in stress (e.g., drought, increased human disturbance) would result in a much greater fraction of the population experiencing marked elevations of glucocorticosteroid levels in the blood. These increases would in turn promote reproductive failure, increase rates of disease, and trigger facultative behavioral patterns leading to animals leaving reserves, and, undoubtedly, increase mortality.

Further investigations of fence lizards reveal potential ways in which elevated adrenocortical responsiveness to stress may influence fitness. Fence lizards infected with the malarial parasite *Plasmodium mexicanum*, have similar baseline levels of corticosterone to uninfected individuals, but their responsiveness to capture and handling is higher (Fig. 5.7; Dunlap & Schall 1995). Baseline levels of testosterone and glucose are lower in infected males, but there is no difference following capture stress (Fig. 5.7). Additionally, males infected with malaria have lower testis mass, less bright nuptial coloration, give fewer courtship displays, and are less aggressive (leading to loss of territories). Females store less fat which is translated into reduced fecundity, or clutch size, the following breeding season (Dunlap & Schall 1995). Experimental implants of corticosterone, to mimic high circulating levels observed during the capture stress protocol, into healthy male fence lizards result in a decrease in testis mass, plasma levels of testosterone, and fat score, but in an increase in blood glucose concentrations. These data suggest that elevated levels of corticosterone, or increased sensitivity to stress, resulting from a natural cause such as malarial infection, directly results in several deleterious morphological, physiological, and behavioral changes, and reduced overall fitness. Again we see the potential for the capture stress protocol and measurement of baseline as well as stress levels of corticosterone to monitor wild populations.

Magellanic penguins in Patagonia, Argentina

Colonial species are particularly vulnerable to disruptive environmental events. Many individuals of a population converge on a small area, e.g., a

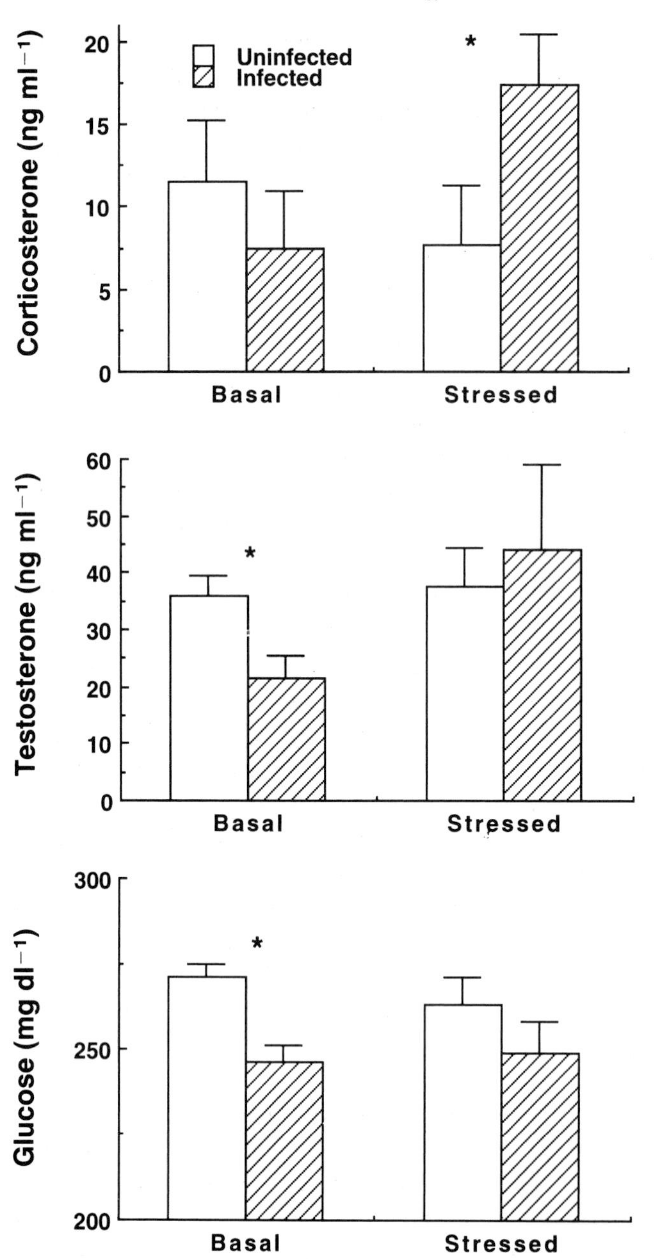

Fig. 5.7. Plasma concentrations of hormones and glucose (mean ± SE) in uninfected and infected Western fence lizards (*Sceloporus occidentalis*). Basal measurements were taken immediately after capture, stressed measurements 1 hour after capture. Asterisks indicate statistically significant differences between infected and uninfected animals. From Dunlap & Schall (1995).

breeding colony, so there is potential for massive catastrophe. Our example here comes from a breeding colony of Magellanic penguins (*Spheniscus magellanicus*) breeding at a coastal site in the Atlantic province of Chubut in Argentina. In recent years this colony has had a chronic problem with oil spills from heavy tanker traffic offshore (e.g., Boersma 1987; Gandini *et al.* 1994). In 1991 there was a substantial oil spill near the colony at a time when investigations on the reproductive endocrinology of the colony were being conducted (Fowler 1993). This unfortunate disaster nevertheless provided a unique opportunity to assess the effects of various degrees of oiling on hormone profiles, as well as to monitor the effects of washing and rehabilitation of oiled birds.

After the spill, heavily oiled penguins were not seen in the breeding colony. Numerous birds with only light oil contamination did move into the colony, although most did not breed. In both males and females, plasma levels of testosterone, and its biologically active metabolite 5-alpha-dihydrotestosterone, are lower in oiled than in non-oiled birds (Fig. 5.8). Circulating levels of luteinizing hormone (LH), a pituitary protein hormone that regulates the secretion of sex steroids, are suppressed only in oiled females. Oiling in males apparently has no effect on LH (Fig. 5.9). Plasma levels of estradiol, a sex steroid that is important for the expression of female sexual behavior as well as the synthesis of yolk for eggs, is markedly reduced in oiled birds (Fig. 5.9; Fowler *et al.* 1995). The general reduction of sex steroid levels in oiled birds is consistent with observations that few pairs with oiled partners established nesting territories and laid eggs. Clearly, even very light oiling, with as little as 5% of body surface area affected, is sufficient to reduce reproductive success. In an area where oil spills are a chronic problem, there is the potential for continuous reduction of breeding success. For an endangered or threatened population, such effects could severely impair recovery, or directly cause further decreases in otherwise healthy small populations.

Baseline plasma levels of corticosterone are higher in oiled females but not in oiled males (Fig. 5.10; Fowler *et al.* 1995). Given the well-known effects of stress-induced rises of corticosterone on reproduction (see above), we expected both sexes to have high levels of glucocorticosteroids. Whether males respond to oil through a corticosterone-independent mechanism, or are more resistant to oil-produced stresses, remains to be determined.

It is now common for volunteers to attempt to rehabilitate heavily oiled birds by washing in detergent and then holding the birds captive until the

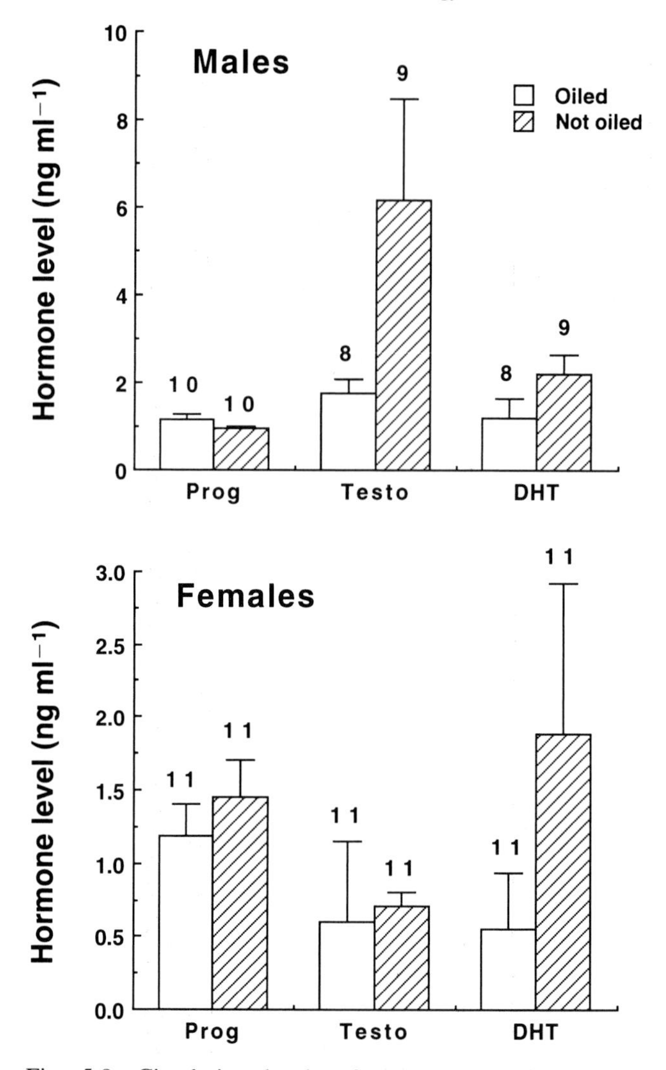

Fig. 5.8. Circulating levels of progesterone (Prog), testosterone (Testo) and 5-alpha-dihydrotestosterone (DHT) in oiled and non-oiled male and female Magellanic penguins (*Spheniscus magellanicus*). Bars are means (± SE), and numbers at tops of bars denote sample sizes. From Fowler *et al.* (1995).

plumage becomes waterproof by preening. Samples collected from heavily oiled, captive penguins as well as from individuals that had been cleaned during the week preceding blood sampling, revealed significantly higher levels of corticosterone in blood (Fig. 5.10). These data suggest

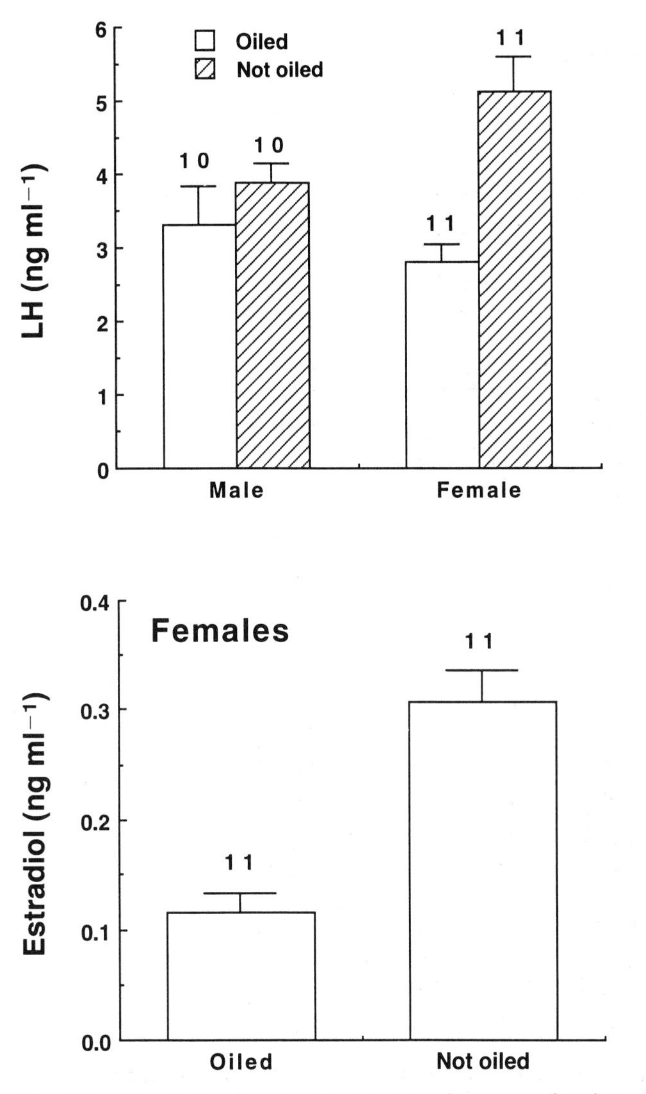

Fig. 5.9. Circulating levels of luteinizing hormone (LH) in male and female, and estradiol−17β in female, Magellanic penguins (*Spheniscus magellanicus*) in relation to oiling. Bars are means ± SE. Numbers at top of bars denote sample sizes. From Fowler *et al.* (1995).

that the act of cleaning is even more stressful and should be born in mind when dealing with endangered populations. Hormone profiles during recovery would be particularly useful (i.e., do they return to normal profiles of non-oiled birds?).

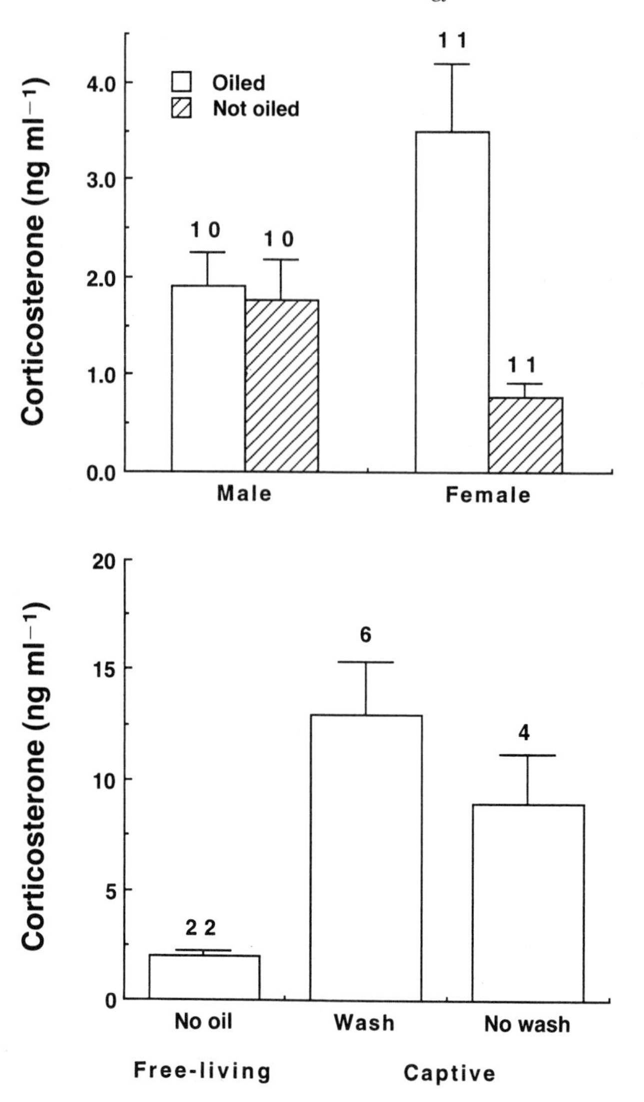

Fig. 5.10. (a) Circulating levels of corticosterone in oiled and non-oiled male and female Magellanic penguins (*Spheniscus magellanicus*). (b) Levels of corticosterone in relation to washing in captives. Bars are means (± SE), and numbers at tops of bars denote sample sizes. From Fowler *et al.* (1995).

Conclusions

An organized, efficient protocol is described as a way of assessing responses to modifying factors (i.e., environmental modifications that are potentially stressful). Measurement of circulating levels of corticosterone, a glucocorticosteroid secreted by the adrenal gland in response to a variety of unpredictable and disruptive environmental events, is used as an indicator of stress. This response is also known as the 'adrenocortical response to stress' and is common virtually throughout the vertebrate classes. The protocol is an organized way of obtaining the maximum information on vulnerability of a population to environmental stressors as well as a direct measure of the degree of stress at the time when samples are collected. Classical field techniques using entirely non-invasive methods such as censusing and observation may take several years to reach the same conclusions. The protocol is designed to give unequivocal information and may be particularly useful when recovery biologists need to establish a program for endangered or threatened populations for which there is little, or no, background information. Note also that these protocols can be modified to fit all vertebrate classes, thus raising the possibility of applying the techniques to many species within a threatened ecosystem.

The elevation of plasma glucocorticosteroid levels is common to a wide spectrum of stressor stimuli, including capture and handling. In birds, capture and restraint results in a rapid increase of corticosterone, usually within 5–10 minutes, and reaches a maximum within 30–60 minutes after capture. We have developed a technique called the capture stress protocol in which birds are captured and very small (20–30 μl) samples of blood collected as soon as possible after capture. The birds are then held in cloth bags for 60 minutes and further samples collected at 5, 10, 30 and 60 minutes after capture for measurement of changing corticosterone levels. Data in the field from species representing several vertebrate classes indicate that sensitivity to stress may change with season and reproductive state, as well as body condition of individuals. Collection of capture stress series in free-living animals and captive populations can tell us a great deal about sensitivity to stress and current stress conditions. Adrenocortical responses are sensitive to human disturbance as well as to natural environmental conditions, disease, exposure to pollution, and other potentially stressful situations.

Several field studies of marked individuals now indicate that capture and blood sampling does not harm the animal. Individuals are re-sighted in the field and appear to breed and behave normally after sampling. Future developments using fecal and urine analysis will reduce the invasive nature of this approach even further. Any investigator beginning this type of work should pay attention to these issues. Individuals must be monitored not just for survival but also for normal behavior following sampling. When working in the field with threatened or endangered species, it is particularly important that sound judgment be used concerning the timing of the sampling and whether local conditions (e.g., severe weather) constitute an additional factor that may be detrimental to the individual after release.

Interpretation of the patterns of glucocorticosteroid levels following the capture stress protocol must also be made with caution. Stress series should be collected as a 'standard' from populations under benign conditions for comparison with further series during times when the population may be threatened. If individuals in a population, for example, show a very rapid and marked elevation of glucocorticosteroids during the capture stress protocol, they may 'over react' to an otherwise relatively minor stress in the natural environment. The result could be deleterious, because such functions as reproduction may be disrupted unnecessarily. This may be especially critical if the breeding season is short and there are few possibilities to breed, and less so if the breeding season is long with many opportunities to initiate a breeding attempt.

We feel that the techniques described will be highly useful additions to the arsenal of methods used to monitor threatened and endangered populations. Although rises in glucocorticosteroids have wide ranging effects on physiology, they also have important ramifications for behavior in several ways. They can promote survival ('facultative behavioral and physiological patterns') in the face of stress – obviously an advantage. If susceptibility to stress is increased by pollution or habitat modification, however, then a relatively minor stress in the future, which normally would *not* have triggered a facultative behavioral and physiological pattern, will become deleterious. Clearly, there is an urgent need to apply these approaches to a broad spectrum of vertebrate species in the field. In this way we can rigorously test and further characterize the adrenocortical responses to stress. It will be then possible to develop guidelines with a heightened awareness and accuracy for monitoring free-living populations.

Literature cited

Astheimer, L. B., W. A. Buttemer, and J. C. Wingfield. 1992. Interactions of corticosterone with feeding, activity and metabolism in passerine birds. *Ornis Scandinavica*. **23**: 355–365.

Axelrod, J., and T. D. Reisine. 1984. Stress hormones: their interaction and regulation. *Science* **224**: 452–459.

Boersma, P. D. 1987. Penguins oiled in Argentina. *Science* **236**: 135.

Buttemer, W. A., L. A. Astheimer, and J. C. Wingfield. 1991. The effect of corticosterone on standard metabolic rates of small passerines. *Journal of Comparative Physiology B* **161**: 427–431.

Chester-Jones, I., D. Bellamy, D. K. O. Chan, B. K. Follett, I. W. Henderson, J. G. Phillips, and R. S. Snart. 1972. Biological actions of steroid hormones in nonmammalian vertebrates. Pages 414–480 in D. R. Idler, editor. *Steroids in non-mammalian vertebrates*. Academic Press, New York.

Creel, S., N. Creel, and S. L. Monfort. 1996. Social stress and costs of dominance. *Nature* (in press).

Creel, S., N. Creel, D. E. Wildt, and S. L. Monfort. 1992. Behavioral and endocrine mechanisms of reproductive suppression in Serengeti dwarf mongooses. *Animal Behaviour* **43**: 231–245.

DeNardo, D. F., and B. Sinervo. 1994. Effects of steroid hormone interaction on activity and home range size of male lizards. *Hormones and Behavior* **28**: 273–287.

Dunlap, K. D., and J. J. Schall. 1995. Hormonal alterations and reproductive inhibition in male fence lizards (*Sceloporus occidentalis*) infected with the malarial parasite *Plasmodium mexicanum. Physiological Zoology* **68**: 608–621.

Dunlap, K. D., and J. C. Wingfield. 1995. External and internal influences on indices of physiological stress. 1. Seasonal and population variation in adrenocortical secretion of free-living lizards, *Sceloporous occidentalis*. *Journal of Experimental Zoology* **271**: 36–46.

Elkins, N. 1983. *Weather and bird behaviour*. Poyser, Calton, UK.

Fowler, G. S. 1993. Ecological and endocrinological aspects of long-term pair bonds in the Magellanic penguin (*Spheniscus magellanicus*). Ph.D. Thesis, University of Washington.

Fowler, G. S., J. C. Wingfield, and P. D. Boersma. 1995. Hormonal and reproductive effects of low levels of petroleum fouling in Magellanic penguins (*Spheniscus magellanicus*). *Auk* **112**: 382–389.

Gandini, P., P. D. Boersma, E. Frere, M. Gandini, T. Holik, and V. Liechtschein. 1994. Magellanic penguins (*Spheniscus magellanicus*) are affected by chronic oil along the coast of Chubut, Argentina. *Auk* **111**: 20–27.

Gessamen, J. A., and G. L. Worthen. 1982. *The effect of weather on avian mortality*. Utah State University Printing Services, Logan, Utah.

Gorbman, A., W. W. Dickhoff, S. R. Vigna, N. B. Clark, and C. L. Ralph. 1983. *Comparative endocrinology*. Wiley, New York.

Gray, J. M., D. Yarian, and M. Ramenofsky. 1990. Corticosterone, foraging behavior, and metabolism in dark-eyed juncos, *Junco hyemalis*. *General and Comparative Endocrinology* **79**: 375–384.

Greenberg, N., and J. C. Wingfield. 1987. Stress and reproduction: reciprocal relationships. Pages 461–503 in D. O. Norris and R. E. Jones, editors. *Reproductive endocrinology of fishes, amphibians and reptiles*. Plenum Press, New York.

Holmes, W. N., and J. G. Phillips. 1976. The adrenal cortex of birds. Pages 293–420 in I. Chester-Jones and I. Henderson, editors. *General and comparative endocrinology of the adrenal cortex*. Academic Press, New York.

Honey, P. K. 1990. Avian flight muscle, pectoralis major, as a reserve of proteins and amino acids. M.S. Thesis, University of Washington.

Hose, J. E., and L. J. Guillette. 1995. Defining the role of pollutants in the disruption of reproduction in wildlife. *Environmental Health Perspectives* **103**: 87–91.

Iatropoulos, M. J. 1994. Endocrine considerations in toxicologic pathology. *Experimental Toxicology and Pathology* **45**: 391–410.

Idler, D. R., editor. 1972. *Steroids in non-mammalian vertebrates*. Academic Press, New York.

Ishii, S., M. Wada, S. Wakabayashi, H. Sakai, Y. Kubodera, N. Yamaguchi and M. Kikuchi. 1994. Endocrinological studies for artificial breeding of the Japanese ibis, *Nipponia nippon*, an endangered avian species in Asia. *Journal of BioScience* **19**: 491–502.

Kirkpatrick, J. F., and J. W. Turner Jr. 1991. Reversible contraception in non-domestic animals. *Journal of Zoo and Wildlife Medicine* **22**: 392–408.

Leibowitz, S. F., C. R. Roland, L. Hor, and V. Squillari. 1984. Noradrenergic feeding via the paraventricular nucleus is dependent upon circulating corticosterone. *Physiology and Behavior* **32**: 857–864.

Munck, A., P. M. Guyre, and N. J. Holbrook. 1984. Physiological functions of glucocorticoids in stress and their relation to pharmacological actions. *Endocrine Reviews* **5**: 25–44.

Sapolsky, R. 1983. Endocrine aspects of social instability in the olive baboon, *Papio anubis*. *American Journal of Primatology* **5**: 365–379.

Sapolsky, R. 1987. Stress, social status, and reproductive physiology in free-living baboons. Pages 291–322 in D. Crews, editor. *Psychobiology of reproductive behavior: An evolutionary perspective*. Prentice Hall, Englewood Cliffs, New Jersey.

Seal, U. S. 1991. Fertility control as a tool for regulating captive and free-ranging wildlife. *Journal of Zoo and Wildlife Medicine* **22**: 1–5.

Selye, H. 1971. *Hormones and resistance*. Springer, Berlin.

Silverin, B. 1986. Corticosterone-binding proteins and behavioral effects of high plasma levels of corticosterone during the breeding period. *General and Comparative Endocrinology* **64**: 67–74.

Smith, G. T., J. C. Wingfield, and R. R. Veit. 1994. Adrenocortical response to stress in the common diving-petrel, *Pelecanoides urinatrix*. *Physiological Zoology* **67**: 526–537.

Wakabayashi, S., M. Kikuchi, M. Wada, H. Sakai, and S. Ishii. 1992. Induction of ovarian growth and ovulation by administration of a chicken gonadotrophin preparation to Japanese quail kept under a short-day regimen. *British Poultry Science* **33**: 847–858.

Wasser, S. K. 1995. Costs of conception in baboons. *Nature* **376**: 219–220.

Wingfield, J. C. 1983. Environmental and endocrine control of reproduction: an ecological approach. Pages 205–288 in S.-I. Mikami, and M. Wada, editors. *Avian endocrinology: Environmental and ecological aspects*. Japanese Scientific Societies Press, Tokyo, and Springer-Verlag, Berlin.

Wingfield, J. C. 1984. Effects of weather on reproduction. *Journal of Experimental Zoology* **232**: 589–594.

Wingfield, J. C. 1988. Changes in reproductive function of free-living birds in direct response to environmental perturbations. Pages 121–148 in M. H. Stetson, editor. *Processing of environmental information in vertebrates*. Springer, Berlin.

Wingfield, J. C. 1994. Modulation of the adrenocortical response to stress in birds. Pages 520–528 in K. G. Davey, R. E. Peter, and S. S. Tobe, editors. *Perspectives in comparative endocrinology*. National Research Council Canada, Ottawa.

Wingfield, J. C., and D. S. Farner. 1975. The determination of five steroids in avian plasma by radioimmunoassay and competitive protein binding. *Steroids* **26**: 311–327.

Wingfield, J. C. and D. S. Farner. 1976. Avian endocrinology – field investigations and methods. *Condor* **78**: 570–573.

Wingfield, J. C. and D. S. Farner. 1978. The endocrinology of a naturally breeding population of the white-crowned sparrow (*Zonotrichia leucophrys pugetensis*). *Physiological Zoology* **51**: 188–205.

Wingfield, J. C., and B. Silverin. 1986. Effects of corticosterone on territorial behavior of free-living male song sparrows, *Melospiza melodia*. *Hormones and Behavior* **20**: 405–417.

Wingfield, J. C., and G. J. Kenagy. 1991. Natural regulation of reproductive cycles. Pages 181–241 in M. Schreibman and R. E. Jones, editors. *Vertebrate Endocrinology: Fundamentals and biomedical implications*. Academic Press, New York.

Wingfield, J. C., H. Schwabl, and P. W. Mattocks Jr. 1990. Endocrine mechanisms of migration. Pages 232–256 in E. Gwinner, editor. *Bird migration*. Springer, Berlin.

Wingfield, J. C., P. Deviche, S. Sharbaugh, L. B. Astheimer, R. Holberton, R. Suydam, and K. Hunt. 1994*a*. Seasonal changes of the adrenocortical responses to stress in redpolls, *Acanthis flammea*, in Alaska. *Journal of Experimental Zoology* **270**: 372–380.

Wingfield, J. C., R. Suydam, and K. Hunt. 1994*b*. Adrenocortical responses to stress in snow buntings and Lapland longspurs at Barrow, Alaska. *Comparative Biochemistry and Physiology* **108**: 299–306.

Wingfield, J. C., K. M. O'Reilly, and L. B. Astheimer. 1995. Ecological bases of the modulation of adrenocortical responses to stress in Arctic birds. *American Zoologist* **35**: 285–294.

:r 6

/ation and the ontogeny of behavior

:CLEAN

Airline pilots are not born. They are made. Some people will make better pilots than others because of variation in attributes such as sight, coordination, perceptual abilities in three dimensions, motivation, and so on. In principle, most of us could learn to fly a plane, just as most of us do learn to drive a car.

Why is there variation in the attributes needed to learn a complex skill? Seligman (1970) argued a heretical view for psychologists of the time: that as a function of their evolutionary history, animals are born differentially prepared to cope with life problems. Modern biologists find it extraordinary that anyone could believe otherwise. Realistically, there will be genetically based variation in most or even all of the particular skills that one calls upon when learning something new. In the above list several attributes clearly vary genetically. Some people are born blind or nearly so. Some people are naturally clumsy; others are poised and controlled. Such variation is the fodder of theoretical analyses in animal behavior such as optimality theory. It is also fundamental to the notion of ecological psychology, a discipline in which researchers study psychological issues (such as perception, cognition, and learning) in their natural context (Johnston & Pietrewicz 1985; Mace 1989). Krall & Misof (1995) provide a recent analysis of the behavioral genetics perspective.

This chapter, however, is about the 'nurture' in nature and nurture, and the importance of nurture to conservation biology. I use the anachronistic terminology purposely to flag my primary intent, which is to focus on the flexibility of behavioral development and to ignore underlying constraints. Nature provides the building blocks from which nurture creates individual organisms appropriately adjusted to their environment. It is that potential for adjustment that conservationists need to understand. Throughout the chapter I offer ideas and examples linking nurture to issues in conservation, with an emphasis on experimental research on endangered species. I provide background on the psychological and biological perspectives of behavioral development and point to controversial questions for the future.

Definition and perspective

My focus is the development of behavior. Perhaps the simplest defining characteristic of behavior is 'responsive (usually directed) movement' (Real 1993, p. 414). Thus, I am concerned with the development of an ability to respond in a directed manner (presumably to a stimulus).

Let's put this notion in context. If I perceive a predator, then a variety of response options will be available to me (run away, hide, attack, give an alarm call, etc.). What I actually do will depend on many factors, including things like context (do I have young nearby?), perceived threat (is the predator hunting?), appropriateness (has it detected me?), and so on. My tendency to do any of those things will depend first, on my basic physical skills (is there an alarm call in my repertoire?) and second, on my experience (what happened when I alarm-called last time?). The more experience I have had, the more likely I am to do something that works effectively in the current context. That experience in responding to the predator stimulus will be gained during development.

Of course, I will likely add the current event to my prior experience, thereby hopefully responding even more effectively next time. With respect to dealing with predators, learning throughout life is likely, although most learning will still occur when an individual is young and inexperienced. Many behaviors, however, develop within a narrow time frame, after which they become set, or *canalized*. Well-known examples are young salmon returning to where they were born, having imprinted on the chemical cue provided by the river; or birds giving the song pattern characteristic of the area in which they live.

Behavioral development does not take place in a vacuum. It takes place in the environmental context in which the animal lives. If it lives in a natural and unmodified environment, then it should develop the suite of behaviors normally exhibited by its species through evolutionary time. If it lives in a modified environment it can still develop those behaviors, but some may be adjusted to suit the new conditions. The extent of such adjustments will depend to a large extent on the degree of environmental modification. If the modifications are extreme, so too will be the adjustments. At some point, the required adjustments will be too extreme for the population to sustain and behaviors may even be expressed inappropriately (my favorite example is a bittern (*Botaurus poiciloptilus*) seen standing rock-steady among the non-existent reeds of an over-grazed meadow). Conservation biologists are only beginning to appreciate that adjustments to behavior are a necessary consequence of environmental modification. More specifically, these adjustments are a consequence of the modified environment in which the individual develops.

The best understanding of how environmental changes affect the behavior of free-living animals will come from an experimental approach. Unfortunately with respect to the important focus of this volume on free-living animals, much of the experimental research on behavioral development is necessarily carried out under controlled conditions, often in the laboratory (or in captivity). I hope the reader can accept my treating captivity as a highly modified environment in which many natural processes still occur.

Animal cognition

Animal cognition involves the notion that animals use "internal mental operations in generating specific behavioral responses to sets of external stimuli" (Real, 1993, p. 413). What this really means is that it is now acceptable to talk about animals perceiving objects, processing information, making decisions, learning new skills, solving problems, forming rules and concepts, and so on (Terrace 1984; Yoerg & Kamil 1991; Griffin 1992).

A cognitive animal is a thinking animal. It seems reasonable to expect that as an individual gains in experience (i.e., develops), its thinking will change. For example, young animals may not perceive combinations of stimuli as distinct elements in the way that adults do; rather, they are likely to perceive the combination as a distinct entity from the component parts (Spear & Kucharski 1984). Similarly, infant humans tend to perceive identical objects seen in different places as different objects (Bower 1979), or occasionally the reverse may occur (e.g., all bald men are Uncle Harry; Carey 1992).

There are two implications. First, if no experience is gained with a particular stimulus, then no thoughts with respect to that stimulus will ever be considered. In other words the animal will not develop or refine its skills with respect to that stimulus. Second, the most effective ways of dealing with particular stimuli are unlikely to be those exhibited by a naive animal when first encountering those stimuli. Naivety is a characteristic of young animals generally, and also of any animal in a novel (or recently modified) environment.

The development of behavior

Any developing animal must integrate an extraordinary array of components in order to express appropriate behavior. To return to the airline pilot analogy for a moment, that pilot will integrate sensory input from sight, air movement (presumably via the sense of feel or touch), magnetism (the compass), air pressure (for altitude), and balance (is the plane level?). Actual control of the craft is via finely tuned neuromuscular control developed originally while the pilot was a child, then refined during specialist pilot training. An important component of training for commercial pilots involves the ability to suppress or control 'panic' responses, so that appropriate behavior is expressed at times of crisis. Lesser mortals sitting in the body of the plane have an intrinsic faith in that training procedure. We know it works because planes do not routinely crash.

The same principle applies to other animals, or at least to animals whose behavior is modifiable by experience. During development, they acquire behavioral patterns that work. Otherwise they would not survive. What are the components that underlie the development of any complex behavior pattern in animals? Here, I give an overview of fundamental principles in the development of behavior, using examples to explore the relevance of those principles to conservation. I use development in a rather general way to include the acquisition of new behaviors, as well as improving behaviors that are already available. I start with a psychological perspective.

The psychological perspective

[I begin with an introduction to learning theory. Readers may prefer to jump straight to the section titled 'Relevance to conservation' and refer to this section as needed.]

Psychologists emphasize learning and memory. Through the first two-thirds of this century, psychological thinking about animal behavior was dominated by the behaviorists, whose roots can be traced to Pavlov's principle of reinforcement, and Thorndike's law of effect (Mackintosh 1994). Since the 1960s and especially recently, however, focus has moved to notions of animal mentality, or cognition (Griffin 1992). The principle here involves three components (Real 1993). First, animals obtain sensory information and encode it in a form accessible to brain functions. Second, mechanisms for processing the stored information are available. Third, the information is used to generate representations about the environment that are used to make decisions about behavioral responses.

The disagreements between behaviorism and cognitive science are of little practical consequence in the conservation context. Several attempts have now been made to integrate the behaviorist and cognitive perspectives (Roitblatt *et al.* 1984; Pearce 1987; Ristau 1991; Mackintosh 1994). Here, my aims are to provide a summary of the basic components of learning theory, to make some comments about animal memory, and to put these ideas into a conservation context.

Conditioning

Pavlovian, or classical, conditioning involves an animal learning predictive relations between a signal (the conditioned stimulus, e.g., a tone) and an important component of the world (the unconditioned stimulus, e.g., food). Thus, Pavlov's dogs learned (or were conditioned) to salivate on hearing a tone, because the tone had been paired with the appearance of food and dogs naturally salivate when they encounter food. Using the language of cognition: as a result of their experiences the dogs stored a representation in memory of a link between the tone (of no original interest) and food (of fundamental evolutionary significance), so that the tone generated an expectation of food, and the appropriate response of salivation. One can imagine wild animals learning about their environment in just this way. Rufous hummingbirds (*Selasphorus rufus*), for example, quickly learn to associate a signal with a food reward (Sutherland & Gass 1995).

Operant, or instrumental, conditioning has a somewhat different structure. Here, if a behavior has favorable consequences, the probability of the act being repeated is increased. If the behavior has negative consequences, then the probability of the act decreases. In essence, the animal operates on its environment to generate a desirable consequence. No conditioned stimulus is required. Rather, the animal, by its behavior, is instrumental in generating an outcome. As the animal interacts with its environment, it learns to link particular behaviors with particular outcomes through a reward process ('rewards' can be negative) that began with random behavioral events. In the example above, the hummingbirds went on to predict the spatial distribution of future food availability and use of their artificial flower environment changed accordingly.

Thus, the hummingbirds exhibited classical conditioning by learning to associate signal and reward, and instrumental conditioning by learning to predict profitable food availability. The language used by Sutherland & Gass, however, was more typically cognitive: '. . . hummingbirds use coarse-grained memories of the spatial patterning of energy availability to guide their foraging . . . [although] . . . they quickly modify their memory of places on the basis of new information.' (p. 1273)

Embryonic animals are responsive to both kinds of conditioning from before birth or hatch (reviewed in Gandelman 1992). Experimental research on the long-term memory of embryonic animals (including humans) has given confusing results and questions about this ability remain unresolved (Moscovitch 1985). It may be that each case or species needs to be assessed in context. Embryonic rats injected intraperitoneally with a variety of substances demonstrated preferences for, or aversions to, those substances (as appropriate) for some time after birth. Classical conditioning was effective on embryo chicks from 15 days of age (hatch is at 21 days), although post-hatch conditioning was short-lived. Embryonic anurans develop the ability to recognize kin at a very early age (reviewed in Waldman 1991), and may carry that skill through metamorphosis (Blaustein *et al.* 1984). Thus, it may be possible to establish dietary or other behavioral preferences from a very early age, and perhaps reinforce those preferences after birth/hatch.

Learning theory has spawned a mind-boggling array of concepts and terms, usually couched in a language so technical that the writing is impervious to an untrained reader. Frequently used concepts include processes such as: (1) habituation (the animal learns not to respond to a stimulus, presumably because repeated exposures have proven the stimulus to be harmless or at least tolerable); (2) extinction (non-linkage between the conditioned stimulus and the unconditioned stimulus will result in the animal ceasing to exhibit the conditioned response to the conditioned stimulus); (3) retention performance (remembering and forgetting); and (4) discrimination learning (distinguishing between two, usually similar, stimuli). Also principles such as: (1) reinforcement schedules (reward and act may only be linked some of the time); (2) latency (an amount of time between two things, e.g., the act and the reward); and (3) stimulus generalization (the tendency to respond to a stimulus that has at least some of the characteristics of another stimulus to which a response is currently given).

Relevance to conservation

The critical point for conservation biologists is that the various forms of conditioning describe real processes or principles that influence the development of behavior in animals and can potentially be used to influence behavioral outcomes in a management context. In biological terms, the processes are adaptive both in the ultimate sense of having been selected for and in the proximate sense of generating appropriate behavior in the current context. The principles provide methodological options. Most importantly, behavior will be modified by these processes and principles whether or not the researcher or manager is aware of them or is attempting to use them.

When placed in a highly modified environment, animals can be conditioned to do bizarre or unnatural things, presumably because they have few alternatives. Thus, a pigeon in a Skinner box can be conditioned to turn around 100 times before pecking a key to obtain a food reward. The process of conditioning is straightforward and involves the researcher rewarding first one turn, then two turns, and so on. Perhaps not surprisingly, pigeons in the wild simply refuse to develop such aberrant behavior because they have alternatives (Cohen & Block 1991). Wild animals are sensible about obtaining food, even if they sometimes behave a little less than optimally; the pigeon can go elsewhere in search of food, and it does. It need not concern us here that some of the research carried out by psychologists in artificial environments is less than realistic; realism was not a relevant concern in the experiments. The issue of what constitutes realistic, or appropriate, behavior is, however, of central concern in conservation biology.

The ability to manipulate behavior is a powerful tool that can certainly be used to generate appropriate behavior in animals once one has decided on that behavior. It can also be used, however, to generate inappropriate behavior, frequently because the researcher or manager is not even aware that conditioning has occurred. Also possible is the use of 'inappropriate' behavior to achieve a desirable management outcome, if only for a short time. I give two examples of inappropriate behavior, then return to the question of appropriate behavior.

The first example is an anecdote, but it demonstrates the power of conditioning and the potential for generating inappropriate behavior. Rufous hare-wallabies (*Lagorchestes hirsutus*; Fig. 6.1) are currently extinct on the Australian mainland and are being bred for reintroduction at a captive-rearing facility in Alice Springs, Northern Territory (Gibson

Fig. 6.1. Management practices for rufous hare-wallabies may inadvertently have conditioned animals to run only short distances after disturbance, a behavior that is likely to be inappropriate if disturbed by a fox. (Photograph by Conservation Commission of the Northern Territory, Australia.)

et al. 1995). The animals are held in pens that are 10–15 m long and are captured monthly for weighing, inspection of reproductive state, etc. Capture involves disturbing them from their day-beds, after which they run around the pen and into hand nets. The process is straightforward and effective. The wallabies, however, can only run a maximum of 15 m after disturbance before they encounter the end of the pen. G. Lundie-Jenkins (personal communication) discovered that after release into the wild, disturbed wallabies still ran a meager 15–20 m before stopping, even though the pen wall was no longer there. Such behavior would be extremely inappropriate if disturbance was by a predator such as a Dingo (*Canis familiaris*) or fox (*Vulpes vulpes*).

The second example is a spin-off from management operations designed to eradicate or control vertebrate 'pest' species. In any poisoning operation there will always be a few individuals that either simply refuse to touch baits, or eat a small amount of bait and become sick ('bait-shyness'). There is an extensive literature on conditioned food-aversion (see Hall 1994), and it seems that animals can develop a life-long aversion to particular foods from just one unpleasant experience, even if there is a latency of several hours between ingesting the food and becoming ill.

These results are hardly surprising to humans, most of whom have had an unpleasant experience with food of one sort or another. An interesting management option is suggested. If a problem animal, such as a predator, can be conditioned to avoid prey targeted for protection for even a short time, the prey might be given enough release from predation pressure to shift from negative to positive survival and/or reproduction measures (Gustavson *et al.*, 1974, Avery & Decker 1994). Gustavson *et al.* conditioned Coyotes (*Canis latrans*) to avoid potential prey (lambs and rabbits) by linking killing and consumption of the prey to illness, using LiCl. Avery & Decker fed chemical-injected quail eggs to crows (*Corvus ossifragus*). In both cases the predators subsequently avoided the food source, at least for a while. Here, an 'inappropriate' conditioned response is used to achieve a desirable management objective as a result of stimulus generalization (that lamb/egg was bad = all lambs/eggs are bad).

The wallaby and conditioned aversion examples can be put into the language of experimental psychology. Classical conditioning involves the formation of an association between a stimulus (the taste of food – a lamb/egg) and a response (being sick). In this case eating lambs/eggs is aversively conditioned and the conditioned stimulus (lamb/egg – shape) produces avoidance behavior because of its association with the unconditioned stimulus (food). Instrumental conditioning involves the animal adjusting its interaction with the environment towards behavior suited to that environment. Wallabies running large distances after being disturbed are negatively reinforced (they crash into a wall), thus they learn to run shorter distances.

Attempts to use conditioning positively in conservation programs are beginning to appear in the literature. Food placed in trees, for example, encourages primates to remain in the trees (Box 1991; it also conditions them to avoid the ground, which tends to be a dangerous place for primates). Early reintroductions of Golden-lion tamarins (*Leontopithecus rosalia*) were accompanied by costly and labor-intensive training (conditioning) before release (Kleiman *et al.* 1991). It turned out that the training offered short-term advantages only, and training carried out after release was just as effective, although it produced some animals with a lifelong dependency on human provisions. Wild-reared offspring of reintroduced animals developed motor and coordination skills much more rapidly than zoo-raised individuals, and the management team came to regard these wild-reared animals as the first generation from which a successful population might establish. By developing in a truly wild environment, tamarins learned (or were conditioned) to use that environment effectively. The primary objective of management shifted from conditioning zoo-raised animals to survive successfully in a wild state, to helping zoo-raised animals to breed successfully in the wild. Although in this example the original conditioning experiments were unsuccessful with respect to the long-term aim, they were an important component of a research and management path that was ultimately successful.

Several attempts have now been made to condition animals to fear predators (e.g., *Colinus virginianus*, Ellis *et al.* 1977; *Grus americana*, Anonymous 1993; *Mustela nigripes*, Miller *et al.* 1994; *Lagorchestes hirsutus*, McLean *et al.* 1995, 1996; *Culaea inconstans*, Chivers *et al.* 1995). In terms of developing changes in behavior towards the predator, these studies seem to be uniformly successful. In terms of generating improved survival, most have either been unsuccessful, or the question was not tested. Clearly, conservation biologists are becoming aware of the potential for using conditioning as a way of generating (or developing) appropriate behavior in animals subject to management.

Memory

Learning about something implies an ability to retain information that has been acquired. Research on memory in animals by psychologists has focussed on issues such as mechanisms of storage, encoding and retrieval, the nature of memory representations, spatial abilities, and the nature of forgetting. Access to this literature is available in Gordon & Klein (1994), although much of it will be of little interest to conservation biologists. There are some critical issues, however, with which conservation biologists do need to concern themselves. In particular, what is the importance of context when skills important for survival are being acquired (and stored)? How long will these memory representations last? Is there a relationship between training intensity and/or number of training events and learning? I will make just a few comments, based primarily on summaries in Spear (1984), Spear & Kucharski (1984), and Gordon & Klein (1994).

Context is important during the acquisition of information. First, very young animals tend to learn more effectively when at home (presumably a safe environment where there are no distractions or stress) than when in a novel environment. Second, skills learned in one context will be expressed most effectively in that context, and movement away from the context results in poorer retention and/or expression. It makes little difference if the skill was acquired by classical or instrumental techniques, and the context change can be as simple as transfer of the trained animal from one room to another slightly different one. If multiple responses are learned in different contexts, then each response tends to be expressed most effectively in the context in which it was learned. Other factors, such as time of day, may also influence the expression of responses – decrements similar to those found for a change in living environment can be obtained by testing at different times of day.

Lifelong food aversions can develop from a single experience (Shettleworth 1994). Thus, it is possible for animals to acquire permanent changes in behavior (i.e., acquire a permanent memory) from experiences that are essentially instantaneous. More usually, animals require considerable experience with a problem or task before they can do it consistently (e.g., Sutherland & Gass 1995). Also, complex tasks requiring complex performance require practice – which, of course, is why airline pilots are required to fly a minimum number of hours annually to maintain their license. Thus, young animals will tend to be clumsy foragers relative to adults (e.g., Wunderle 1991; see Spear 1984 for a general review of learning by young animals), and animals that have learned a new skill will tend to forget that skill over time (e.g., McLean *et al.* 1996). Some extraordinary examples of long-term memory of complex information, however, have been identified (e.g., Balda & Kamil 1989). Clearly, there is huge variability in the ability of animals to develop and retain complex behaviors. Also clearly, the abilities they have will tend to be ecologically relevant. Thus, Balda & Kamil (1989) found that the ability of three food-storing bird species to remember where food had been stored was correlated with their dependence on those stores during resource-limited times of the year.

Finally, it is even possible to hinder the development of particular skills at a later age if animals get the wrong kind of training, or if they have the right kind of training but are too young (examples in Spear 1984).

The implications of these results on context and memory for conservation programs where animals are required to adjust to new environments are enormous.

Biological perspective

Adaptation and adjustment

Biologists emphasize the adaptiveness of behavior. The myriad patterns of behavior to be found in nature all reflect a lengthy history of natural selection: the outcome seen today represents a 'best fit' situation (see Dupre 1987, Reeve & Sherman 1993, or any textbook on evolution for general discussion of this principle). Put another way, the currently most frequently used solution to any problem will be the most effective of the options available to the species (or population) in evolutionary time.

Unfortunately, adaptation is a slippery concept. In recent years, biologists have backed away from the notion that all things are a consequence of selection (an extreme position attacked vigorously by Gould & Lewontin 1979), and the debate has shifted ground to a discussion of "levels of selection" (Sherman 1988; Via *et al.* 1995). For conservation biologists, the notion that animals are adapted to their environment is best viewed as a framework that constrains options for rehabilitation, restoration, protection, translocation, or whatever is the current objective. Put more positively, a primary conservation objective should be to allow selection to continue operating on target organisms adapted to reasonably natural conditions.

Biologists use the notion of adaptation in a second way. For physiologists, adaptation means adjustment (Schmidt-Nielson 1979). The tolerances of winter-acclimatized fish, for example, will cover a lower temperature range than the tolerances of the same fish acclimatized to summer conditions. Some winter conditions may be lethal for a summer-acclimatized fish, and vice versa. The point is that individual animals are strongly influenced by the environment in which they live, sometimes in ways that are difficult to detect (e.g., at the level of cell chemistry). Put another way, currently the animal will be adapted to a subset of the tolerances with which it is potentially able to cope. Placing a well-adjusted (or adapted) animal into a new environment (for example, by changing that environment; there is no need to move the animal) without giving it the opportunity to adjust could result in some tolerances being exceeded. The coping skills are simply not available, even though the animal may be capable of developing them.

The notion of adaptation to a modified environment encapsulates both of these notions of adaptation (captivity is a convenient example, although a similar argument could be developed for, say, a fragmented population). Even in a successful captive population, some individuals will not breed or will have a high proportion of failed attempts. Perhaps they are unable to cope with the stress of constant association with humans, or there are missing components in the diet to which they are unusually sensitive. Despite the best efforts of well-meaning managers, these individuals will be selected against because they are not contributing to the future gene pool (Frankham 1995). The effect is to create genetic lineages that are well adapted to the current environment (captivity) and probably poorly adapted to the original environment (where a stress response to humans was appropriate, and the missing components of the diet were readily available).

Added to this genetic change is the problem that individuals within the captive population experience a limited array of components of a less modified environment. They are protected from predators, are not challenged by parasites, are not shown where to obtain essential dietary components by their parents, and so on. In the physiological sense, one does not need to actually experience 32 °C to be able to cope with it after adaptation (or acclimatization) to summer conditions. There is a useful behavioral analogy here. Growing up in an environment containing predators is likely to result in behavior appropriate to coping with predators (such as being continuously watchful or remaining near to cover). Even if a new predator shows up, the individual's general behavior patterns incorporate at least some potential for coping with the new problem because of prior adaptation to a similar problem. The new problem comes within current tolerances.

Earliest behaviors

Behavior begins before hatch or birth. In Atlantic salmon (*Salmo salar*), the following behaviors may all be observed prior to hatching: muscle contractions (including body flexion and heart beats), movements of fins, jaw and gills, movement away from tactile stimuli, response to light and gravity, and swimming movements (Fig. 8.1 in Goodenough *et al.* 1993). Although it is a significant (and potentially traumatic) event for a developing embryo, hatching is simply another in a stream of new experiences. In fact, behavioral abilities at hatch vary considerably with the life history pattern of the species of interest. Put another way, within classes of animals, hatching occurs at a relatively random moment during embryogenesis. Thus, hatchling birds range from highly altricial passerines (with minimal sensory abilities, unable to thermoregulate, and little muscular control) to highly precocial megapodes (that are able to fly and fend for themselves entirely) (Starck 1993). Possibly larger extremes can be found among the mammals, in part because of differences between marsupials and eutherians.

Some behavior can be influenced by experiences prior to birth. The influence of testosterone during embryogenesis on the sexual behavior of rats as adults is described below. Vom Saal (1984), however, identified a more subtle effect called the intrauterine position phenomenon. Embryonic male mice secrete higher titers of testosterone than embryonic females. Mice are polytocous (give birth to multiple young); thus, within the uterus, embryos of each sex can be subjected to differing hormonal influences from their neighboring sibs. A male (or female) can have two, one or no females as neighbors. Vom Saal determined the sex of the intrauterine neighbors for a sample of mice, then raised the mice to adulthood. He tested them by placing a newborn pup in their cage, and found that adult males with two male neighbors as embryos were strongly parental and weakly infanticidal, whereas males with two females as neighbors were the reverse. Such results indicate that marked behavioral variability can arise from subtle influences in the early experiences of individual animals. From a conservation perspective, the results suggest the potential for large behavioral shifts arising from supposedly minor environmental change (e.g., human application of a hormonally based chemical could adjust the proportion of potentially infanticidal males in the local mouse population). They also offer the potential for useful manipulations, for example, to solve management problems (e.g., a higher proportion of 'parental' males might be desirable).

Cultural evolution: Social transmission of information

Biologists have long debated the question of inheritance of acquired characteristics. The original version of this concept, frequently portrayed as Lamarckian giraffes stretching their necks to reach ever higher into the trees, has been thoroughly debunked (e.g., Williams 1966). However, behavioral versions are acceptable and may even be essential to the maintenance of the social complexity seen in many vertebrate species (Avital & Jablonka 1994). Rather than inheritance of genes, Dawkins (1976) talks of the inheritance of memes. A meme is a culturally derived behavioral trait (akin to a tradition in human culture) that is transferred across generations behaviorally (e.g., by teaching; see Caro & Hauser 1992). Relative to genetically based traits, memes are easy to acquire (they can appear and spread through a population in one generation), easy to modify (mistakes in transfer represent cultural mutations), and are easily lost. Thus, in general, memes evolve much more rapidly than genes.

Many complex social systems in animals are structured, in part, around memes, and memes can remain quite stable across generations (e.g., song in Indigo buntings (*Passerina cyanea*); Payne *et al.* 1981). In contrast, Humpback whale (*Megaptera novaeangliae*) songs change annually, although all individuals track the changes in the same way (Payne & Payne 1985). Presumably, it is the orchestration and matching of song patterns that is important to the whales; consistency of structure is unimportant.

Rats develop dietary preferences from smelling odors on the mouths of other group members (work of Galef and coworkers, reviewed in Shettleworth 1994). Chickens (Evans & Marler 1991), blackbirds (*Turdus merula* , Vieth *et al.* 1980), and monkeys (Mineka & Cook 1993) learn to recognize predators by watching others give alarm calls or mobbing behavior. Behaviors that are either useful or essential for survival can be obtained from members of one's social group.

Clearly, conservation biologists need to be aware that traits essential for survival in an original environment may have been lost by cultural mutation. The mutation could be a result of small population effects: the randomly chosen lucky few who survived did not happen to carry the trait with them. Perhaps the population was established by young animals that were isolated from the parent population before they acquired the trait. Perhaps an essential factor maintaining the trait has disappeared from the environment in which the population currently exists.

Given that a species has the ability to inherit culturally acquired traits, many options become available to biologists. If a known trait has been lost, it may be possible to recreate the conditions under which it was acquired in order to re-establish the trait in the population. Even more challenging is the possibility of providing animals with new skills for coping with new problems (e.g., see the Takahe/Pukeko example below).

Sensitive periods and imprinting: the development of preferences

Behaviors may be acquired more easily or more quickly during some periods of development than during others. The extreme case was termed "imprinting" by Lorenz (1935, in Goodenough *et al.* 1993), who borrowed the term from embryology. Lorenz's now classic example was of newly hatched geese forming a filial attachment. The attachment developed instantly between the hatchling and the first reasonably large moving object it encountered (under natural conditions such an attachment is appropriate because that object ought to be a parent). Lorenz, and subsequently many others, went on to show that filial imprinting occurred during a highly constrained "sensitive period" lasting just a few hours, and the imprinting event was virtually instantaneous. The many arguments about whether imprinting is a unique form of learning, how instantaneous it really is, whether its effectiveness can be adjusted with further experience (see Bolhuis 1991), or whether it is a form of classical conditioning (Shettleworth 1994, and see below), need not concern us here.

Essentially, imprinting means the development of a preference (Bolhuis 1991). Rarely, that preference may be so absolute as to preclude future behavioral options, as in a few well-publicized cases of inappropriate sexual imprinting (e.g., Whooping cranes (*Grus americana*); Kepler 1977). More usually, the preference is not exclusive and may even be over-ridden by future experience.

Sensitive periods, and analogues of imprinting, have been identified in many circumstances. To list just a few: (1) Hormones: the future expression of appropriate sexual behavior in rats depends on whether or not the embryonic brain develops under the influence of testosterone (a full description is provided by Alcock 1993). (2) Sexual imprinting: male Zebra finches (*Taenopygia guttata*) will prefer female Bengalese finches (*Lonchura striata*) to female Zebra finches as mates if they were raised to independence by Bengalese finches (Immelman 1969, in Goodenough *et al.* 1993). (3) Kin recognition: at hatching, toads (*Bufo americana*) form chemical preferences that become resistant to change after two weeks (Waldman 1991). (4) Generalized learning: rats exhibit maximum sensitivity to learning about their environment for three weeks immediately after weaning day 22 (Forgays & Read 1962, in Gandelman 1992).

Imprinting is now well-enough known that conservation biologists are likely to be aware of potential problems arising in management situations, particularly with respect to mate choice. The usual approach, however, is to minimize the possibility of mal-imprinting, either on the assumption that there might be a problem, or because experimental work designed to investigate just how significant the problem might be is difficult. Also, little or no attempt may be made to develop a full understanding of the sensitive periods of the target species.

In captive situations, for example, problems with sexual imprinting are now routinely solved by hand-raising using puppets and minimizing contact with humans. This approach can be time-consuming and expensive, may be applied for considerably longer than is necessary to avoid the problem, and frequently avoids both filial and sexual imprinting on humans although the former might not be a problem. Also, the compensating solution of creating situations in which the mal-imprinted animals are forced to mate appropriately (usually because the inappropriate but preferred sexual partner is not available) can still result in effective breeding. Lastly, imprinting can be used as a potentially effective management tool to create preferences that solve conservation problems, even if the preferences are 'unnatural' in some sense.

To give some examples: in two programs in New Zealand, Black robins (*Petroica traversi*) were cross-fostered to Pied tits (*P. macrocephala*; Butler & Merton 1992), and Black stilts (*Himantopus novaezealandiae*) were fostered to Pied stilts (*H. himantopus*; Pierce 1984). In both cases, some individuals of the endangered species eventually mated with (or attempted to mate with) individuals of the fostering species. The cause may not have been inappropriate sexual imprinting because these pairings may have resulted from there being no mate of the appropriate species available. That issue aside, the solution was simple if drastic. Tits were eradicated on a second island, all tit-raised robins were transferred there, and they paired normally. A second (and preferred) solution involved fostering older nestlings back to black robins after they had been incubated and partially raised by tits; these individuals generally did not develop inappropriate sexual preferences. For the stilts, the "second island" option did not exist and the pied mate of a black bird was removed wherever possible. As more black mates became available, they tended to be accepted by the individuals that had initially mated inappropriately (D. Murray & C. Reed, personal communication).

Immelman's male Zebra finches eventually mated normally with female Zebra finches if given these as the only option, although they never entirely lost their preference for female Bengalese finches if these were offered (Goodenough *et al.* 1993). In contrast, hand-raised Whooping cranes remained resolute in their preference for humans as sexual partners, despite imaginative attempts to convince them otherwise (Kepler 1977). Clearly, inappropriate sexual imprinting can be a significant problem and should be avoided if possible, but it is not an absolute phenomenon and may be a relatively minor issue for many species.

In an attempt at habitat imprinting, Lovegrove (1992) encouraged Saddlebacks (*Philesturnus carunculatus*) to choose boxes for roosting and

Fig. 6.2. Takahe were cross-fostered to closely related Pukeko in an attempt to encourage young Takahe to develop predator–response behaviors that are effective for Pukeko. (Photograph by the author.)

nesting. Saddlebacks roost and nest in natural holes where they are easily taken by rats (*Rattus* spp.); however, boxes can be protected from rats. Lovegrove's source population lived on a rat-free island with few natural holes and many boxes. Thus, most birds were raised in boxes, and at least some appeared to prefer to roost in boxes. Lovegrove translocated two groups to a second island on which *R. norvegicus* were common: source individuals that preferentially used boxes, and source individuals that rarely used boxes. As predicted, box-using birds did have higher survival on the second island. Unfortunately, box-using may not have been a successfully imprinted trait because similar proportions of the two groups did and did not use boxes on the second island (i.e., the translocation was a success, although the experiment may have failed).

Bunin & Jamieson (1996, personal communication) went one step further. They argued that the issue of sexual imprinting should be subordinated to the possibility of learning successful predator-coping skills. They cross-fostered Takahe (*Porphyrio mantelli*) to Pukeko (*P. porphyrio*) with the specific objective of generating improved predator-coping skills in the cross-fostered Takahe (Fig. 6.2). Pukeko cope well with an array of predators including mustelids, cats, dogs, and hawks. Despite their large size, takahe appear to be particularly susceptible to mustelid predation. Unfortunately, this innovative experimental approach did not get beyond the pilot study due to high levels of infertility in takahe eggs, and it is not known if sexual preference problems would have arisen.

Lastly, Hulet *et al.* (1987) achieved what Bunin & Jamieson could not, although using rather more prosaic species. By bonding lambs to cattle, they found that the lambs obtained protection from Coyotes. Mixed-species groups are an acknowledged mechanism for individuals to obtain protection from predation (Diamond 1981), although such groupings have rarely been considered in a conservation or management framework.

A great deal of experimental work on sensitive periods and imprinting has now been conducted, but on only a small number of species (mostly chickens and ducks, see Bolhuis 1991 for review). The experimental designs developed in these programs could potentially be used to address similar issues with endangered species (or closely related surrogates) in order to gain a better understanding of the timing and scale of the problems to be addressed. Conservation biologists could be generating a great deal of unnecessary work for themselves if they simply assume that imprinting will be a problem. They may also eliminate management options that could be both cost-effective and easily implemented, if they do not exploit sensitive periods that the target animal may have. Frequently with endangered species, the chosen research design will depend more on logistics than on currently available knowledge, and it may be necessary to assume that any unknown imprinting problems can be dealt with later, if and when they arise. On the other hand, imprinting certainly needs to be considered in the design of any management program and there is a clear need for detailed research on imprinting covering a broader taxonomic base than a few bird species.

Issues for the future

The first step in any attempt to integrate behavioral issues into conservation programs is recognition of the important behavioral questions. Outbreeding, for example, is routinely a primary objective in management programs. The managers preference for outbreeding, however, could conflict with the animals' own mating preferences to the point where animals simply refuse to mate. The solution could be as simple as ensuring that a choice of mates is available, even if the outbreeding objective is somewhat compromised. That solution, however, might never be implemented if questions about mate choice are not asked. Alternatively, mating preferences may be manipulable using imaginative rearing procedures.

In many cases, even if there is an awareness that behavioral questions need to be asked, there may be logistical or financial constraints on asking them. Thus, conservation programs are routinely beset by small sample sizes, lack of resources, inexperienced personnel, and a need for decisions to be made now (in the absence of essential information) rather than after desirable experiments have been conducted. Thus, in the early 1970s when Black robins were reduced to five birds (three males, two females), the decision was made to transfer the entire population to a new island. It was not known if robins would survive the process of transfer, so the best possible experiment was performed: the unmated male was transferred ($N = 1$). The 'experiment' was successful (he survived), and a pair was transferred the next day with the unfortunate consequence that the female abandoned her long-time mate for the 'resident' male in the new site. After transfer, the second female similarly abandoned her mate for the (now single) mate of the first female (Butler & Merton 1992). Recent experiments with birds indicate that divorce rates will be high after translocation (Armstrong *et al.* 1995), and in retrospect the Black robin result was unsurprising. At the time the management team was rightly concerned that their procedures had disrupted two breeding pairs.

Clearly, inappropriate or unwanted behavioral development should be avoided wherever possible. Animals should live in environments that are as close to natural as possible to ensure development of behaviors appropriate to survival in the wild. Conditioning procedures generating inappropriate behavior should be recognized and minimized, or compensating conditioning should be incorporated into the management regimen. In the hare-wallaby example it might have been possible to create a lengthy runway around the pens and occasionally chase animals around it without catching them. When in the wild they might have run much further when disturbed. One does not always know if such procedures will have the desired effect, but at least they give animals opportunities, experiences and skills that seem appropriate.

On the other hand, inappropriate developmental contexts can occasionally be of immense value in a management program. Thus, cross-fostering has been used in crisis situations for a variety of bird species (e.g., Black robins, Black stilts, Whooping cranes, Takahe; references above). Alarm calls of several different species were used as stimuli during attempts to teach Black stilts to recognize predators by Hume

(1995). Cross-fostering in rats and mice leads to changes in corticosterone levels and in behavior (review in Gandelman 1992). Cross-fostering can be a useful tool for manipulating specific aspects of behavior, or for achieving particular management objectives.

Modern animal ethics legislation can be directly contradictory to desirable conservation outcomes. As noted by Kleiman *et al.* (1991), animal welfare is assumed to be enhanced if stress is minimized, health is maximized, and environments are close to natural. Yet wild-living animals are subject to stressful events such as food shortages, internal parasites, extremes of weather, and interactions with predators. Protecting managed animals from these difficulties will result in populations with limited experience and poor coping skills. Subjecting them to stress or extremes may be counter to animal ethics legislation, or arouse the wrath of animal welfare organizations. For example, many of the experiments conducted in the past by experimental psychologists have used mild electric shock as a conditioning stimulus, and mild shock is extremely effective in a broad array of experimental situations. Yet modern ethics committees are unlikely to approve experiments using mild shock, nor would they be likely to approve Gustavson *et al.*'s (1974) work with Coyotes, lambs, and rabbits. The objectives of the legislation are laudable, but in a conservation framework they may need to be balanced against the life or death situation of species in crisis.

Conservation biologists must do good science, but they are frequently restricted by problems such as small sample sizes, constraints on experimental design imposed by management requirements, or limited time. It may therefore be even more difficult to conduct convincing experimental research than is usual in science. We can live with that, although we may have to spend more time on advocacy than scientists are accustomed to doing, some research will necessarily be carried out on surrogate species, and we may have to seek imaginative ways of designing experiments and analyzing data.

Perhaps most importantly, conservation biologists need to be aware that behavior is a fundamental component of the population structure and demography of any species. Animals modify their behavior as a result of experience, whether or not we understand the background to those modifications, and whether or not we have attempted to manipulate those experiences. The flexibility of behavioral development provides a powerful tool for species management, but can be a dangerous weapon if ignored or misused. We entrust our lives to airline pilots because we trust the learning environment in which they developed. To some extent at least, we control the learning environment of managed species. Our role involves assisting them to become successful pilots.

Literature cited

Anonymous. 1993. Regional news reports. *Endangered Species Technical Bulletin* **18**: 2 11.

Alcock, J. 1993. *Animal behavior*. Fifth edition. Sinauer Associates, Sunderland, Massachusetts.

Armstrong, D. P., T. G. Lovegrove, D. G. Allen, and J. L. Craig. 1995. Composition of founder groups for bird translocations: does familiarity matter? Pages 105–112 in M. Serena, editor. *Reintroduction biology of Australian and New Zealand fauna*. Surrey Beatty and Sons, Chipping Norton, Australia.

Avery, M. L., and D. G. Decker. 1994. Responses of captive fish crows to eggs treated with chemical repellents. *Journal of Wildlife Management* **58**: 261–266.

Avital, E., and E. Jablonka. 1994. Social learning and the evolution of behaviour. *Animal Behaviour* **48**: 1195–1199.

Balda, R. P., and A. C. Kamil. 1989. A comparative study of cache recovery by three corvid species. *Animal Behaviour* **38**: 486–495.

Blaustein, A. R., R. K. O'Hara, and D. H. Olson. 1984. Kin recognition behaviour is present after metamorphosis in *Rana cascadae* tadpoles. *Animal Behaviour* **32**: 445–450.

Bolhuis, J. J. 1991. Mechanisms of avian imprinting: a review. *Biological Reviews of the Cambridge Philosophical Society* **66**: 303–345

Bower, T. G. R. 1979. *Human development*. W. H. Freeman, San Francisco.

Box, H. 1991. *Primate responses to environmental change*. Chapman and Hall, London.

Bunin, J. S., and I. G. Jamieson. 1996. Response to a model predator by two species of New Zealand rail. *Conservation Biology*, in press.

Butler, D, and D. Merton. 1992. *The black robin*. Oxford University Press, Auckland.

Carey, S. 1992. Becoming a face expert. *Philosophical Transactions of the Royal Society of London, series B* **335**: 95–103.

Caro, T. M., and M. D. Hauser. 1992. Is there teaching in nonhuman animals? *Quarterly Review of Biology* **67**: 151–174.

Chivers, D. P., G. E. Brown, and J. F. Smith. 1995. Acquired recognition of chemical stimuli from pike, *Esox lucius*, by brook sticklebacks, *Culaea inconstans* (Osteichthyes, Gasterosteidae). *Ethology* **99**: 234–242.

Cohen, P. S., and M. Block. 1991. Replacement of laboratory animals in an introductory-level psychology laboratory. *Human Innovations and Alternatives* **5**: 221–225.

Dawkins, R. 1976. *The selfish gene*. Oxford University Press, Oxford.

Diamond, J. 1981. Mixed-species foraging groups. *Nature* **292**: 408–409.

Dupre, J. 1987. *The latest on the best*. MIT Press, Massachusetts.

Ellis, D. H., S. J. Dobrott, and J. G. Goodwin. 1977. Reintroduction techniques for masked bobwhites. Pages 345–354 in S. A. Temple, editor. *Endangered birds: Management techniques for preserving threatened species*. University of Wisconsin Press, Wisconsin.

Evans, C. S., and P. Marler. 1991. On the use of video images as social stimuli in birds: audience effects on alarm calling. *Animal Behaviour* **41**: 17–26.

Frankham, R. 1995. Genetic management of captive populations for reintroduction. Pages 31–34 in M. Serena, editor. *Reintroduction biology of Australian and New Zealand fauna*. Surrey Beatty and Sons, Chipping Norton, Australia.

Gandelman R. 1992. *Psychobiology of behavioral development*. Oxford University Press, Oxford.

Gibson, D. F., K. A. Johnson, D. G. Langford, J. R. Cole, D. E. Clarke, and Wollowra community. 1995. The rufous hare-wallaby *Lagorchestes hirsutus*: a history of experimental reintroduction in the Tanami Desert, Northern Territory. Pages 171–176 in M. Serena, editor. *Reintroduction biology of Australian and New Zealand fauna*. Surrey Beatty and Sons, Chipping Norton, Australia.

Goodenough, J., B. McGuire, and R. Wallace. 1993. *Perspectives on animal behavior*. John Wiley and Sons, New York.

Gordon, W. C., and R. L. Klein. 1994. Animal memory: the effects of context change on retention performance. Pages 255–279 in N. J. Mackintosh, editor. *Animal learning and cognition*. Academic Press, San Diego.

Gould, S. J., and R. C. Lewontin. 1979. The spandrels of San Marco and the Panglossian paradigm: a critique of the adaptationist programme. *Proceedings of the Royal Society of London* **205**: 581–598.

Griffin, D. R. 1992. *Animal minds*. Chicago University Press, Chicago.

Gustavson, C. R., J. Garcia, W. G. Hawkins, and K. R. Rusiniak. 1974. Coyote predation control by aversive conditioning. *Science* **184**: 581–583.

Hall, G. 1994. Pavlovian conditioning: laws of association. Pages 15–43 in N. J. Mackintosh, editor. *Animal learning and cognition*. Academic Press, San Diego.

Hulet, C. V., D. M. Anderson, J. N. Smith, and W. L. Shupe. 1987. Bonding of sheep to cattle as an effective technique for predation control. *Applied Animal Behavior Science* **22**: 261–267.

Hume, D. K. 1995. Anti-predator training: An experimental approach in reintroduction biology. Unpubl. M.S. Thesis, University of Canterbury, Christchurch, New Zealand.

Johnston, T. D., and A. T. Pietrewicz. 1985. *Issues in the ecological study of learning*. Lawrence Erlbaum Assoc., New Jersey.

Kepler, C. B. 1977. Captive propagation of whooping cranes: a behavioral approach. Pages 231–241 in S. A. Temple, editor. *Endangered birds: Management techniques for preserving threatened species*. University of Wisconsin Press, Wisconsin.

Kleiman, D. G., B. B. Beck, J. M. Dietz, and L. A. Dietz. 1991. Costs of a re-introduction and criteria for success: accounting and accountability in the golden lion tamarin conservation program. *Symposium of the Zoological Society of London* **62**: 125–142.

Krall, P., and K. Misof. 1995. Constraints on the evolution of organismic structures and their consequences for adaptation of cognitive mechanisms. *Perspectives in Ethology* **11**: 149–190.

Lovegrove, T. 1992. Effects of introduced predators on the saddleback. Unpublished Ph.D. Thesis, University of Auckland, New Zealand.

Mace, W. M. 1989. Editorial. *Ecological Psychology* **1**: 1.

Mackintosh, N. J. 1994. *Animal learning and cognition*. Academic Press, San Diego.

McLean, I. G., G. Lundie-Jenkins, and P. J. Jarman. 1995. Teaching captive rufous hare-wallabies to recognise predators. Pages 177–182 in M. Serena, editor. *Reintroduction biology of Australian and New Zealand fauna.* Surrey Beatty and Sons, Chipping Norton, Australia.

McLean, I. G., G. Lundie-Jenkins, and P. J. Jarman. 1996. Teaching an endangered mammal to recognise predators. *Biological Conservation* **56**: 51–62.

Miller, B., D. Biggens, L. Hanebury, and A. Vargas. 1994. Reintroduction of the black-footed ferret (*Mustela nigripes*). Pages 455–464 in P. J. S. Olney, G. M. Mace, and A. T. C. Feistner, editors. *Creative conservation: Interactive management of wild and captive animals.* Chapman and Hall, London.

Mineka, S., and M. Cook. 1993. Mechanisms involved in the observational conditioning of fear. *Journal of Experimental Psychology: General* **122**: 23–38.

Moscovitch, M. 1985. Memory from infancy to old age: implications for theories of normal and pathological memory. *Annals of New York Academy of Sciences* **444**: 78–96.

Pearce, J. M. 1987. *An introduction to animal cognition.* Lawrence Erlbaum Associates, London.

Pierce, R. 1984. Plumage, morphology and hybridisation of New Zealand stilts *Himantopus* spp. *Notornis* **31**: 106–130.

Payne, K., and R. S. Payne. 1985. Large scale changes over 19 years in songs of humpback whales in Bermuda. *Zeitschrift für Tierpsychologie* **46**: 298–305.

Payne, R. B., W. L. Thompson, K. L. Fiala, and L. L. Sweany. 1981. Local song traditions in indigo buntings: cultural transmission of behaviour patterns across generations. *Behaviour* **77**: 199–221.

Real, L. A. 1993. Towards a cognitive ecology. *Trends in Ecology and Evolution* **8**: 413–417.

Reeve, H. K., and P. W. Sherman. 1993. Adaptation and the goals of evolutionary research. *Quarterly Review of Biology* **68**: 1–32.

Ristau, C. A. 1991. *Cognitive ethology: The minds of other animals.* Lawrence Erlbaum Associates, New Jersey.

Roitblatt, H. L., T. G. Bever, and H. S. Terrace. 1984. *Animal cognition.* Lawrence Erlbaum Associates, New Jersey.

Schmidt-Nielsen, K. 1979. *Animal Physiology: Adaptation and environment.* Cambridge University Press, Cambridge.

Seligman, M. E. 1970. On the generality of the laws of learning. *Psychological Review* **77**: 406–418.

Sherman, P. W. 1988. The levels of analysis. *Animal Behaviour* **36**: 616–619.

Shettleworth, S. J. 1994. Biological approaches to the study of learning. Pages 185–219 in N. J. Mackintosh, editor. *Animal learning and cognition.* Academic Press, San Diego.

Spear, N. E. 1984. Ecologically determined dispositions control the ontogeny of learning and memory. Pages 326–358 in R. Kail and N. E. Spear, editors. *Comparative perspectives on the development of memory.* Lawrence Erlbaum Associates, New Jersey.

Spear, N. E., and D. Kucharski. 1984. Ontogenetic differences in stimulus selection during conditioning. Pages 227–252 in R. Kail and N. E. Spear, editors. *Comparative perspectives on the development of memory.* Lawrence Erlbaum Associates, New Jersey.

Starck, J. M. 1993. Evolution of avian ontogenies. *Current Ornithology* **10**: 275–366.
Sutherland, G. D., and C. L. Gass. 1995. Learning and remembering of spatial patterns by hummingbirds. *Animal Behaviour* **50**: 1273–1286.
Terrace, H. S. 1984. Cognition in animals and humans. Pages 7–28 in H. L. Roitblatt, T. G. Bever, and H. S. Terrace, editors. *Animal cognition*. Lawrence Erlbaum Associates, New Jersey.
Via, S., R. Gomulkiewicz, G. De Jong, S. M. Scheiner, C. D. Schlichting, and P. H van Tienderen. 1995. Adaptive phenotypic plasticity: consensus and controversy. *Trends in Ecology and Evolution* **10**: 212–217.
Vieth, W., E. Curio, and E. Ulrich. 1980. The adaptive significance of avian mobbing. III. Cultural transmission of enemy recognition in blackbirds: cross species tutoring and properties of learning. *Animal Behaviour* **28**: 1217–1229,
Vom Saal, F. 1984. Proximate and ultimate causes of infanticide and parental behavior in male house mice. Pages 401–424 in G. Hausfater and S. Blaffer Hrdy, editors. Infanticide: *Comparative and evolutionary processes*. Aldine, New York.
Waldman, B. 1991. Kin recognition in amphibians. Pages 162–219 in P. G. Hepper, editor. *Kin recognition*. Cambridge University Press, Cambridge.
Williams, G. C. 1966. *Adaptation and natural selection*. Princeton University Press, Princeton.
Wunderle, J. M. 1991. Age-specific foraging proficiency in birds. *Current Ornithology* **8**: 273–324.
Yoerg, S. I., and A. C. Kamil. 1991. Integrating cognitive ethology with cognitive psychology. Pages 273–289 in C. A. Ristau, editor. *Cognitive ethology: The minds of other animals*. Lawrence Erlbaum Associates, New Jersey.

Chapter 7

Hatching asynchrony in parrots: Boon or bane for sustainable use?

SCOTT H. STOLESON & STEVEN R. BEISSINGER

Recently biologists have become increasingly alarmed over the accelerating loss of biodiversity, particularly in tropical regions. Academic scientists, including most behaviorists, are now trying more than ever to apply their seemingly irrelevant knowledge and expertise to real world conservation problems.

Among a broad range of strategies proposed to conserve biodiversity (Soulé 1991) is a growing interest in the pragmatic approach that stresses the value of maintaining healthy ecosystems and populations. Conservation efforts may be most effective when organisms and ecosystems 'pay their own way,' either through the value of the goods and services provided by an ecosystem, or by utilizing particular organisms with economic value in a sustainable manner (e.g., Fearnside 1989; Robinson & Redford 1991*a*; Balick & Mendelsohn 1992). Sustainable use is one conservation strategy where the need for basic behavioral and ecological research has been particularly recognized (Lubchenco *et al.* 1991; Mangel *et al.* 1993; Meyer & Helfman 1993).

In many cases a behaviorist's evolutionary perspective, using multiple working hypotheses, can provide unique insights into effective management options (Arcese *et al.*, Chapter 3). We present a case study showing how theory-based behavioral studies can be not only useful, but essential, for developing appropriate and effective conservation strategies. Specifically, we discuss how studies of incubation behavior in parrots can provide useful information for the design and implementation of sustainable harvest programs. Many parrots in the world are highly valued in the international pet trade, so a program of sustainable harvest may be a feasible conservation option for them (Thomsen & Brautigam 1991; Beissinger & Bucher 1992*a*). Unfortunately, knowledge of the basic biology of these birds lags behind the need for such information to establish sustainable levels of harvest and to increase the productivity of managed populations. Studies of parrot behavior can help to bridge that gap.

The current conservation crisis of parrots

Status of parrot populations

Parrots (family Psittacidae) are one of the most threatened groups of birds in the world (Beissinger & Snyder 1992). Approximately one-third of the 140 New World species of macaws, parrots, and parakeets are considered to be at risk of extinction (Collar & Juniper 1992). These threatened species are concentrated in relatively small areas of the Neotropics, primarily in the Andean highlands, Atlantic forests of Brazil, and islands of the Caribbean (Collar & Juniper 1992). With few exceptions, populations of almost all New World parrot species are thought to be declining (Collar & Juniper 1992).

The status of Old World parrots is much less comprehensively known. In Australia many forest species are known to be endangered or declining, although several non-forest species have increased dramatically with the expansion of industrialized agriculture (Long 1984; Joseph 1988). Relatively few parrots of Africa or mainland Asia are currently considered endangered, but this may reflect a lack of information more than healthy population levels (Collar & Stuart 1985; Collar *et al.* 1994). A high proportion of species inhabiting islands are threatened or declining, in part because many occupy very restricted ranges (e.g., Lambert 1985; Taylor 1985; Merritt *et al.* 1986; Rinke 1989; Evans 1991; Robinet *et al.* 1995).

Causes for declines in parrot populations

Parrot populations are declining for two primary reasons: habitat destruction and direct exploitation for the wild bird trade (Fig. 7.1). As with most threatened organisms, habitat destruction is an important factor reducing population sizes for many parrot species. Tropical habitats, particularly forests, are being destroyed through logging, conversion to agriculture, and urbanization. Burgeoning human populations and problems of poverty and inadequate land tenure have accelerated these losses. Parrot species inhabiting montane forests have been especially hard hit because highland regions in the tropics are the most favored areas for

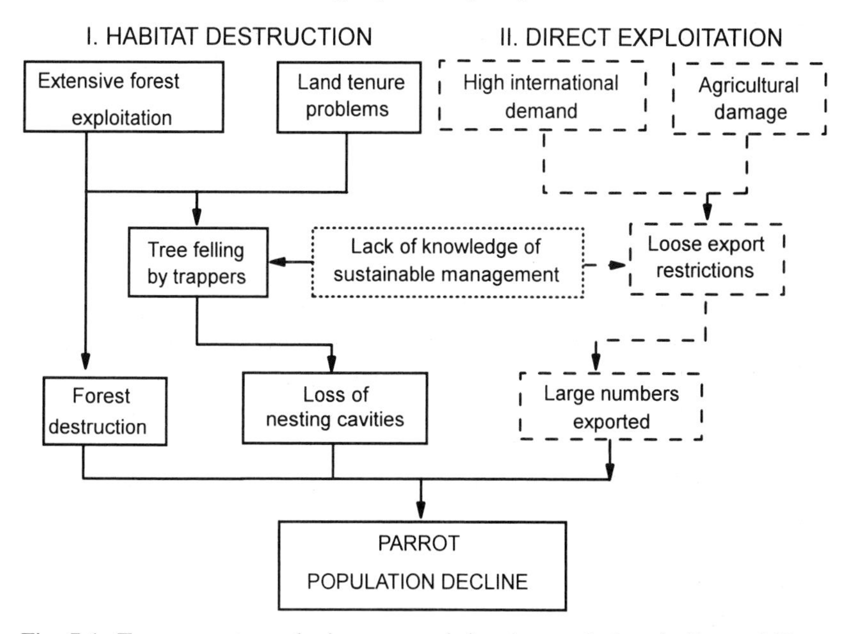

Fig. 7.1. Two separate paths have caused drastic population declines of Neotropical parrots: habitat destruction and direct exploitation for the pet trade. Providing the information needed for sustainable management potentially addresses both paths. Adapted from Beissinger & Bucher (1992*a*).

settlement and agriculture (Houghton *et al.* 1991). Habitat loss is thought to be wholly or partially responsible for population declines in over two-thirds of all parrot species (Collar & Stuart 1985; Collar & Juniper 1992). In Australia, where the capture and exportation of wild parrots was banned in the 1950s, habitat loss resulting from agricultural practices is the single most important cause of population declines (Joseph 1988; Joseph *et al.* 1991; Saunders 1991).

Outside of Australia, parrots bear the additional burden of being exploited directly for the pet trade. Large numbers of wild caught birds are transported from developing tropical countries to developed countries each year. Over 250 000 psittacines were exported annually from mainland Neotropical countries between 1982 and 1986 (Thomsen & Brautigam 1991). The impact of harvesting on wild populations is greater than export figures suggest because pre-export mortality may exceed 50% of individuals captured (Iñigo-Elias & Ramos 1991). In most countries the volume of non-export, internal trade is also high (Beehler 1985; Desenne & Strahl 1991; Thomsen & Brautigam 1991). Exploitation for the pet trade affects over half of all parrot species (Smiet 1985; Jorgenson & Thomsen 1987; Thomsen & Mulliken 1992).

Additional factors are a major cause of mortality in some psittacine populations. Island populations have been hard hit by introduced predators (e.g., Beggs & Wilson 1991; Gnam & Rockwell 1991; Robinet *et al.* 1995). Hunting of parrots for food or plumes by indigenous peoples can be a significant source of mortality in remote areas (Beehler 1985; Redford & Robinson 1987).

The two processes of habitat loss and direct exploitation represent two distinct avenues for population declines in parrots (Fig. 7.1; Beissinger & Bucher 1992*a*). Conservation solutions that address one suite of problems fail to affect the other. For example, while habitat protection may be necessary for supporting parrot populations, it is insufficient as long as economic incentives for capturing wild birds exist. On the other hand, an import ban to reduce the demand for wild birds would do nothing to alleviate the rate of forest destruction in the tropics (Beissinger & Bucher 1992*a*).

Life history and behavior of parrots

The decline of parrot populations from habitat loss and direct exploitation is exacerbated by certain characteristics of the life history and behavior of parrots. Parrots have low reproductive rates because of slow maturation, small clutch sizes in larger species, and a high proportion of non-breeders (e.g., Lanning & Shiflett 1983; Smith & Saunders 1986; Gnam & Rockwell 1991; Munn 1992; Navarro *et al.* 1992). In most populations only a small proportion of potential breeders actually attempt to breed in any given year. For example, in the Puerto Rican parrot (*Amazona vitatta*), non-breeders comprise 57% of the population (Snyder *et al.* 1987). In the Manú Biosphere Reserve, Perú, only 10 to 20% of adult mated pairs of macaws (*Ara* spp.) nest in any given year (Munn 1992). In our own studies of the Green-rumped parrotlet (*Forpus passerinus*), an average of 37% of the banded adults seen did not breed in any given year (unpublished data). This figure certainly underestimates the actual proportion of non-breeders in our study population because our banding and resighting efforts have focused on breeding birds. The net result of having a large proportion of non-breeders is that the effective size of any parrot population is actually much smaller than its census size. Also, parrots exhibit asynchronous hatching, which frequently results in the mortality of the smallest young (Rowley 1990; Beissinger & Waltman 1991).

Almost all species are secondary cavity nesters, and therefore nest sites may be limited (Snyder 1977; Forshaw 1989). In many species of hole-nesting birds, the size of the breeding population is constrained by the availability of nest sites (Newton 1994). Local people frequently destroy nesting trees to gain access to nestlings for the pet trade. This activity compounds the loss of nest sites because of habitat destruction.

Conservation alternatives for parrots

Three general conservation strategies based on biological factors have been suggested as appropriate for parrots (Beissinger & Snyder 1992): habitat protection, captive breeding, and sustainable harvesting for the pet trade.

Habitat protection

Habitat protection is crucial for maintaining any wild population. Protecting essential habitat for parrots has the advantage of protecting other organisms that share the habitat. Many parrot species, however, do not require pristine habitat, and can thrive in disturbed areas (e.g., Snyder *et al.* 1987; Bucher 1992; Waltman & Beissinger 1992; Wiley *et al.* 1992). More importantly, habitat protection alone does not address the problem of direct exploitation (Redford 1992). Thus, habitat protection is necessary but not sufficient for conserving many parrot species.

Captive breeding

Ideally captive breeding programs could provide parrots for reintroductions or to augment existing populations; however, many problems interfere with its implementation and use (Derrickson & Snyder 1992; Wiley *et al.* 1992). To start a captive population, individuals must be removed from the wild, exacerbating the problems inherent in small populations. Captive breeding programs are very expensive compared with *in situ* methods; channeling scarce funds into captive breeding may pre-empt more cost-effective *in situ* techniques (Snyder *et al.* 1996). Once implemented, captive breeding programs for parrots face numerous and significant problems, including obtaining consistent reproduction, avoiding progressive domestication, controlling disease, reintroducing captive-raised birds into the wild, and maintaining financial and administrative support (Dobson & May 1986; Derrickson & Snyder 1992).

Captive breeding for the purpose of conserving threatened parrots has had little success in the past and is unlikely to prove successful at rescuing wild populations of parrots in the future. Captive programs ignore the ultimate causes of population declines in parrots: habitat loss and exploitation for the bird trade. Captive breeding, therefore, is not a useful tool for parrot conservation, except in a few extreme cases in which a species is in imminent danger of extinction, such as the Puerto Rican parrot or Spix's macaw (Snyder *et al.* 1987; Juniper & Yamashita 1991). Aviculturalists have argued that captive breeding may help to alleviate the pressures on wild populations by supplying birds for the pet trade. The costs of captive production far exceed the costs of harvesting and transporting wild birds, however (Chubb 1992). Captive-bred birds are unlikely to compete successfully in the marketplace as long as there are wild birds to be caught (Thomsen & Brautigam 1991; Derrickson & Snyder 1992).

Sustainable harvesting

Proponents have promoted sustainable use as a way to integrate conservation priorities with social and economic necessities (Reid 1989; Lubchenco *et al.* 1991), although the definition and even the attainability of sustainable use are debated (Simon 1989; Ludwig *et al.* 1993; Mangel *et al.* 1993). Here we use the following definition: sustainable harvesting refers to the continued persistence and replenishment of a resource despite utilization (Beissinger & Bucher 1992*a*). The conservation function of sustainable harvesting of parrots is to provide the pet trade with an abundant, dependable source of birds, while simultaneously providing local people with economic incentives to maintain healthy populations of parrots and their habitats. Thus, sustainable utilization of parrots may address both of the ultimate causes of declining parrot populations (Fig. 7.1).

Before sustainable harvesting programs can be implemented, substantial information is required about population structure, factors that limit reproduction, and habitat requirements (Beissinger & Bucher 1992*a*,*b*). Long-term studies with marked individuals are required to collect this information. Once determined, factors that limit populations may be manipulated to increase the productivity of a population.

Models for sustainable harvesting

In a sustained yield model, the harvest rate used is the annual increment to the population such that the net change in the population size after harvesting is zero. Typically, the harvest levels used are near the maximum sustained yield, which is that level of harvest from the population size where the population growth rate is maximized. In traditional natural resource management of timber or fisheries, the maximum sustained yield is usually achieved when populations are at about half of their carrying capacity, assuming a logistic growth pattern for the population (Fig. 7.2*a*).

Traditional maximum sustained yield models are not really appropriate for parrots because often the models do not differentiate between individuals of different age classes. Most managed species are harvested as adults. With parrots, however, nestlings are preferable to adults for harvesting because: (1) chicks make better pets than wild-caught adults; (2) chicks bring higher market prices than adults; and (3) chicks have a lower reproductive value to the population than adults (Fisher 1930). If only nestlings are to be harvested, then the harvestable portion of the population is maximized when the population is near its carrying capacity (Fig. 7.2*b*). Logically, the more birds there are in the population, the more harvestable young they can produce. Thus, a program of sustainable harvesting of parrots would not only reduce pressure from the pet trade on wild populations, but also provide incentives to maintain robust population sizes.

In species such as parrots where we lack sufficient biological knowledge to establish sustainable harvest rates, the use of a Conservative Sustained Harvest model is warranted (Beissinger & Bucher 1992*a*,*b*). This approach suggests that when a population is increasing or stable, then any management programs that increase productivity would lead to a harvestable surplus (Fig. 7.3). By managing factors that limit population growth, productivity may be increased by: (1) increasing the number of breeders within the population; (2) increasing the proportion of breeders that successfully fledge young; or (3) increasing the number of young fledged per successful nest (Beissinger & Bucher 1992*a*,*b*). The number of breeders in a population may be increased by providing supplemental food or nest sites (boxes). The proportion of successful breeders can be increased by adding or predator-proofing nest boxes. Fledging success may be improved by adding nest boxes (to decrease the incidence of infanticide), providing supplemental food, or forcing double clutching

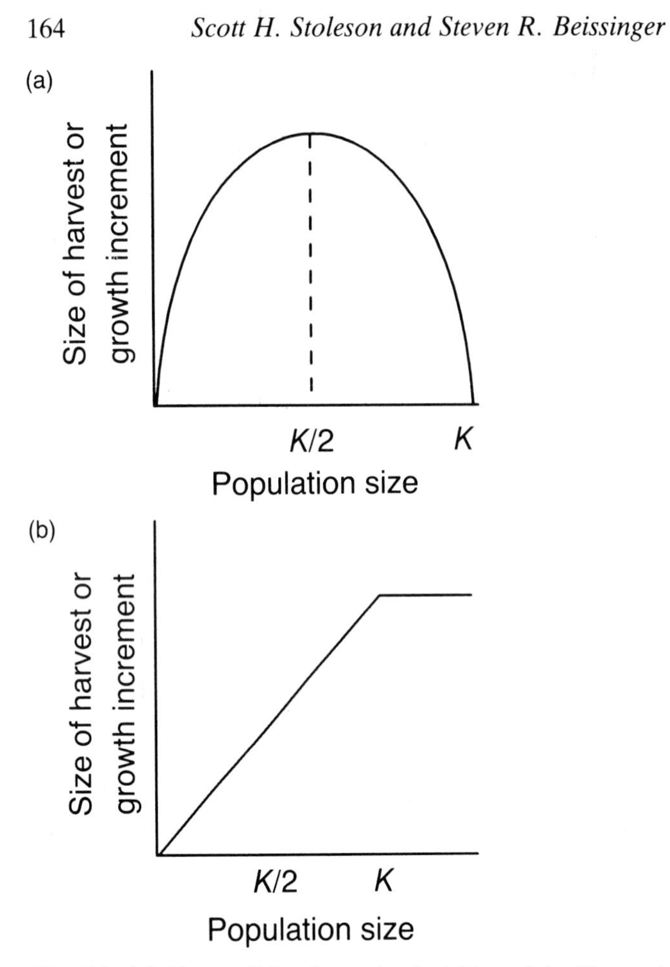

Fig. 7.2. (*a*) The traditional sustained yield model of harvest as a function of population size. The maximum sustained yield occurs at approximately $K/2$ when population growth is logistic. (*b*) A hypothetical sustained yield model for harvesting parrot nestlings. Harvest size = annual growth increment as a function of population size. The maximum sustained yield occurs near K. Adapted from Beissinger & Bucher 1992*b*.

(Beissinger & Bucher 1992*b*). Manipulating the degree of hatching asynchrony can increase fledging success in some cases, but an understanding of the functional basis of this behavior is required.

Hatching asynchrony

Many birds delay the initiation of incubation until the last egg in a clutch has been laid, causing all eggs to develop and hatch synchronously. If

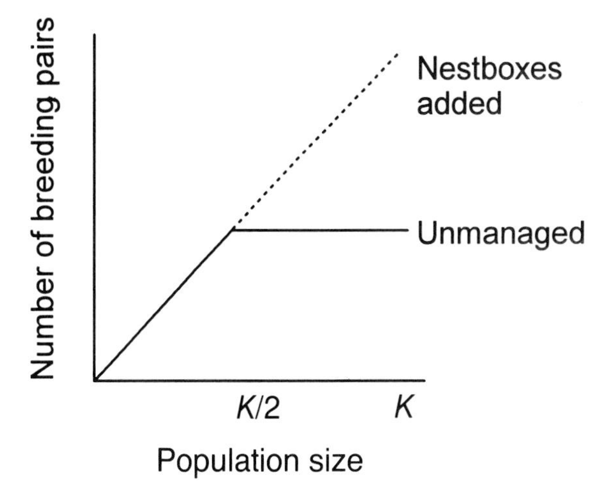

Fig. 7.3. Effect of adding nest sites on the number of breeding pairs when nest sites are limiting, based on the Conservative Harvest Model. Adapted from Beissinger & Bucher (1992*b*).

parents initiate incubation before the last egg is laid, the result is a staggered hatching of offspring. Asynchronous hatching produces chicks of different sizes in a nest, and frequently results in the mortality of the smallest nestlings (Lack 1947; Stokland & Amundsen 1988; Beissinger & Waltman 1991). Beissinger & Bucher (1992*a*,*b*) suggested that because these late-hatched chicks have low probabilities of survival, they may be harvested without adversely affecting the population.

Hatching asynchrony may be applied to sustainable use in other ways as well. The early onset of incubation and resulting asynchronous hatching may serve a number of different functions that reflect different constraints on reproductive success (Stoleson & Beissinger 1995). By identifying the function(s) of hatching asynchrony in a particular species through experimental studies, these constraints on reproduction can be identified. Management actions can then be formulated that address or correct for those particular constraints. Thus, by understanding the function of early incubation behavior, the behavioral ecologist can identify ways to effectively increase productivity in a population.

At least 17 hypotheses have been proposed to explain hatching asynchrony in birds (reviewed in Stoleson & Beissinger 1995). Four categories of hypotheses can be described based on the nature of the selective forces or constraints influencing the decision of when to incubate, and the period(s) in the nesting cycle when those forces or

Table 7.1 *Hypotheses for the evolution of hatching asynchrony*

Evolutionary significance of asynchrony	Constraint or selective force	Critical period in the nesting cycles	Hypothesis
Non-adaptive	Physiological	Laying	Hormone
		Laying	Energy constraints
Adaptive	Environment	Laying	Egg viability
		Laying	Egg protection
		Laying	Limited breeding opportunities
	Behavior	Laying	Brood parasitism
	Parental efficiency	Nestling	Peak load reduction
		Nestling	Dietary diversity
		Nestling	Larder
	Behavior	Nestling	Sibling rivalry reduction
		Nestling	Sex ratio manipulation
	Environmental	Nestling	Brood reduction
	Variation	Nestling	Hurry-up
		Nestling	Insurance
	Behavior	Laying/brooding	Sexual conflict
	Predation	Laying/fledgling	Nest failure
		Laying/brooding	Adult predation

For each hypothesis, the adaptive significance of asynchrony, the nature of the selective force, and the critical period in the nesting cycle are indicated. Adopted from Stoleson & Beissinger (1995).

constraints are relevant (Table 7.1). Two hypotheses suggest there is no adaptive function to either early incubation or the resulting hatching asynchrony; rather, parents are physiologically constrained energetically or hormonally to initiate incubation before the last egg is laid. Four hypotheses posit an adaptive function to the early onset of incubation. The resulting hatching asynchrony is considered incidental, and any offspring mortality resulting from asynchrony would be maladaptive. Early incubation may function to protect eggs or nest sites from unfavorable ambient temperatures, predators, conspecifics, or brood parasites. Eight hypotheses are premised on the idea that asynchronous hatching and the resulting developmental asymmetries among nestlings have an adaptive function. Asynchronous hatching may increase parental efficiency in feeding, or enable parents to deal with stochasticity of food resources or hatching success. Finally, three hypotheses suggest that the timing of the incubation and fledging periods are a response to differential predation through the nesting cycle, or are part of a strategy to increase male investment by lengthening the time females are on the nest.

Differentiating between these groups of hypotheses to determine the function of hatching asynchrony in any given taxon requires answering several questions. Does early incubation serve to protect the eggs, create an asynchronous hatching pattern, or something else? Does the mortality of small offspring have a function, as with the Brood Reduction Hypothesis, or is it merely incidental? Identifying the function of early incubation and resultant hatching asynchrony can help to identify limits to reproduction. In turn, these limits can indicate potential management actions for increasing productivity, and designing a valid, sustainable harvest program. For example, by experimentally identifying asynchrony as an adaptation to adjust brood size to unpredictable food resources (e.g., Wiebe & Bortolotti 1994), a behaviorist would know that providing supplemental food would be an effective strategy to increase productivity. In this case, it is unlikely that manipulating the degree of asynchrony would have a beneficial effect. This process is better explained through the use of several examples.

Applying hypotheses for hatching asynchrony to sustainable use of parrots

Macaws and the insurance function of asynchrony

Macaws (e.g., *Anodorhynchus* spp., *Ara* spp., *Cyanopsitta* sp.) are among the largest parrots and the most prized in the pet trade. Most species are considered endangered, and some are on the verge of extinction, primarily because of persecution for the exotic bird trade (Yamashita 1987; Collar & Juniper 1992; Munn 1992). Macaws share a number of reproductive traits. They generally lay clutches of 2 to 3 eggs. The smallest chick in a brood rarely survives, even when there is a surplus of food. However, the smallest chick survives if older sibs do not hatch or die very young (Lanning 1991; Munn 1992). Thus, asynchrony in macaws appears to pertain, at least partially, to an insurance function of the smallest chicks.

Understanding the insurance function of asynchrony in macaws has several uses. Because parents rarely raise smallest young, strategies that attempt to reduce offspring mortality may be ineffectual and are unlikely to increase productivity. From a conservation perspective, when the older chicks survive these surplus young can be harvested directly with little or no impact on the population. They represent potential productivity that would otherwise be lost.

Charles Munn and his associates in Manú, Perú, have experimentally harvested the last-hatched young in several macaw species. Chicks were removed from nests at a young age and then hand-raised. They were kept in the area as tourist attractions, however, and never marketed as an economic resource (Munn *et al.* 1991; Munn 1992).

The idea of harvesting small young with low reproductive value has been applied to other species as well. Gyrfalcons (*Falco rusticolus*) are highly coveted by falconers, and like the large macaws, command high prices in the international bird market. Similarly, chicks hatch asynchronously, and the smallest chick functions as insurance (Clum & Cade 1994). The Canadian government has experimented with programs that permitted the harvest of smallest chicks from selected nest sites (Mossop & Hayes 1982). These programs proved biologically and economically feasible, but have been discontinued due to political problems (Clum & Cade 1994). A similar experimental harvest of Prairie falcon (*F. mexicanus*) nestlings resulted in little effect on population parameters or size and appeared to be sustainable over a seven year period (Conway *et al.* 1995).

Parrotlets and the maintenance of egg viability

We have researched the breeding biology and demography of the Green-rumped parrotlet in the llanos of Venezuela. This species is among the smallest of the New World parrots, and one of the few species that may be increasing in number (Forshaw 1989). This species is also unusual because females typically lay large clutches (average = 7, range = 4 to 10 eggs) that hatch completely asynchronously. A clutch of eight eggs may take up to 14 days to hatch (Beissinger & Waltman 1991). Partial brood loss is frequent in larger broods, and the smallest one or two chicks frequently die.

We have tested experimentally a number of hypotheses to determine the function or functions of early incubation that produces extreme hatching asynchrony in this species. We used synchronization experiments (Forbes 1994) to test hypotheses based on the premise that an asynchronous hatching pattern confers some tangible benefit to parents. To summarize the results, we found that broods manipulated to hatch relatively synchronously enjoyed greater reproductive success than natural asynchronous broods, as measured by number and survivorship of offspring (Fig. 7.4). In addition, parents at synchronized broods incurred no detectable costs in terms of effort, survivorship, or impact on future reproduction (Stoleson and Beissinger, in review). Thus, asynchronous hatching in this species appears to serve no obvious adaptive function for

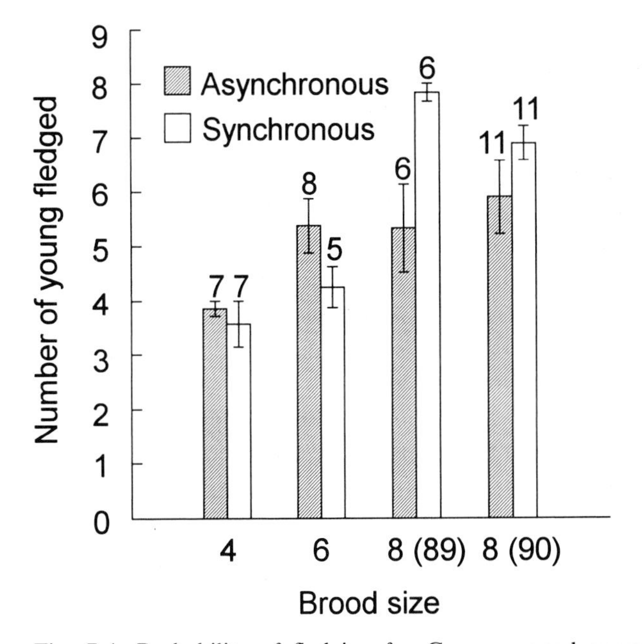

Fig. 7.4. Probability of fledging for Green-rumped parrotlet nestlings from experimental synchronous and asynchronous broods as a function of brood size. Bars indicate one standard error around the mean, and sample sizes appear above each bar. Values exclude depredated nests. Brood size (F=19.5, P<0.001) and the interaction of brood size and asynchrony (F=8.1, P=0.001) affected the number of young fledged (two-way ANOVA). Large synchronous broods produced more young than asynchronous broods in 1989 (t=4.3, P=0.003) and marginally so in 1990 (t=1.5, P=0.15).

the survival of chicks, and offspring mortality appears to be maladaptive. For parrotlets, therefore, any management techniques that address the problem of offspring mortality associated with asynchronous hatching may increase reproductive productivity. Our experiments specifically demonstrated that reproductive success can be increased by manipulating eggs or young to create broods with reduced degrees of asynchrony (Fig. 7.4).

When we found no evidence to support an adaptive function to asynchronous hatching patterns in the parrotlet, we began testing whether the early onset of incubation confers benefits to parents by ensuring the survival of early embryos. One hypothesis suggests that early incubation functions to maintain the viability of first-laid eggs (the Egg Viability Hypothesis; Arnold *et al.* 1987; Veiga 1992). Embryos begin to develop when eggs are heated above physiological zero (the temperature below which no embryological development occurs, approximately 26–28°C; Drent 1973; Webb 1987). As embryos develop, they become sensitive to changes in temperatures. Prolonged or repeated exposure to temperatures above physiological zero yet below normal incubation levels (34 to 38°C) results in abnormal development or death (Wilson 1991; Deeming & Ferguson 1992). In areas where ambient temperatures regularly exceed physiological zero, embryos may begin to develop in the absence of incubation. Thus, parents may be obliged to begin continuous incubation to maintain the hatchability of first laid eggs.

We tested this hypothesis by removing freshly laid eggs from nest boxes and isolating them in an empty but otherwise identical nest box. After being held for various lengths of time to simulate delays in the onset of incubation, these eggs were placed under incubating females and were incubated to completion. We found that compared with control eggs in recipient nests that were not exposed to ambient temperatures, experimental eggs showed a significant reduction in hatchability with just one day of exposure, and a rapid decline in hatchability with greater exposure times (Fig. 7.5). Thus, there appears to be a physiological basis for the early onset of incubation in the parrotlet.

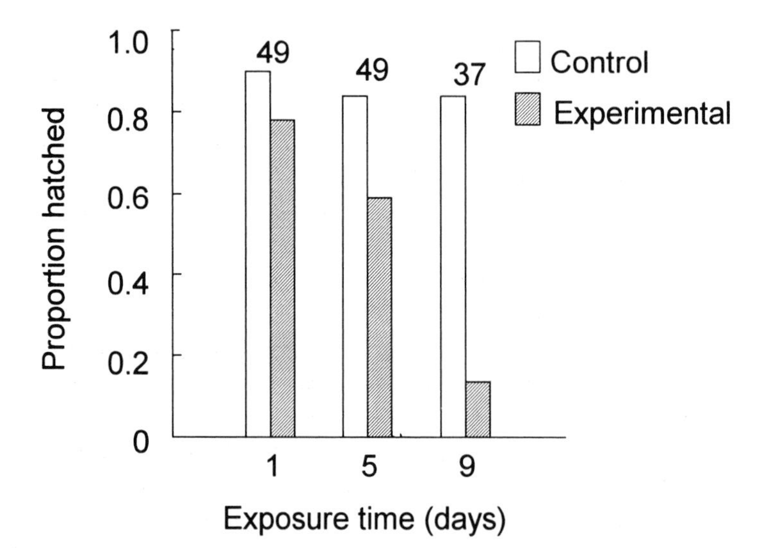

Fig. 7.5. Proportion of experimental parrotlet eggs that hatched in relation to exposure to ambient temperatures prior to incubation for periods of one, five, and nine days, compared with untreated controls. Sample sizes are the number of matched pairs of experimental and control eggs. The probability of hatching did not differ significantly between control and experimental eggs after one day exposure ($\chi^2_1=2.7$, $P=0.10$), but did differ after five days exposure ($\chi^2_1=7.2$, $P=0.007$) and nine days exposure ($\chi^2_1=36.6$, $P<0.001$).

Early incubation to protect limited nest sites

When nest sites are limited in number, parents may be forced to protect the site by initiating incubation early (the Limited Breeding Opportunities Hypothesis; Beissinger & Waltman 1991), particularly in species such as parrots that do not defend general-purpose territories. In our population of parrotlets, behavioral evidence including fights, egg destruction, and infanticide indicate intense competition for nest sites (personal observations). The hypothesis that the early onset of incubation functions to protect eggs from takeover may be tested by placing several parrotlet eggs in empty nest boxes for a period of three days to simulate laying without incubating. Behaviors of individual birds at the site would be observed, and the fate of the eggs followed during the test. Our preliminary results suggest there is strong selective pressure to protect eggs from destruction by conspecifics.

In the parrotlet, the availability of nesting sites appears to be a limiting factor constraining population growth. Although it is difficult to document a change in the number of breeding birds due to the use of nest boxes without carefully structured experiments, evidence suggests the productivity of our study population may have increased considerably since boxes were placed. Parrotlets breed naturally in hollow trees or limbs, termitoriums, or hollowed fence posts (Forshaw 1989; Waltman & Beissinger 1992). At Hato Masaguaral in Venezuela, artificial nest boxes hung on fence posts rapidly became the most used nesting site (Beissinger & Bucher 1992*a*). The number of nesting attempts in boxes increased from 58 in 40 boxes in 1988 to 181 in 105 boxes in 1994 (Table 7.2). During the same period, no more than 11 nesting attempts per year were recorded in fence posts within the study site. Reproductive success was slightly but not significantly greater (Student's *t*-test, t = 0.87, DF = 15.1, $P = 0.40$) in successful nest boxes (4.5 ±1.9 fledglings, N = 571 nests) than the successful natural sites (4.1 ± 1.6 fledglings, N = 15 nests). Nests in boxes were more likely ($\chi^2 = 9.83$, DF = 1, P <0.002) to fledge some young (60%, N = 971) than were nests in fenceposts from 1985 to 1994 (34%, N = 38). Thus, the addition of nest boxes has increased productivity in parrotlets through an increase in the number of breeders and by increasing reproductive success. Between 1988 and 1994, a total of 2562 parrotlets fledged from nest boxes at this site (Table 7.2). During the same period the number of birds known to have fledged from natural sites was 62, but may have been as high as 95. Increases in productivity through management actions such as these

Table 7.2 *Use and success of nest boxes and natural cavities by* Forpus passerinus *in the Llanos of Venezuela 1988–1994*

	Year						
Nest use characteristic	1988	1989	1990	1991	1992	1993	1994
No. of nest boxes	40	100	100	100	100	100	105
No. of nesting attempts in boxes	58	119	153	171	173	146	181
No. of successful attempts in boxes	38	56	105	118	108	75	111
No. of young fledged from boxes	156	260	469	483	465	295	507
No. of attempts in natural sites	5	3	3	2	6	6	11
No. of young fledged from natural sites	8	5	5	4	11	10	17

represent potential for harvesting according to the Conservative Harvest Model (Fig. 7.3).

When behavioral studies indicate that nest sites are limiting, the use of nest boxes can allow a greater proportion of birds to breed, and thereby increase productivity (Toland & Elder 1987; Beissinger & Bucher 1992*b*). Nest boxes have been employed in the management of other parrots (Snyder 1977; Rowley 1990; Munn *et al.* 1991; Mawson & Long 1994), as well as other cavity-nesting species (e.g., Marti *et al.* 1979; Johnson 1987; Caine & Marion 1991; Petty *et al.* 1994).

These examples show that early incubation and asynchrony can have different functions in related species or even in a single species. Simple generalizations across taxa may be inappropriate. Only by experimentally testing appropriate hypotheses can the function(s) of asynchrony in a particular species be determined, and the corresponding constraint on reproduction identified. Once identified, these constraints can be addressed or manipulated to increase productivity and create a harvestable surplus of young. Determining the function of asynchrony in different parrot species is just one area where behavioral studies are needed to guide conservation efforts.

Areas for future research

Parrots are highly social and intelligent animals and display a variety of complex behaviors. Despite increasing interest in wild parrots by both behavioral ecologists and conservation biologists, many aspects of their biology and behavior remain enigmatic. Parrot behavior presents tremendous opportunities for behavioral research that has immediate applicability to conservation.

Aside from our own work, virtually nothing is known of the function or consequences of hatching asynchrony in parrots. A number of species are known to be only partially asynchronous (e.g., Navarro & Bucher 1990; Rowley 1990). Why such variation exists is unclear. In some species, such as the Puerto Rican parrot (Snyder *et al.* 1987), last-hatched nestlings do not suffer reduced survival in spite of pronounced size asymmetries. Further behavioral studies are needed to know if these young somehow differ in quality from older sibs, and how the population might be impacted by their harvest.

Much needs to be learned of mating and reproductive behaviors of parrots. Many of the parrots in the genus *Amazona* will not use nest boxes (Snyder *et al.* 1987). A study of the criteria used in cavity selection in these species is urgently needed and would have immediate application to conservation. The question of why only a fraction of adults in most populations actually breed in any given year also remains unanswered. This issue provides a unique opportunity for research that combines demography, life-history theory, and social behavior. An understanding of this phenomenon might lead to the development of management actions to increase the proportion of breeders, and thus increase the productivity of a population.

The complex social systems and high intelligence of parrots offer particular challenges for conservation. Young birds are slow to learn essential skills such as foraging and predator-avoidance, and appear to require much social contact to successfully acquire these behaviors (e.g., Diamond & Bond 1991; Snyder *et al.* 1994). Although learning and social facilitation in parrots may not be relevant to sustainable use, the success of reintroduction programs may depend on an understanding of these behaviors. For example, recent efforts to reintroduce the Thick-billed parrot (*Rhynchopsitta pachyrhyncha*) in Arizona were hampered by the use of inexperienced birds (Snyder *et al.* 1994).

Some intriguing observations have been published from studies of captive parrots. For example, parent Budgerigars (*Melopsittacus undulatus*) preferentially allocated food to young based on sex or size (Stamps *et al.* 1985, 1987). In several species, pairs were more likely to breed when exposed to certain stimuli, including access to multiple potential nest sites (Shields *et al.* 1989; Millam *et al.* 1995). These results call for behavioral studies of wild birds to verify if these phenomena occur in the wild or are artifacts of captivity.

Conclusion

We brought a theoretically based, multiple hypothesis approach to the question of why parent parrotlets initiate incubation early and consequently hatch their eggs asynchronously. This approach allowed us to test and discriminate among the bewildering number of factors that potentially constrain productivity in parrots by influencing parental incubation behavior. Our research into the constraints on parrot reproduction offers potential insight and direction for population-level management.

Sustainable harvesting of parrots is feasible because the trade in wild birds confers many species with economic value. Harvesting has been tried or proposed for many commercially valuable taxa, including other birds (Feekes 1991), reptiles (e.g., Thorbjarnarson 1991; Werner 1991), mammals (Robinson & Redford 1991*b*), and plants (Olmsted & Alvarez-Buylla 1995). The concept of sustainable harvesting is probably not an option for the majority of species without recognized economic value. Behavioral studies dealing with the characteristics affecting productivity, survivorship, and other aspects of population demography and regulation, however, can be crucial to developing successful management and recovery strategies other than for sustainable use. As more species become endangered, the need for the expertise of animal behaviorists will become increasingly urgent.

Once a sustainable harvesting program is established, a variety of social, political, and economic problems can impede its viability (Beissinger & Bucher 1992*a*). Problems may arise with the smuggling of birds on the black market, laundering of illegal birds through legal programs, over-harvesting (especially of large species with low reproductive rates and high market value), and poaching of nestlings from sustained harvest programs. Some of these potential problems can be ameliorated through the development of a reliable marking system to distinguish legal from illegal birds.

Ultimately, the success of sustainable harvest programs depends upon their ability to compete economically with captive-reared birds and with unsustainably harvested birds. Effective regulation of the wild bird trade in importing countries can help to promote the development of sustainable harvest programs. In 1992, the United States passed the Exotic Wild Bird Conservation Act. This Act prohibits the importation of any species listed in the CITES Appendices I, II, III (including all parrots), except for captive-reared birds and those produced by sustainable ranching programs. Although not perfect, it represents a first attempt in dealing with the problem of continued exploitation by reducing the demand for wild birds and by providing incentives for captive-breeding and sustainable harvest programs.

Conservation problems cannot be solved by scientific solutions exclusively. Biological research can promote an understanding of the systems and organisms we wish to save, but many conservation problems are ultimately political and social. As Meyer & Helfman (1993: p. 570) stated in a discussion of the relevance of ecology to conservation, 'Ecological research is necessary but not sufficient. A better understanding of global ecology will be to no avail without the political will to implement the changes dictated by that understanding.' Even when relevant, behavioral studies may serve no conservation purpose if they are not actively integrated into the policy-making process.

We began the research described here as a behavioral ecology study of the factors that affect the onset of incubation in a bird with extreme hatching asynchrony, knowing only that parrots were endangered, extensively traded, and poorly known. We had no idea that our work would lead to the first detailed study of the potential for sustainable use of parrots. Only when we observed the incredible response of the parrotlet population to our nest boxes did we fully recognize the importance of our work to conservation. Rather than write up our observations for the scientific community to interpret and use, we invested additional time to obtain extra data of direct relevance to conservation. Instead of limiting our audience to fellow behavioral ecologists, we found a much larger community of conservationists and policy-makers hungry for scientists to help them determine if they could wisely incorporate sustainable use into legislation, and an enthusiastic audience when one of us (SRB) testified before the US Congress. To apply behavioral research to conservation involves trade-offs, but the rewards can be gratifying, significant, and long-lasting.

Literature cited

Arnold, T. W., F. C. Rohwer, and T. Armstrong. 1987. Egg viability, nest predation, and the adaptive significance of clutch size in prairie ducks. *American Naturalist* **130**: 643–653.

Balick, M. J., and R. Mendelsohn. 1992. Assessing the economic value of traditional medicines from tropical rain forests. *Conservation Biology* **6**: 128–130.

Beehler, B. 1985. Conservation of New Guinea rainforest birds. Pages 233–247 in A. W. Diamond and T. E. Lovejoy, editors. *Conservation of tropical forest birds*. ICBP Technical Publication 4. ICBP, Cambridge.

Beggs, J. R., and P. R. Wilson. 1991. The Kaka *Nestor meridionalis*, a New Zealand parrot endangered by introduced wasps and mammals. *Biological Conservation* **56**: 23–38.

Beissinger, S. R., and E. H. Bucher. 1992*a*. Can sustainable harvesting conserve parrots? *BioScience* **42**: 164–173.

Beissinger, S. R., and E. H. Bucher. 1992*b*. Sustainable harvesting of parrots for conservation. Pages 73–115 in S. R. Beissinger and N. F. R. Snyder, editors. *New World parrots in crisis: Solutions from conservation biology*. Smithsonian Institution Press, Washington, DC.

Beissinger, S. R., and N. F. R. Snyder, editors. 1992. *New world parrots in crisis: solutions from conservation biology*. Smithsonian Institution Press, Washington, DC.

Beissinger, S. R., and J. R. Waltman. 1991. Extraordinary clutch size and hatching asynchrony of a neotropical parrot. *Auk* **108**: 863–871.

Bucher, E. H. 1992. Neotropical parrots as agricultural pests. Pages 201–219 in S. R. Beissinger and N. F. R. Snyder, editors. *New World parrots in crisis: Solutions from conservation biology*. Smithsonian Institution Press, Washington, DC.

Caine, L. A., and W. R. Marion. 1991. Artificial addition of snags and nestboxes to slash pine plantations. *Journal of Field Ornithology* **62**: 97–106.

Chubb, S. L. 1992. The role of private aviculture in the conservation of neotropical psittacines. Pages 117–131 in S. R. Beissinger and N. F. R. Snyder, editors. *New World parrots in crisis: Solutions from conservation biology*. Smithsonian Institution Press, Washington DC.

Clum, N. J., and T. J. Cade. 1994. Gyrfalcon (*Falco rusticolus*). Pages 1–28 in A. Poole and F. Gill, editors. *The birds of North America, No. 114*. The Academy of Natural Sciences, Philadelphia; The American Ornithologists' Union, Washington, DC.

Collar, N. J., M. J. Crosby, and A. J. Stattersfield. 1994. *Birds to watch 2: The world list of threatened birds*. Birdlife International, Cambridge.

Collar, N. J., and A. T. Juniper. 1992. Dimensions and causes of the parrot conservation crisis. Pages 1–24 in S. R. Beissinger and N. F. R. Snyder, editors. *New World parrots in crisis: Solutions from conservation biology*. Smithsonian Institution Press, Washington, DC.

Collar, N. J., and S. N. Stuart. 1985. *Threatened birds of Africa and related islands*. ICBP, Cambridge, IUCN, Gland, Switzerland.

Conway, C. J., S. G. Anderson, D. E. Runde, and D. Abbate. 1995. Effects of experimental nestling harvest on prairie falcons. *Journal of Wildlife Management* **59**: 311–316.

Deeming, D. C., and M. W. J. Ferguson. 1992. Physiological effects of incubation temperature on embryonic development in reptiles and birds. Pages 147–173 in D. C. Deeming and M. W. J. Ferguson, editors. *Egg incubation: Its effects on embryonic development in birds and reptiles*. Cambridge University Press, Cambridge.

Derrickson, S. R., and N. F. R. Snyder. 1992. Potentials and limits of captive breeding in parrot conservation. Pages 133–163 in S. R. Beissinger and N. F. R. Snyder, editors. *New World parrots in crisis: Solutions from conservation biology*. Smithsonian Institution Press, Washington, DC.

Desenne, P. A., and S. D. Strahl. 1991. Trade and conservation status of the family Psittacidae in Venezuela. *Bird Conservation International* **1**: 153–169.

Diamond, J., and A. B. Bond. 1991. Social behavior and the ontogeny of foraging in the Kea (*Nestor notabilis*). *Ethology* **88**: 128–144.

Dobson, A. P., and R. M. May. 1986. Disease and conservation. Pages 345–365 in M. E. Soulé, editor. *Conservation biology*. Sinauer Associates, Sunderland, Massachusetts.

Drent, R. H. 1973. The natural history of incubation. Pages 262–322 in D. S. Farner, editor. *Breeding biology of birds*. National Academy of Science, Washington, DC.

Evans, P. G. 1991. Status and conservation of Imperial and Red-necked Parrots *Amazona imperialis* and *A. arausiaca* on Dominica. *Bird Conservation International* **1**: 11–32.

Fearnside, P. M. 1989. Extractive reserves in Brazilian Amazonia. *BioScience* **39**: 387–393.

Feekes, F. 1991. The Black-bellied Whistling Duck in Mexico – from traditional use to sustainable management. *Biological Conservation* **56**: 123–131.

Fisher, R. A. 1930. *The genetical theory of natural selection*. Oxford University Press, London.

Forbes, L. S. 1994. The good, the bad, and the ugly: Lack's brood reduction hypothesis and experimental design. *Journal of Avian Biology* **25**: 338–343.

Forshaw, J. M. 1989. *Parrots of the world*, third edition. Landsdowne Editions, Willoughby, Australia.

Gnam, R. S., and R. F. Rockwell. 1991. Reproductive potential and output of the Bahama Parrot *Amazona leucocephala bahamensis*. *Ibis* **133**: 400–405.

Houghton, R. A., D. S. Lefkowitz, and D. L. Skole. 1991. Changes in the landscape of Latin America between 1850 and 1985. I. Progressive loss of forests. *Forest Ecology and Management* **38**: 143–172.

Iñigo-Elias, E. E., and M. A. Ramos. 1991. The psittacine trade in Mexico. Pages 380–392 in J. G. Robinson and K. H. Redford, editors. *Neotropical wildlife use and conservation*. University of Chicago Press, Chicago.

Johnson, D. H. 1987. Barred Owls and nest boxes: results of a five-year study in Minnesota. Pages 129–134 in R. W. Nero, R. J. Clark, R. J. Knapton and H. R. Hamre, editors. *Biology and conservation of northern forest owls: symposium proceedings*. USDA Forest Service General Technical Report RM 142.

Jorgenson, A., and J. B. Thomsen. 1987. The world trade in African parrots. TRAFFIC **7**: 9–14.

Joseph, L. 1988. A review of the conservation status of Australian parrots in 1987. *Biological Conservation* **46**: 261–280.

Joseph, L., W. B. Emison, and W. M. Bren. 1991. Critical assessment of the conservation status of the Red-tailed Black-Cockatoos in south-eastern Australia with special reference to nesting requirements. *Emu* **91**: 46–50.

Juniper, A. T., and C. Yamashita. 1991. The habitat and status of Spix's Macaw *Cyanopsitta spixii*. *Bird Conservation International* **1**: 1–9.

Lack, D. 1947. The significance of clutch size. *Ibis* **89**: 302–352.

Lambert, F. 1985. The St. Vincent parrot, an endangered Caribbean bird. *Oryx* **19**: 34–37.

Lanning, D. V. 1991. Distribution and breeding biology of the Red-fronted Macaw. *Wilson Bulletin* **103**: 357–365.

Lanning, D. V., and J. T. Shiflett. 1983. Nesting ecology of Thick-billed Parrots. *Condor* **85**: 66–73.

Long, J. L. 1984. The diets of three species of parrots in the south of Western Australia. *Australian Wildlife Research* **11**: 357–371.

Lubchenco, J., A. M. Olson, L. B. Brubaker, S. R. Carpenter, M. M. Holland, S. P. Hubbell, S. A. Levin, J. A. MacMahon, P. A. Matson, J. M. Melillo, H. A. Mooney, C. H. Peterson, H. R. Pulliam, L. A. Real, P. J. Regal, and P. G. Risser. 1991. The sustainable biosphere initiative: an ecological research agenda. *Ecology* **72**: 371–412.

Ludwig, D., R. Hilborn, and C. Walters. 1993. Uncertainty, resource exploitation, and conservation: lessons from history. *Science* **260**: 17–36.

Mangel, M., R. J. Hofman, E. A. Norse, and J. R. Twiss. 1993. Sustainability and ecological research. *Ecological Applications* **3**: 573–575.

Marti, C. D., P. W. Wagner, and K. W. Denne. 1979. Nest boxes for the management of Barn owls. *Wildlife Society Bulletin* **7**: 145–148.

Mawson, P. R., and J. L. Long. 1994. Size and age parameters of nest trees used by four species of parrot and one species of cockatoo in South-west Australia. *Emu* **94**: 149–155.

Merritt, R. E., P. A. Bell, and V. Laboudallon. 1986. Breeding biology of the Seychelles Black Parrot (*Coracopsis nigra barklyi*). *Wilson Bulletin* **98**: 160–163.

Meyer, J. L., and G. S. Helfman. 1993. The ecological basis of sustainability. *Ecological Applications* **3**: 569–571.

Millam, J. R., B. Kenton, L. Jochim, T. Brownback, and A. T. Brice. 1995. Breeding Orange-winged Amazon parrots in captivity. *Zoo Biology* **14**: 275–284.

Mossop, D. H., and R. Hayes. 1982. The Yukon Territory Gyrfalcon harvest experiment (1974–80). Pages 263–280 in W. N. Ladd and P. F. Schempf, editors. *Proceedings of the symposium and workshop: raptor management and biology in Alaska and western Canada.* US Fish and Wildlife Service, Anchorage.

Munn, C. A. 1992. Macaw biology and ecotourism, or 'when a bird in the bush is worth two in the hand.' Pages 47–72 in S. R. Beissinger and N. F. R. Snyder, editors. *New World parrots in crisis: Solutions from conservation biology*. Smithsonian Institution Press, Washington, DC.

Munn, C. A., D. Blanco, E. Nycander, and D. Ricalde. 1991. Prospects for sustainable use of large macaws in southeastern Perú. Pages 42–47 in J. Clinton-Eitnear, editor. *Proceedings of the first Mesoamerican workshop on the conservation and management of macaws.* Center for the Study of Tropical Birds, San Antonio.

Navarro, J. L., and E. H. Bucher. 1990. Growth of Monk Parakeets. *Wilson Bulletin* **102**: 520–524.

Navarro, J. L., M. B. Martella, and E. H. Bucher. 1992. Breeding season and productivity of Monk Parakeets in Cordoba, Argentina. *Wilson Bulletin* **104**: 413–424.

Newton, I. 1994. The role of nest sites in limiting the numbers of hole-nesting birds: a review. *Biological Conservation* **70**: 265–276.

Olmsted, I., and E. R. Alvarez-Buylla. 1995. Sustainable harvesting of tropical trees: demography and matrix models of two palm species in Mexico. *Ecological Applications* **5**: 484–500.

Petty, S. J., G. Shaw, and D. I. K. Anderson. 1994. Value of nest boxes for population studies and conservation of owls in coniferous forests in Britain. *Journal of Raptor Research* **28**: 134–142.

Redford, K. H. 1992. The empty forest. *BioScience* **42**: 412–422.

Redford, K. H., and J. G. Robinson. 1987. The game of choice: patterns of Indian and colonist hunting in the Neotropics. *American Anthropologist* **88**: 650–667.

Reid, W. V. C. 1989. Sustainable development: lessons from success. *Environment* **31**: 7–35.

Rinke, D. 1989. The reproductive biology of the red shining parrot *Prosopeia tabuensis* on the island of 'Eua, Kingdom of Tonga. *Ibis* **131**: 238–249.

Robinet, O., F. Beugnet, D. Dulieu, and P. Chardonnet. 1995. The Ouvéa parakeet: state of knowledge and conservation status. *Oryx* **29**: 143–150.

Robinson, J. G., and K. H. Redford, editors. 1991*a*. *Neotropical wildlife use and conservation*. University of Chicago Press, Chicago.

Robinson, J. G., and K. H. Redford. 1991*b*. Sustainable harvest of neotropical forest mammals. Pages 415–429 in J. G. Robinson and K. H. Redford, editors. *Neotropical wildlife use and conservation*. University of Chicago Press, Chicago.

Rowley, I. 1990. *Behavioural ecology of the galah* Eolophus roseicapillus *in the wheatbelt of Western Australia*. Surrey Beatty and Sons, Chipping Norton.

Saunders, D. A. 1991. The effect of land clearing on the ecology of Carnaby's Cockatoo and the Inland Red-tailed Black Cockatoo in the wheatbelt of western Australia. *Acta XX Congressus Internationalis Ornithologici*: 658–665.

Shields, K. M., J. T. Yamamoto, and J. R. Millam. 1989. Reproductive behavior and LH levels of Cockatiels (*Nymphicus hollandicus*) associated with photostimulation, nest-box presentation, and degree of mate access. *Hormones and Behavior* **23**: 68–82.

Simon, D. 1989. Sustainable development: theoretical construct or attainable goal? *Environmental Conservation* **16**: 41–48.

Smiet, F. 1985. Notes on the field status and trade of Moluccan parrots. *Biological Conservation* **34**: 181–194.

Smith, G. T., and D. A. Saunders. 1986. Clutch size and productivity in three sympatric species of Cockatoo (Psittaciformes) in the southwest of Australia. *Australian Wildlife Research* **13**: 275–285.

Snyder, N. F. R. 1977. Puerto Rican Parrots and nest-site scarcity. Pages 47–53 in S. A. Temple, editor. *Endangered birds: Management techniques for preserving threatened birds*. University of Wisconsin Press, Madison.

Snyder, N. F. R., S. R. Derrickson, S. R. Beissinger, J. W. Wiley, T. B. Smith, and W. D. Toone. 1996. Limitations of captive breeding in endangered species recovery. *Conservation Biology* **10**: 338–348.

Snyder, N. F. R., S. E. Koenig, J. Koschmann, H. A. Snyder, and T. B. Johnson. 1994. Thick-billed Parrot releases in Arizona. *Condor* **96**: 845–862.

Snyder, N. F. R., J. W. Wiley, and C. B. Kepler. 1987. *The parrots of Luquillo: Natural history and conservation of the Puerto Rican Parrot*. Western Foundation of Vertebrate Zoology, Los Angeles.

Soulé, M. E. 1991. Conservation tactics for a constant crisis. *Science* **253**: 744–750.

Stamps, J., A. Clark, P. Arrowood, and B. Kus. 1985. Parent-offspring conflict in Budgerigars. *Behaviour* **94**: 1–40.

Stamps, J., A. Clark, B. Kus, and P. Arrowood. 1987. The effects of parent and offspring gender on food allocation in Budgerigars. *Behaviour* **101**: 177–199.

Stokland, J. N., and T. Amundsen. 1988. Initial size hierarchy in broods of the Shag: relative significance of egg size and hatching asynchrony. *Auk* **105**: 308–315.

Stoleson, S. H., and S. R. Beissinger. 1995. Hatching asynchrony and the onset of incubation in birds, revisited: When is the critical period? Pages 191–270 in D. M. Power, editor. *Current Ornithology*. Volume 12. Plenum Press, New York.

Taylor, R. H. 1985. Status, habits, and conservation of *Cyanoramphus* parakeets in the New Zealand region. Pages 195–211 in P. J. Moors, editor. *Conservation of island birds: Case studies for the management of threatened island species*. ICBP Technical Publication 3. ICBP, Cambridge.

Thomsen, J. B., and A. Brautigam. 1991. Sustainable use of neotropical parrots. Pages 359–379 in J. G. Robinson and K. H. Redford, editors. *Neotropical wildlife use and conservation*. University of Chicago Press, Chicago.

Thomsen, J. B., and T. A. Mulliken. 1992. Trade in neotropical psittacines and its conservation implications. Pages 221–239 in S. R. Beissinger and N. F. R. Snyder, editors. *New World parrots in crisis: Solutions from conservation biology*. Smithsonian Institution Press, Washington, DC.

Thorbjarnarson, J. B. 1991. An analysis of the spectacled caiman (*Caiman crocodilus*) harvest program in Venezuela. Pages 217–235 in J. G. Robinson and K. H. Redford, editors. *Neotropical wildlife use and conservation*. University of Chicago Press, Chicago.

Toland, B. R., and W. H. Elder. 1987. Influence of nest-box placement and density on abundance and productivity of American Kestrels in central Missouri. *Wilson Bulletin* **99**: 712–718.

Veiga, J. P. 1992. Hatching asynchrony in the House sparrow: a test of the egg-viability hypothesis. *American Naturalist* **139**: 669–675.

Waltman, J. R., and S. R. Beissinger. 1992. Breeding behavior of the Green-rumped Parrotlet. *Wilson Bulletin* **104**: 65–84.

Webb, D. R. 1987. Thermal tolerance of avian embryos: a review. *Condor* **89**: 874–898.

Werner, D. I. 1991. The rational use of green iguanas. Pages 181–201 in J. G. Robinson and K. H. Redford, editors. *Neotropical wildlife use and conservation*. University of Chicago Press, Chicago.

Wiebe, K. L., and G. R. Bortolotti. 1994. Food supply and hatching spans of birds: energy constraints or facultative manipulation? *Ecology* **75**: 813–823.

Wiley, J. W., N. F. R. Snyder, and R. S. Gnam. 1992. Reintroduction as a conservation strategy for parrots. Pages 165–200 in S. R. Beissinger and N. F. R. Snyder, editors. *New World parrots in crisis: Solutions from conservation biology*. Smithsonian Institution Press, Washington DC.

Wilson, H. R. 1991. Physiological requirements of the developing embryo: temperature and turning. Pages 145–156 in S. G. Tullett, editor. *Avian incubation. Poultry science symposium 22*. Butterworth-Heineman, London.

Yamashita, C. 1987. Field observations and comments on the Indigo Macaw (*Anodorhynchus leari*), a highly endangered species from northeastern Brazil. *Wilson Bulletin* **99**: 280–282.

Chapter 8

Behavioral variation: A valuable but neglected biodiversity

RICHARD BUCHHOLZ AND JANINE R. CLEMMONS

> *Science linked to human purpose is a compass: a way to gauge directions when sailing beyond the maps . . . but the prudent voyager uses all the instruments available, profiting from their individual virtues.*
>
> *Kai N. Lee, Compass and Gyroscope*

Variation in animal characteristics is used to classify and monitor biological diversity and to establish conservation priorities. Given the magnitude of the extinction crisis, it is understandable that conservationists and policy-makers have attempted to simplify the task of conservation prioritization by adopting traditional systematic methods and taxonomic classifications for organizing biological diversity into conservation units. Unfortunately, behavioral diversity, which is important for the survival of species and contributes to human welfare, is unlikely to be preserved by current schemes of taxonomic classification alone. In this chapter we explore how behavior can be better incorporated into conservation systematics. It is not our intention to discard the use of systematics in conservation, nor do we belittle the contributions of traditional systematics and systematists, past and present, to the protection of nature (Savage 1995). We too lament the deplorable and dangerous decline of taxonomic expertise in universities and museums (Cotterill 1995; Simpson & Cracraft 1995). Nevertheless, for the future, we support a more critical examination of the sole use of theoretical systematics for prioritizing conservation problems. The failure of biologists to agree on a single definition of biodiversity or to identify a single method for measuring biodiversity is indicative of the need to entertain organismal complexity and conservation problems with approaches of great breadth. Along these lines Feldså (1995) has called for "good interaction – a broad-mindedness – all the way from 'pure science' to 'pure application'." Reliance on any single definition or measurement will fail to achieve conservation biology's goals. We must strive to develop and implement a multifaceted conservation strategy that includes the assessment of behavior patterns to identify, evaluate, and protect biodiversity in all of its forms.

Not all measures of diversity have equal and immediate value for enhancing population viability. The neglect of behavioral diversity in classification schemes ultimately results in an underestimation of the extinction crisis, and in some cases, a misdirection of conservation efforts. To begin our exploration of this problem we review the limitations of traditional systematic measures of biodiversity. Next we discuss how the inclusion of behavioral diversity into estimates of global biological diversity would impact our estimates of species loss. And finally, we demonstrate how the behavioral adaptations likely to be ignored by classical systematics are nevertheless crucial to conservation practice and politics.

Problems of using traditional taxonomy for conservation

In this section we explore briefly some basic tenets of the systematic sciences that underlie traditional taxonomy and conservation. Although we recognize that systematics, taxonomy, and classification represent separate endeavors often conducted by different scientists, in practice these disciplines are combined to generate frameworks assumed to represent the evolutionary history of animals (sometimes incorrectly; Harvey & Pagel 1991, p. 53). The specific conservation implications of this practice are raised later in our chapter. For now we explore in general terms how traditional taxonomy, broadly defined, performs two tasks of interest to conservation efforts. The first task is the identification of species. The second task involves arranging these species into nested subsets on the basis of similarity.

Species concepts and conservation units

Units of preservation have always been a contentious topic in conservation biology. While some propose that the only economical approach to conservation is to protect large areas encompassing entire landscapes or ecosystems (Franklin, 1993; Beissinger, Chapter 2), other workers have suggested that we must target evolutionary hot spots (Myers 1988), keystone species (Terborgh 1986), overall genetic heterozygosity (Ralls & Ballou 1983), or heterozygosity at specific loci (e.g., the MHC locus, Hughes 1991; or rare alleles, Miller 1995) for conservation. Despite the range of alternatives, the legal unit of most conservation efforts in developed countries remains the species (e.g., O'Brien 1994). Debates over what constitutes a species, however, are a chronic feature of evolutionary biology (e.g., Coyne *et al.* 1988) with obvious impact on the taxonomic sciences. Currently, at least three species concepts hold sway: (1) the classical biological species (Mayr 1963); (2) the recognition species (Paterson 1985), and (3) the genetically cohesive species (Temple 1989). In essence each concept attempts to describe the factors that maintain the independent evolutionary trajectories of the gene pools of species (Nelson 1989). While academic debates over methods for defining species may seem peripheral to the practical concerns of conservationists, species classifications affect both population-level and community-level conservation efforts.

Population-level implications of species definitions

Legal actions and other protection measures are usually directed at the species level, allowing populations of subspecific status to go extinct (Daugherty *et al.* 1990; Collins 1991). Subspecific extinctions warrant concern because many 'species' actually consist of complexes of reproductively isolated subspecies. The discovery of reproductive incompatibility among 'conspecific' populations (e.g., Ayala *et al.* 1974) raises doubts about the effective population size, and thus, long-term viability of a species complex and brings into question the wisdom and efficacy of attempts to reduce the loss of heterozygosity by translocating individuals between populations (Templeton *et al.* 1986; Woodruff 1989).

Species are not absolutes. They are subjective classes that reflect a set of relationships specified by the particular type of characters chosen by, or available to, the taxonomist (Johnson 1994). In practice the species characteristics examined to determine specific status will depend greatly on the practitioner's underlying species concept. The taxon is likely to be bounced back and forth between species and subspecies designation if different types of characters provide conflicting information (e.g., Avise & Nelson 1989). It may surprise some readers that species and other taxa are more likely to be defined by consensus than by irrefutable evidence. To emphasize this point, Soulé (1986, p. 3) recounts a popular saying among taxonomists: "a species was 'good' (real) when enough systematists agreed that it was good."

No single species concept has generated a standard set of rules to satisfactorily define the conspecific status of a group of related taxa (Diamond 1986; May 1995; O'Donnell *et al.* 1995). Consequently, adherents to each concept often employ rather arbitrary criteria for identifying conspecifics. Genetically defined species serve as an example of the problem. Bacteria are deemed the same species if they share 70%

or more of their DNA sequences, whereas humans and chimpanzees are separate species, yet differ in DNA by less than 2%. Similarly, the sequence differences that exist between bacterial species are equivalent to those that exist between mammals and fishes (May 1995). Furthermore, the degree of genetic similarity may be dependent on the molecular technique employed: species of bacteria that show less than 1% difference in RNA sequencing may exhibit very low DNA reassociation in DNA hybridization experiments (O'Donnell *et al.* 1995), indicating strong dissimilarity in DNA sequences.

In conclusion, it is clear that species-level designations are highly subjective, despite claims by systematists that species are the only genuine evolutionary unit (Brooks & McLennan 1991). An event in 19th century American history provides a humorous parallel. After the discovery of gold in California, many first time prospectors left their urban homes in the east to make their fortunes in the west. When they arrived at their destination, they were known to ask naively, "Where is the best place to find gold?" A common reply to the innocents was said to be "Son, gold is where you find it." The conservationist faces frustrations parallel to those of the pioneering gold prospectors because of legal and typological misconceptions about species (Meffe & Carroll 1994).

Community-level implications of species definitions

In addition to its role in deciding the fate of populations, the species is also used to allocate conservation attention on a regional scale. Species numbers and abundance are used to construct measures of species richness and evenness, and their products, indices of diversity (Hayek 1994). These diversity values may be used in a number of ways: for example, to estimate diversity within larger ecological contexts (Whittaker 1972), or to assign weighted values to species to emphasize taxonomic distinctness ('taxic diversity,' Vane-Wright *et al.* 1991). Other conservation methods elaborate upon this basic theme. Information about species abundance, for example, can prove to be a sensitive indicator of pollution or disturbance (Magurran 1988). As a result, species concepts and the operational definitions they beget influence which populations and communities are targeted for conservation protection.

Once the limits of a species have been defined by traditional taxonomy, the species is classified into a framework of genera, families, and orders. Classification dictates, in part, how unique or 'special' a species is in the eyes of both scientists and decision-makers (Avise 1989; Avise & Nelson 1991). The threatened status of monotypic taxa cause greater concern among conservationists than the decline of more speciose taxa (Daugherty *et al.* 1990; Stiassny 1994). In the following section we first examine how taxonomic methods may affect the classification of species; then, we explore how behavioral variation is an important though misunderstood and neglected character in traditional taxonomy.

Brief review of systematic schools and their methods

Futuyma (1979) describes systematics as "the study of the historical evolutionary and genetic relationships among organisms, and of their phenotypic similarities and differences." Systematics may take advantage of multiple lines of biological evidence (e.g., genetics and osteology), and may consider non-biological information (Stiassny 1994), such as tectonic movement (Carson & Clague 1995). For the most part, modern systematists attempt to reconstruct the genealogical history of a group of species (but see below). The relationship of these organisms is represented as a branching tree called a phylogeny. In its simplest incarnation the phylogeny shows which pairs of species have the most recent common ancestors. In addition, recent advances have allowed the estimation of the timing of species bifurcation (Harvey & Pagel 1991, p. 51). Taxonomists use systematic information to name and arrange biological units into an organizational scheme called a taxonomy. As discussed above, conservation efforts can be legally dependent on how a population has been classified in a taxonomy.

Taxonomists fall into one of three major schools: phenetics, cladistics, or evolutionary taxonomy (Harvey & Pagel 1991). Students of these schools differ in how they treat character similarities and differences within and between species groups. Pheneticists construct taxonomies based on overall phenotypic similarity, ignoring the source of similarity in appearance. Statistical clustering techniques are used to decide which species are distinctly different in phenotype from other clustered groups. This approach does not consider that the similarities between groups may be the result of convergent evolution (similar selective pressures resulting in analogous adaptations) rather than shared ancestry (homologous characters inherited from a common ancestor).

In contrast, cladists attempt to use phylogenetic branching as the basis for classifying organisms rather than overall similarity. A group of species is designated as a clade with common ancestry if member species share uniquely derived characters that are not shared with taxa outside the group (Funk 1995). Relationships within the group are determined by parsimonious arrangement of taxa such that the fewest number of evolutionary events are required to explain the distribution of character states. Cladistics relies on comparison with an outgroup to establish the ancestral character state (for example, present or absent), and thereby the polarity of character changes within the clade. The conservation of primitive outgroups should be of great practical concern to cladists (Stiassny 1994).

Similarly to cladists and unlike pheneticists, evolutionary taxonomists only use characters that are identical by descent to establish taxonomic affinities. Evolutionary taxonomists differ from cladists in that they consider character differences to be more important to classification than character similarities. For example, dinosaurs (Reptilia) are classified separately from birds (Aves) mainly because they do not have feathers (Harvey & Pagel 1991).

Conservation implications of traditional systematics and taxonomies

Although the majority of systematists probably favor a cladistic approach to the construction of taxonomies, this is a fairly recent and still controversial development (Meffe & Carroll 1994). The majority of commonly accepted classification schemes probably do not represent phylogenetic branching patterns (Harvey & Pagel 1991). This situation is of significance to conservation because phenetic and evolutionary taxonomic approaches tend to produce two types of artificial taxa: paraphyletic and polyphyletic taxa. Paraphyletic groups are missing closely related species and are the result of placing unrelated species under the same higher level of classification elsewhere, i.e. creating polyphyletic groups. In fact, as many as 50% of the classes in the Phylum Chordata may be polyphyletic taxa (Felsenstein 1985, p. 7). Hence, most classifications currently available for determining phylogenetic uniqueness, and thus the level of conservation attention legally warranted, are flawed.

The recent reorganization of the taxonomy of the francolins (partridge-like birds of Africa and Asia) by Crowe *et al.* (1992) provides an example of the importance for considering the problem of artificial taxa in conservation. As with most galliform birds, francolins differ unremarkably in their osteology, and hence all 41 species of this group have been considered congeners in the genus *Francolinus* for many years. Given the speciose nature of this one group of mostly nondescript birds, one can imagine how the threatened status of several of the francolins has not caused very great alarm nor conservation attention (WPA 1985; Johnsgard 1988). As a result of their classification, individual francolin species in this largest genus in the avian order Galliformes are not perceived as being very unique. This is evidenced by Johnsgard (1988), who often dubs francolin species 'forms,' a moniker not used for other taxa in his monograph on quails and partridges.

Crowe *et al.* (1992) re-evaluated the systematics of this group by considering behavioral, morphological, and genetic characters in his analysis. He provides convincing evidence that the single clade of francolins currently recognized is paraphyletic and may consist of four distinct genera. Speciose taxa are considered less 'valuable' than less polytypic groups. Thus, after re-classification, the rare and endangered species of francolins take on new light; now they would be treated as being unique and valuable objects of conservation attention under prominent systems of conservation triage (US Congress 1987, p. 111). Francolin systematics illustrate the conservation dangers of artificial taxa in our established schemes of classification and points to the value of behavioral characters in revealing phylogenetic relationships.

This brief review of the complications of species designation and classification raises some doubt about current methodologies for deciding conservation units. If conservation biology is to save the evolutionary diversity of organisms (Soulé 1985; Meffe & Carroll 1994), the unmodified use of traditional systematic classifications to decide conservation units must be re-examined more closely. Even if we could miraculously re-classify organisms along phylogenetic lines overnight, and thereby eliminate misleading artificial classifications, there remains an additional problem. How do we ensure that conservation taxonomies make equitable use of a wide range of systematic characters, including behavior?

Search for appropriate taxonomic characters

Despite the fact that scientific progress has always relied on tacit agreement among scholars, many find the taxonomic reliance on consensus to be somewhat 'unscientific.' There is an underlying hope that a single class of characters will be found to reflect unequivocally the taxonomic status and detailed ancestry of all species. In addition, systematists often have personal preferences for specific types of characters. Most recently, the development of molecular techniques for determining genealogical associations was heralded as the systematist's panacea (Williams *et al.* 1991). It was thought that differences in DNA could be translated directly into phylogenetic patterns, thus enabling us to bypass the homoplasious obfuscation of phenotype (i.e., character similarities due to convergent evolution, not shared ancestry). So far, genetic characters have been unable to provide all the answers to our systematic questions.

In spite of the vast improvements in molecular technology over the past decade, genetic systematists agree that genetic characters simply supplement an old repertoire of traditional measures of relatedness (Woodruff 1989). The ebb and flow of debate over 'good' characters for systematics, reveals much about how historical biases in training create long-lasting misconceptions. Ignorance of behavioral biology has resulted in an almost complete avoidance of the use of behavioral variation in systematics. In this next section we first review the historical factors that have led to the exclusion of behavioral considerations in taxonomic classification, and then explain why a behavior pattern is just as valid a systematic character as a DNA sequence, a muscle attachment site, or another more conventional character.

A reaffirmation of ethology in systematics

Systematic neglect: lability, plasticity and the variation of behavioral characters

Brooks & McLennan (1991, p. 5) review the reasons for the 'eclipse of history in ethology,' and we base our historical discussion on their review. After reading their review it seems reasonable to conclude that what actually happened was that ethology became 'eclipsed' in systematics, rather than the other way around. As the authors point out, an explicitly phylogenetic framework was inherent in ethology since its inception (e.g., Heinroth 1911). Debates over the ability to detect homology in behavior patterns, however, led to polarized opinions about the operational value of behavior in reconstructing phylogenies (Brooks & McLennan, p. 6). For example, Atz (1970; in Brooks & McLennan 1991, p. 7) concluded that behavior is too "functional" and "cannot be traced" among ancestors, and thus its use is "fraught with danger." Despite the fact that behavioral phylogenies mirrored morphological ones and prominent pleas for the inclusion of behavioral characters (Mayr 1958, and Alexander 1962, in Brooks & McLennan 1991), behavioral information was relegated to the role of systematic "anecdote" (Alexander 1962, in Brooks & McLennan 1991).

Brooks & McLennan (1991, p. 7) conclude that the dismissal of behavior for use in phylogenetic reconstruction was based not on the inability to detect homology in behavior, but rather on the simplistic nature of early attempts in comparison with the older, more sophisticated quantitative methodologies available for morphological characters. Despite the fact that behavioral traits have continued to show value in sorting out phylogenetic problems (Crowe *et al.* 1992; Löhrl & Thaler 1992; Wenzel 1992) left unresolved by the analysis of other characters, there remains an ingrained cultural avoidance of behavior by many systematists.

De Queiroz & Wimberger (1993) claim that two misconceptions about behavior still predominate in systematic circles: (1) behavior cannot be homologized, and (2) behavior is too evolutionarily labile to be a reliable indicator of relationships. These authors admit that behavioral characters may be more difficult to compare as homologues, but remark that the precision with which behavioral phylogenies have mirrored morphologic ones suggests that behavioral characters can be effectively homologized (e.g., Langtimm & Dewsbury 1991; Kennedy *et al.* 1996). As evidence they show that the levels of homoplasy of behavioral and morphologic characters in systematic data sets for a broad range of taxa did not differ significantly within or between taxa. De Queiroz & Wimberger (1993) caution that this is not necessarily true for all behavioral characters, but holds for the ones chosen for inclusion in the data sets they examined. These results confirm that behavior can be homologized effectively.

Behavioral characters are also assumed to be more evolutionarily labile than morphological ones because behavior is thought to be particularly susceptible to natural selection (Atz 1970, in Brooks & McLennan 1991). This was not the case, however, with the behavioral characters in the data sets De Queiroz & Wimberger (1993) examined, nor did Kennedy *et al.* (1996) find any evidence of differential lability of behavioral and non-behavioral characters. In addition, field studies, of fish for instance (reviewed by Magurran *et al.* 1995), have found that behavior is not necessarily more responsive to selection pressures than other types of characters. Ironically, at the same time that behavioral characters are confirmed to be no more labile than other types of characters, characters previously thought to be relatively untouched by natural selection (e.g., mtDNA) are coming under suspicion (Ballard & Kreitman 1995). Again, it would seem that there is little evidence that behavioral characters are evolutionarily any different than other systematic characters.

Intraspecific and intraindividual variation in behavior has also been cited as a reason to avoid using behavioral characters in systematics. De Queiroz & Wimberger (1993), however, allay these fears by pointing out that "countless examples exist of such plasticity in morphological characters," including such things as inducible jaw morphologies in fishes and exercise or altitude dependent physiologies of humans. Thus, behavioral plasticity can be treated in systematics in much the same way as morphological plasticity: avoid plastic characters or come up with ways of incorporating flexible character states into the data set. Finally, plasticity of behavior should not be taken as further evidence of the evolutionary lability of behavior: plasticity is caused by environmental differences, rather than heritable differences that might be evolutionarily labile. The influence of learning on behavioral character states also has been of great concern to systematists. Wenzel (1992), however, dismisses this view by showing that the structural details of the learning process are what must be used as characters, not that which is learned. As an example he suggests that we focus not on the imprinting cues of neonate animals *per se*, but instead homologize when and how they imprint. If one were attempting to homologize song learning in birds, for example, the timing of the critical imprinting period and the mode of imprinting would be used as characters, rather than characteristics of the song type learned. The parameters of learning can serve as effective systematic characters.

Just as genetics is discovered not to be a systematist's panacea, neither should behavioral characters be their anathema. Despite the renewed validation of behavioral characters for systematics, conservationists must consider the fact that behavioral traits have not been used to assign taxonomic ranking to the vast majority of animal populations. Consequently, functionally valuable variation in behavior is not likely to be protected under present classification schemes. Phyletic schools of taxonomic methodology and the traditional biases against behavioral characters in the construction of taxonomic classifications should generate concern that species that do not exhibit marked variation in morphology or molecular genetics will be relegated to the extinction dustbin in plans to prioritize species conservation efforts, no matter how unique the behavioral variation they exhibit. Behavioral variation must be assessed on par with other characters if we are genuine in our goal to preserve a representative cross-section of the evolutionary uniqueness of Earth's taxa.

Hidden extinctions

Behavioral variation is not only useful in elucidating the phylogenetic uniqueness of taxa, it can also be used to uncover hidden species that might otherwise be lost in the conservation shuffle. The depth of the extinction crisis is plumbed by authors who cite alarming rates of species loss. These depressing numbers, ranging from as few as one species per day to as many as 50 000 species (Myers 1983; Mann & Plummer 1992), are usually accompanied by a standard platitude declaring the rate of loss is speculative because many invertebrates have not been described by science and remote regions remain unexplored by biologists. Indeed, taxonomic experts have found a plenitude of new insects in the crown of a single neotropical tree, newly revealed fish species are common place, and recent explorations have discovered mammals and birds previously undescribed (Raven & Wilson 1994). Unfortunately, the elation with which biologists welcome the discovery of new species is tempered by the realization that the apparent omnipresence of undescribed species means that extinction rates are not just uncertain but probably are higher than we realize. In addition to the oft cited real and opportunity costs to human welfare (Myers 1983), the loss of undescribed species means that fascinating examples of evolutionary adaptation and accident are vanishing before we have the opportunity to witness and explore them (Wilson 1984).

What would be the quantitative impact of including behavioral characters in traditional systematic assessments of diversity loss? At first glance it seems that there would be relatively minor impact on the numbers of unique, large, diurnal vertebrates that have been classified. Even for such well-known species, however, there are many behavior patterns that have been missed by decades worth of research. For example, until relatively recently most passerine bird species have been assumed to be monogamous. Detailed behavioral observations and genetic parentage studies of nestlings, however, have shown that extra pair copulations can be quite common (see Parker & Waite, Chapter 10), and mating systems can vacillate between monogamy and polygamy depending on environmental conditions (Komdeur & Deerenberg, Chapter 11). It seems that diurnal vertebrate taxa can be very familiar, while their behavior remains relatively unknown. One of us (Buchholz 1995), for example, discovered that male Yellow-knobbed curassows (*Crax daubentoni*; a large neotropical galliform bird) commonly use a 'booming' display behavior that was thought to be absent in this keystone species (Delacour & Amadon

1973; Strahl & Silva 1989; Silva & Strahl 1991). Findings such as these do not have earth-shattering impact on our estimates of higher order diversity, but they do suggest that we could very easily miss behaviorally unique populations of vertebrates that are under foot or that fly past our noses. Species or subspecies that are similar or identical in appearance may differ markedly in the behavioral rules (e.g., neophobia, Greenberg 1983; or migratory direction or restlessness, Berthold *et al.* 1992) they use to maximize fitness. Discoveries of different karyotypes in morphologically similar primates and ungulates (Templeton *et al.* 1986; Ryder 1986; Ryder *et al.* 1988; Benirschke & Kumamoto 1991) provide an analogous situation where evolutionarily separate populations were assumed to be identical on the basis of only one type of character.

It is perhaps not as surprising that behavior continues to be an important character for recognizing taxonomic diversity among animals cryptic in activity or size. Baptista & Gaunt (Chapter 9) describe studies of the vocalizations of nocturnal primates that suggest as many as 40 undescribed species may be classified by a few morpho-types under the present system. Vocalizations of drab, crepuscular ground birds called tinamous (Tinamiformes) also suggest more species diversity than is apparent to the eye (Blake 1977; Meyer de Schauensee & Phelps 1978, p. 6). Tinamou systematics is of great relevance to conservation because these species exhibit a relatively uncommon mating system, polyandry, they are probably important seed dispersers in neotropical forests (Erard *et al.* 1991), and they seem to be particularly prone to extinction in fragmented forests (Terborgh 1974). Hidden vertebrate species are surprising and garner media attention but probably represent only a tiny fraction of species yet to be described. Most new species will be found among the invertebrate phyla.

Close examination of the reproductive behaviors of invertebrates has revealed multiple species where only one was recognized previously. The diversity of the speciose Hawaiian drosophilids are most apparent in their behavior (Kaneshiro 1988; Hoikkala *et al.* 1994). Similarly, Dingle *et al.* (Chapter 4) review the behaviorally isolated species discovered among Hawaiian crickets. Perhaps the discovery of behavioral species is no more important and alarming than among the freshwater mussels. Based on historical collections of bivalve shells, we know that these animals have suffered widespread extinctions in North America (Williams & Mulvey 1994), and probably elsewhere. Closer behavioral attention to the remaining taxa will likely show that those with similar shells may have different reproductive behavior patterns. Hartfield & Butler (1995), for example, described remarkable minnow-like lures that some mussels use to infect host fish with their glochidia. Further studies of the means by which mussels infect their preferred host fish no doubt will generate examples of hidden bivalve species (O'Brian *et al.* 1995).

Another place we can expect to find more hidden species is in the realm of disease organisms. The rapid generation time and the continuing arms race between host and parasite mean that variation is likely to be generated more quickly here than in other taxa (Hamilton & Zuk 1982). Host preferences of parasites and their vectors (Dingle *et al.* Chapter 4), and variation in the circadian cycling of parasites within the host will be of great assistance in establishing the unique identities of protozoal disease organisms that appear identical to the microscopist (Atkinson & van Riper 1991, p. 24).

These few studies suggest that a few new species of diurnal vertebrates will be detected by their behavior in the next decade, but that the numbers of cryptic vertebrate and invertebrate species that might be revealed by behavioral study may reach several orders of magnitude greater than we now recognize. This statement might make the faint of heart back away from the whole extinction issue with the opinion that it is even more hopeless than previously imagined. Instead, we suggest that the presence of cryptic, behavioral species represents an opportunity for behavioral biologists to exercise their curiosity and professional skills for the good of conservation. These hidden species are of academic interest because their unstudied biology may make us question the ethologic, evolutionary, and ecological generalities we have concluded from more easily observed species (Buchholz 1992; Stiassny 1994). Also, recently discovered species surprise and excite the layman. Hidden species demonstrate to the public how rampant habitat destruction has outpaced science's ability to explore our natural heritage. No doubt the discovery of new species, whether in suburbia or remote wilderness, will increase pressure on governments to establish wildlife corridors and green spaces in developed areas, and protect undisturbed ecosystems from habitat fragmentation.

Conservation needs multiple measures of diversity

To this point we have reviewed some of the general pitfalls of using traditional taxonomy for conservation, and specifically we show how an ignorance of behavioral biology has probably biased our views of taxonomic uniqueness, rarity, and species loss – all important to national and international conservation strategies. These problems require careful contemplation by conservationists. As others before us (Daugherty *et al.* 1990; Collins 1991; Meffe & Carroll 1994), our critique of conservation taxonomy provides no easy solutions for conservation policy-makers; but it does provide straightforward directives for animal behaviorists and systematists. Behavior is a vital clue to the diversity of organisms. Behavior is no worse or better at revealing evolutionary history than are other more typical characters. It is an additional tool that should be incorporated in attempts to fully describe the identity and relatedness of species. We think the value of behavior, however, extends past its use as a systematic and taxonomic tool.

The adoption of traditional taxonomies and species concepts for conservation was an effective way to jump-start the arduous and daunting process of saving the evolutionary diversity, and future, of life on Earth. As stated at the outset, we believe systematists have been, and will continue to be, essential to the conservation task. There are, however, measures of diversity that are unimportant to the theoretical systematist in academia, but that are crucial to the conservation challenge. In the rest of our chapter, we explore how animal behaviorists and conservationists need to extend beyond traditional taxonomies to demonstrate the necessity and value of protecting local populations.

Fostering a functional unit of animal diversity

As explored earlier, species units do not lend themselves very readily to conservation. Despite our well-intended typological wishes, species rarely present themselves as the clearly distinct entities we would like them to be (Mallet 1995). The debate over species concepts that so invigorates evolutionary biologists is a source of great frustration to conservation managers and policy-makers. If animal species are things that do not interbreed in nature, what do we do with Red wolves (*Canis rufus*) and Coyotes (*Canis latrans*)? These canids are obviously different in their appearance and natural history, but their molecular genetics show signs of significant hybridization (Wayne & Jenks 1991). How different must two taxa be to be classified as separate species, and which characters should be used to determine phylogenetic distances (Cronin 1993)? What if reserves are set up specifically to save a morphological subspecies and subsequent measures of genetic distance reveal the protected taxa to be genetically very similar to the unprotected taxa (Gavin & May 1988)? How genetically or morphologically different is not different enough to be worth saving? What conservation measures should be taken for taxa that previously were geographically or temporally isolated and begin hybridizing because of anthropogenic habitat alteration (Hepp *et al.* 1988; Kodric-Brown 1989; Rhymer *et al.* 1994; Polziehn *et al.* 1995)? Clearly a single answer, i.e., a single species concept or a single measurement of diversity, will not hold for the range of biodiversity (O'Hara 1993) nor for the conservation dilemmas with which we find ourselves confronted. Conservationists and policy-makers should not back away from the diverse ways in which organisms display variation and the multitude of ways in which this variation can be valued. The present situation may be confusing, but it also holds promise for developing a better set of approaches to determining conservation units. The taxonomically based conservation units currently in use clearly are inadequate for describing, and thus for preserving, the diversity inherent in locally adapted populations of animals.

The non-systematic value of behavioral diversity

The emphasis on traditional systematics has resulted in a blind eye to properties of local populations of animals that are crucial to conservation (Furlow & Armijo-Prewitt 1995). Mallet (1995) framed this quite eloquently: "We are much more interested in conserving actual . . . diversity than in structuring conservation around a nebulous taxonomic level about which, in the past, there has been so much disagreement." When decision-makers use established taxonomies or those based on only one type of character to allocate protected status, they miss properties that are essential to population viability and for generating public support for conservation. Current protection schemes rarely consider the functional value of animal adaptations, especially behavioral ones.

It is generally accepted that genetic diversity is at the root of species' abilities to adapt to new environments, cope with pathogens, and provide a buffer against other deleterious consequences of homozygosity (May 1995). It may seem reasonable, therefore, to base populational conservation strategies on maximizing genetic diversity. This strategy ultimately will be ineffective for two reasons: (1) the relationship between genotypic variation and phenotypic expression is nonlinear, and (2) the degree of genotypic difference, or for that matter, phenotypic difference, is not necessarily evidence of essential local adaptation.

Genomic diversity versus specific adaptation

Genetic adaptation One problem with a genetic emphasis on weighting the value of biological diversity is the fact that a few genotypic differences can make disproportionately large or important phenotypic differences. The consequence is that methods used to screen diversity by genetic differences (Magurran *et al.* 1995), or for that matter morphological ones (Chan 1995), alone may fail to detect small behavioral differences that nevertheless have critical survival consequences. To illustrate, there are numerous examples in which the behavior of animals serves as a first line of defense against disease (Hart 1990) or predation (Magurran *et al.* 1995). The nest maintenance behavior of honey bees (*Apis mellifera*) is one example of a critical behavior that is controlled by only two genes. Honey bees have experienced dramatic population declines of up to 85% (Nabhan 1996) in North America, in large part because of introduced pests that infest the hives. The behavior patterns necessary to successfully uncover and dispose of bee larvae suffering from the highly infectious American foul brood disease are controlled by two alleles at each of two loci (Rothenbuhler 1964). Thus, the success of the colony, and the persistence of these keystone pollinators in certain habitats, is largely dependent on intraspecific differences that would be virtually undetectable by molecular genetics approaches.

Studies of the feeding habits of coastal and inland garter snakes (*Thamnophis elegans*; Arnold 1980; Ayres & Arnold 1983) provide another example of the value of specific loci. Only coastal populations feed on slugs. Inland populations of this snake coexist with blood-sucking leeches that have an appearance similar to the slugs and may harm the snake if ingested. The inland populations exhibit heritable leech avoidance behavior while the coastal populations do not. If the coastal snakes were exterminated, the unwillingness of inland snakes to feed on slugs may make reintroduction of these animals into coastal areas difficult and costly. If inland snakes are extinguished, translocated coastal snakes would perhaps suffer high mortality from leech ingestion. Minor genetic differences between the two populations represent important local adaptations of import to conservation.

Similarly, populations of fishes can exhibit heritable adaptations against specific predators (reviewed by Magurran *et al.* 1995) that may be unlikely to be recognized or detected by regular molecular means (Gervai & Csányi 1985). Adaptations against interspecific nest parasites may also be genetically variable (Mark & Stutchbury 1994) between populations. Even when forms are recognized as being different in some way, for example morphologically, the possible presence of heritable psychological adaptations that complement morphology are not considered. In a classic study Wecker (1963) demonstrated that a meadow-dwelling subspecies of deer mouse (*P. leucopus bairdii*) has a genetically determined preference for open habitats, which better suit its light pelage and short tail and legs. Given the impact of habitat structure on light environments and predation risk, and perhaps mating success (Théry & Vehrencamp 1995; Endler, Chapter 14), it is reasonable to expect locally adapted populations to show heritable behavioral preferences for habitats that maximize individual fitness. Simple, genetically-based rules of behavior, perhaps not reflected in general measures of genetic divergence, may enhance the viability of populations.

Population differences in behavior may also lead to the discovery of other, non-behavioral differences. Published correlative data on regional differences in the foraging habits of Black-throated green warblers, for example, led Parrish (1995) to discover population differences in foraging habits and bill and limb morphology that are apparently genetically-based. This has led to a better understanding of the forest habitat of this species in different parts of its range.

Why is it essential to preserve heritable local adaptation, even if it is not reflective of higher order genetic diversity? As populations become reduced and fragmented, the chance of stochastic extinction increases markedly (Lande 1988). It is logical to expect locally adapted populations to maintain higher population sizes, and thus lower risks of extinction, in their native locale than introduced individuals. Likewise, it will be difficult for neighboring, differently adapted, populations to re-colonize areas that require special adaptation. Not every population will demonstrate local adaptation, and it is possible that locally adapted populations can be replaced by individuals from differently adapted populations. It is unlikely that non-adapted populations will be as large and persistently viable as locally-adapted populations. Finally, the diversity of local adaptations that exist over a species' range may be necessary for species to respond to the variable, rapid, and extreme environmental changes caused by humans.

Just as the absence of significant genetic distance should not be used to discard populations, the presence of genetic divergence alone is not evidence of unique value. Magurran *et al.* (1995, p. 182–4) describe how a genetic distance equivalent to over 300 000 years of geographic separation was produced in only 30 generations by founding a population with a single inseminated female. Conant (1988) relates more examples of rapid evolutionary change as a result of translocating few founders. Changes in gene frequencies can occur as a result of non-adaptive factors, such as genetic drift and the founder effect, and do not necessarily warrant conservation dollars.

Animal traditions and learned adaptation Assessments of genetic or morphological variation do not consider the importance of environmentally determined behavior patterns to population survival. The narrow escape from extinction of the endemic Mauritius kestrel (*Falco punctatus*) illustrates how a locally adaptive, learned behavior pattern is essential for species survival. This species, which nested historically in tree cavities, was reduced to a population of two breeding pairs through nest predation by introduced macaques (Temple 1986). Temple observed that the kestrel population rebounded from its precariously low size only after one breeding pair switched to nesting on cliffs that were inaccessible to the primates. The offspring of these pairs also bred on cliffs rather than in the tree holes previously favored by the species. Hence, the learned cliff-nesting habits of this species are now crucial to its survival in an altered landscape. A ground-nesting colony of Snowy egrets (*Egretta thula*) on an island off the coast of Maine, USA, provides a related example (Drennan & Bowman 1993). This colony represents a disjunct range extension of the species that normally nests only in low shrubs or trees unavailable at this site. Evidently, willingness to nest in a new habitat has allowed this unique population to persist for now with few predators. Continued study of this colony will reveal whether fledglings return to breed on similar, species-atypical, colony sites.

A great variety of activities that are critical for population persistence have been found to have a learned component in wild populations, including long distance migration patterns (Temple 1978), selection of spawning grounds (Hasler & Scholz 1983), food preferences (Rowley & Chapman 1986), and daily foraging routes (Helfman & Schultz 1984). Anecdotal evidence from elephant (*Loxodonta africana*) populations in Zambia, where herd matriarchs have been poached for their ivory, suggests that traditional knowledge is crucial to population fitness (ABC News 1996).

Learned behaviors can be important to conservation in somewhat unpredictable ways as well. If conservation biology hopes to perpetuate evolutionary processes, it must also focus on how non-genetically determined behavior seems to drive evolutionary processes (Cavalli-Sforza *et al.* 1982; West-Eberhard 1983; Lott 1991). Learned and other environmentally-induced adaptations are usually not factored into assessments of the worthiness of population protection, but should be.

Hence, if the objective is to conserve the evolutionary potential of populations, the processes of underlying environmentally determined behavior patterns must be conserved as well. Recent attempts at re-introducing captive bred animals have demonstrated the costs and disappointments of attempting to teach animals what they need to know to survive in the wild (Beissinger, Chapter 2; McLean, Chapter 6). The inclusion of crucial learned traditions in conservation prioritization will result in more effective and less costly conservation management in the long run.

Political advocacy and quality of life

Perhaps the strongest arguments for preserving the behavioral attributes of local populations is one of political advocacy and quality of life. Political and social factors are at the root of biological extinctions, and so any conservation strategy that does not address human valuation of other species is certain to fail (Norton 1986). The popularity of nature films, national parks, zoos, aquariums, bird-feeders, and bird, bat, and butterfly houses in North America and elsewhere, attests to the powerful desire of scientists and non-scientists alike to experience the behavior of animals. People become endeared to other species not because of the genetic make-up of those organisms but because of the animals' peculiar and interesting habits, and the realization that other species have characteristics and needs in common with humans. Knowing and understanding the diversity of ways that animals behave under natural conditions may even be of value in resolving social issues in humans (Tinbergen 1976; Emlen 1995).

The greatest agony of extinction is the unrecognized extermination of experience with animals from our lives (Kinch 1996). Yet, governmental funding for studies of behavior that seem useless and expendable to the average politician – for example, the mating behavior of a newt – are often called into question. The public majority, however, often values such information and abhors environmental degradation, especially species extinction. There is little doubt that public interest in animal diversity can be used to impact governmental policy to the benefit of conservation. Anti-environmental US politicians recently suspended their efforts to cut our environmental laws after discovering that most citizens opposed these efforts (Linden 1996). Obtaining the support and the level of public education necessary to conserve diversity worldwide begins with the conservation of diversity at a local level (Hunter & Hutchinson 1994). Traditional measures of evolutionarily significant units alone cannot address this problem and win public support. To the public it likely does not matter that solitary bee species differ very slightly in their morphology, but it probably does matter to them that local bees must pollinate local wild flowers (Dingle *et al.* Chapter 4; Nabhan 1996). It is unimportant to systematics that a common bird species, such as a cardinal (*Cardinalis cardinalis*), sings in my yard – they can be found in disturbed environments in much of the United States – but it does matter that I enjoy listening to them, watching them, and peering into their nests. Public interest in backyard biodiversity is burgeoning, and much of this interest entails animal behavior. The Laboratory of Ornithology at Cornell University, USA, has made important steps towards harnessing public interest in animal behavior, particularly by organizing the behavioral observations of people with backyard bird feeders. This work has resulted in interesting findings on regional differences in the occurrence of eye infections in House finches (*Carpodacus mexicanus*; Dondt 1996) and changes in avian seed preferences with latitude (Rosenberg 1996). Perhaps more importantly, these programs have generated a national group of citizen observers attuned to the animals around them.

The benefit of local, behaviorally-adapted animal populations to humans, need not be so esoteric. Despite their obvious commercial value, neither classification schemes nor genetic assays distinguish the unique drainage-specific populations of anadromous salmon (*Onchorhynchus mykiss*; Reisenbichler *et al.* 1992), nor do they prioritize non-anadromous forms of the same species as separate taxa (Taylor 1995) worthy of unique protection efforts. Only behavior tells the fisheries manager which fish breed where. Native peoples often recognize the value of behaviorally unique populations. For example, the local peoples of central and west Africa place different value on the talking abilities of local populations of Grey parrots (*Psittacus erithacus*, Dändliker 1992).

There are levels of diversity that contribute to the quality of life (personal, professional, commercial) that are not captured by attempts to use systematics to prioritize the value of populations. This is no surprise to the animal behaviorist interested in conservation, nor is it an epiphany to systematists. Oliver Ryder (1996), a zoo geneticist, recently described his latest challenge as "linking concern for wildlife around the world . . . with concern for habitat preservation to ensure species persistence . . . in my own backyard." What is important to realize is that current conservation schemes that use taxonomy to not just organize conservation efforts but to prioritize them, ignore the crucial and valuable interface of the human–non-human experience.

Conclusions

In practice, species concepts attempt to pigeon-hole the continuous variation that accompanies population divergence. While this practice is heuristically productive for exploring the tempos and patterns of evolutionary processes, it does not translate well into attempts to preserve biodiversity (Ryder 1986; Ryder *et al.* 1988). It is necessary to keep in perspective the original purpose of the methods we employ and to acknowledge their limitations as well as virtues. Systematics attempts to organize what evolution has wrought; conservation biology, on the other hand, attempts to ensure that evolution can continue its unpredictable path and that we humans can witness it. The world is not simply dotted with evolutionarily significant, or conservation, units (species, populations, communities, or whatever the current unit in favor). Interactions occur between biological 'units': processes that affect individual survival hold the key to population persistence. Behavior is the most common mode of interaction between an organism and its biotic and abiotic environment. Behavior plays a major role in the processes underlying survival and evolution, so it should come as no surprise that behavioral diversity will be a major determinant of population viability. Sole reliance on flattened museum skins, bleached out alcohol specimens, or the geneticist's gels to determine what is 'worth' saving, will not conserve interesting and valuable behavioral functions performed by animals, nor will it guarantee the retention of locally adaptive action patterns that enhance population viability in the face of stochastic forces. When habitats become fragmented and populations become smaller, population fitness can be directly dependent on the ability of individuals to find mates, food, and protect their young (Lande 1988; Caro & Laurenson 1994).

We propose not a revamping of theoretical systematics, but a reanalysis of methods of conservation taxonomy. First, equitable consideration of characters must be used to construct the framework for conservation that initially prioritizes our global efforts. Second, on a local scale we must weigh the behavioral adaptations of populations more strongly than we have in the past. Behavioral adaptations promote population viability and provide services to human populations, both of which are unlikely to be protected by approaches that merely consider overall genetic or morphological variability.

What we propose is consistent in part with the present academic responsibilities and funding systems of most animal behaviorists. Many in this field already study the unique ways in which individuals and populations are selected upon by their environments, and the ways in which animals affect the environment around them. By collecting these data in a way that reflects population viability, animal behaviorists can begin to make a positive impact in conservation biology (Beissinger, Chapter 2; Ralls, Chapter 15).

In light of the immensity of the conservation task with the species we already recognize, our plea to consider behavioral diversity between and within species may be drowned out. Yet, given the urgency of the task and the immensity of what is being lost, it seems all the more imperative that conservation biologists of every background use all the conservation tools at their disposal, including those provided by behavioral biology.

Literature cited

ABC News. 1996. Deadly game: The Mark and Delia Owens story. 30 March. *Prime Time Live*, New York.

Arnold, S. J. 1980. Behavioral variation in natural populations. II. The inheritance of a feeding response in crosses between geographic races of the garter snake, *Thamnophis elegans*. *Evolution* **35**: 510–515.

Atkinson, C. T., and C. van Riper III. 1991. Pathogenicity and epizootiology of avian haematozoa: *Plasmodium, Leucocytozoon* and *Haemoproteus*. Pages 19–48 in J. E. Loye and M. Zuk, editors. *Bird–parasite interactions*. Oxford University Press, Oxford.

Avise, J. C. 1989. A role for molecular genetics in the recognition and conservation of endangered species. *Trends in Ecology and Evolution* **4**: 279–281.

Avise, J. C., and W. S. Nelson. 1989. Molecular genetic relationships of the extinct dusky seaside sparrow. *Science* **243**: 646–648.

Ayala, F. J., M. L. Tracey, D. Hedgecock, and R. C. Richmond. 1974. Genetic differentiation during the speciation process in *Drosophila*. *Evolution* **28**: 576–592.

Ayres, F. A., and S. J. Arnold. 1983. Behavioural variation in natural populations. IV. Mendelian models and heritability of a feeding response in the garter snake, *Thamnophis elegans*. *Heredity* **51**: 405–413.

Ballard, J. W. O., and M. Kreitman. 1995. Is mitochondrial DNA a strictly neutral marker? *Trends in Ecology and Evolution* **10**: 485–488.

Benirschke, K., and A. T. Kumamoto. 1991. Mammalian cytogenetics and conservation of species. *Journal of Heredity* **82**: 187–191.
Berthold, P., A. J. Helbig, G. Mohr, and U. Querner. 1992. Rapid microevolution of migratory behaviour in a wild bird species. *Nature London* **360**: 668–670.
Blake, E. R. 1977. *Manual of neotropical birds*. University of Chicago Press, Chicago.
Brooks, D. R., and D. A. McLennan. 1991. *Phylogeny, ecology and behavior*. University of Chicago Press, Chicago.
Buchholz, R. 1992. Confusing models with tests in studies of sexual selection: Reply to Jones. *Auk* **109**: 199–201.
Buchholz, R. 1995. The descending whistle display and female visitation rates in the Yellow-knobbed Curassow in Venezuela. *Ornitologia Neotropical* **6**: 27–36.
Caro, T. M., and M. K. Laurenson. 1994. Ecological and genetic factors in conservation: a cautionary tale. *Science* **263**: 485–486.
Carson, H. L., and D. A. Clague. 1995. Geology and biogeography of the Hawaiian islands. Pages 14–29 in W. L. Wagner, and V. A. Funk, editors. *Hawaiian biogeography: Evolution on a hot spot archipelago*. Smithsonian Institution, Washington.
Cavalli-Sforza, L. L., M. W. Feldman, K. H. Chen, and S. M. Dornbusch. 1982. Theory and observation in cultural transmission. *Science* **218**: 19–27.
Chan, K. 1995. Comparative study of winter body composition of resident and migrant grey-breasted silvereyes. *Auk* **112**: 421–428.
Collins, J. T. 1991. Viewpoint: a new taxonomic arrangement for some North American amphibians and reptiles. *Herpetological Review* **22**: 42–43.
Conant, S. 1988. Saving endangered species by translocation: are we tinkering with evolution? *BioScience* **38**: 254–257.
Cotterill, F. P. D. 1995. Systematics, biological knowledge and environmental conservation. *Biodiversity and Conservation* **4**: 183–205.
Coyne, J. A., H. A. Orr, and D. J. Futuyma. 1988. Do we need a new species concept? *Systematic Zoology* **37**: 190–200.
Cronin, M. A. 1993. Mitochondrial DNA in wildlife taxonomy and conservation biology: cautionary notes. *Wildlife Society Bulletin* **21**: 339–348.
Crowe, T. M., E. H. Harley, M. B. Jakutowicz, J. Komen, and A. A. Crowe. 1992. Phylogenetic, taxonomic and biogeographical implications of genetic, morphological, and behavioral variation in francolins (Phasianidae: *Francolinus*). *Auk* **109**: 24–42.
Dändliker, G. 1992. The Grey Parrot in Ghana: a population survey, a contribution to the biology of the species, a study of its commercial exploitation and management recommendations. *CITES Project S30*.
Daugherty, C. H., A. Cree, J. M. Hay, and M. B. Thompson. 1990. Neglected taxonomy and continuing extinctions of tuatara (*Sphenodon*). *Nature* **347**: 177–179.
Delacour, J., and D. Amadon. 1973. *Curassows and related birds*. American Museum of Natural History, New York.
De Queiroz, A., and P. H. Wimberger. 1993. The usefulness of behavior for phylogeny estimation: levels of homoplasy in behavioral and morphological characters. *Evolution* **47**: 46–60.
Diamond, J. 1986. Foreword. Pages v–viii in K. Benirschke, editor. *Primates: The road to self-sustaining populations*. Springer, New York.
Drennan, M. P., and Bowman, R. S. 1993. Ground-nesting snowy egrets in

Maine – a new northernmost breeding record. *American Birds* Fall: 376–377.

Dondt, A. 1996. Finch disease update. *Birdscope* Spring:4–5.

Emlen, S. T. 1995. Can avian biology be useful to the social sciences? *Journal of Avian Biology* **26**: 273–276.

Erard, C., M. Théry, and D. Sabatier. 1991. Régime alimentaire de *Tinamus major* (Tinamidae), *Crax alector* (Cracidae) et *Psophia crepitans* (Psophiidae) en forêt guyanaise. *Gibier Faune Sauvage* **8**: 183–210.

Feldså, J. 1995. Have ornithologists 'slept during class'? On the response of ornithology to the 'Biodiversity Crisis' and 'Biodiversity Convention.' *Journal of Avian Biology* **26**: 89–93.

Felsenstein, J. 1985. Phylogenies and the comparative method. *American Naturalist* **125**: 1–15.

Franklin, J. F. 1993. Preserving biodiversity: Species, ecosystems, or landscapes? *Ecological Applications* **3**: 202–205.

Funk, V. A. 1995. Cladistic methods. Pages 30–38 in W. A. Wagner, and V. A. Funk, editors. *Hawaiian biogeography: Evolution on a hot spot archipelago*. Smithsonian Institution, Washington.

Furlow, F. B., and T. Armijo-Prewitt. 1995. Peripheral populations and range collapse. *Conservation Biology* **9**: 1345.

Futuyma, D. J. 1979. *Evolutionary biology*. Sinauer Associates, Sunderland, Massachusetts.

Gavin, T. A., and B. May. 1988. Taxonomic status and genetic purity of the Columbian white-tailed deer. *Journal of Wildlife Management* **52**: 1–10.

Gervai, J. & V. Csányi. 1985. Behaviour-genetic analysis of the paradise fish (*Macropodus opercularis*). I. Characterization of the behavioural responses of inbred strains in novel environment. A factor analysis. *Behaviour Genetics* **15**: 503–519.

Greenberg, R. 1983. The role of neophobia in determining the degree of foraging specialization in some migrant warblers. *American Naturalist* **122**: 444–453.

Hamilton, W. D., and M. Zuk. 1982. Heritable true fitness and bright birds: a role for parasites? *Science* **218**: 384–387.

Hart, B. 1990. Behavioral adaptations to pathogens and parasites: five strategies. *Neuroscience and Biobehavioral Reviews* **14**: 273–294.

Hartfield, P., and R. S. Butler. 1995. Superconglutinates – the movie. Page 38 in Program and abstracts. *The conservation and management of freshwater mussels II. Initiative for the future*. October 16–18, St. Louis.

Harvey, P. H., and M. D. Pagel. 1991. *The comparative method in evolutionary biology*. Oxford University Press, Oxford.

Hasler, A. D., and A. T. Scholz. 1983. Olfactory imprinting and homing in salmon: investigations into the mechanism of the imprinting process. *Zoophysiology*, Volume 14. D. S. Farner, B. Heinrich, K. Johansen, H. Langer, G. Neuweiler, and D. J. Randall, editors. Springer, Berlin.

Hayek, L. C. 1994. Analysis of amphibian biodiversity data. Pages 207–269 in W. R. Heyer, M. A. Donnelly, R. W. McDiarmid, L. C. Hayek, and M. S. Foster, editors. *Measuring and monitoring biological diversity: Standard methods for amphibians*. Smithsonian Institution Press, Washington.

Heinroth, O. 1911. Beiträge zur Biologie, namentlich Ethologie und Psychologie der Anatiden. Pages 598–702 in *Proceedings of the Fifth International Ornithological Congress*, Berlin, 1910.

Helfman, G. S., and E. T. Schultz. 1984. Social transmission of behavioural traditions in a coral reef fish. *Animal Behaviour* **32**: 379–384.

Hepp, G. R., J. M. Novak, K. T. Scribner, and P. W. Stangel. 1988. Genetic distance and hybridization of black ducks and mallards: a morph of a different color? *Auk* **105**: 804–807.

Hoikkala, A., K. Y. Kaneshiro, and R. R. Hoy. 1994. Courtship songs of the picture-winged *Drosophila planitibia* subgroup species. *Animal Behaviour* **47**: 1363–1374.

Hughes, A. L. 1991. MHC polymorphism and the design of captive breeding programs. *Conservation Biology* **5**: 249–251.

Hunter, M. L., Jr., and A. Hutchinson. 1994. The virtues and shortcomings of parochialism: conserving species that are locally rare, but globally common. *Conservation Biology* **8**: 1163–1165.

Johnsgard, P. A. 1988. *Quails and partridges of the world*. Oxford University Press, Oxford.

Johnson, N. K. 1994. Old-school taxonomy versus modern biosystematics: species-level decisions in *Stelgidopteryx* and *Empidonax*. *Auk* **111**: 773–780.

Kaneshiro, K. Y. 1988. The Hawaiian *Drosophila*. *BioScience* **38**: 258–263.

Kennedy, M., H. G. Spencer, and R. D. Gray. 1996. Hop, step and gape: do the social displays of the Pelecaniformes reflect phylogeny? *Animal Behaviour* **51**: 273–291.

Kinch, J. A. 1996. Ecology forum: love of life. *Nature Conservancy* March/ April:8–9.

Kodric-Brown, A. 1989. Genetic introgression after secondary contact. *Trends in Ecology and Evolution* **4**: 329–330.

Lande, R. 1988. Genetics and demography in biological conservation. *Science* **241**: 1455–1460.

Langtimm, C. A., and D. A. Dewsbury. 1991. Phylogeny and evolution of rodent copulatory behaviour. *Animal Behaviour* **41**: 217–225.

Lee, K. N. 1993. *Compass and gyroscope*. Island Press, Washington.

Linden, E. 1996. G.O.P. hears nature's call. *Time* 4 March: 57.

Löhrl, H., and E. Thaler. 1992. Behavioural traits as an aid to solving taxonomic problems. *Bulletin BOC Centenary Supplement* **112A**: 199–208.

Lott, D. F. 1991. *Intraspecific variation in the social systems of wild vertebrates*. Cambridge University Press, Cambridge.

Magurran, A. E. 1988. *Ecological diversity and its measurement*. Princeton University Press, Princeton.

Magurran, A. E., B. H. Seghers, P. W. Shaw, & G. R. Carvalho. 1995. The behavioral diversity and evolution of Guppy, *Poecilia reticulata*, populations in Trinidad. *Advances in the Study of Behavior* **24**: 155–202.

Mallet, J. A. 1995. A species definition for the Modern Synthesis. *Trends in Ecology and Evolution* **10**: 294–298.

Mann, C. C., and M. L. Plummer. 1992. The butterfly problem. *Atlantic Monthly* **269**(1): 47–70.

Mark, D., and B. J. Stutchbury. 1994. Response of a forest-interior songbird to the threat of cowbird parasitism. *Animal Behaviour* **47**: 275–280.

May, R. M. 1995. Conceptual aspects of the quantification of the extent of biological diversity. Pages 13–20 in D. L. Hawksworth, editor. *Biodiversity: Measurement and estimation*. Chapman and Hall, London.

Mayr, E. 1963. *Animal species and evolution*. Belknap Press of Harvard University Press, Cambridge, Massachusetts.

Meffe, G. K., and C. R. Carroll. 1994. *Principles of conservation biology*. Sinauer Associates, Sunderland, Massachusetts.
Meyer de Schauensee, R., and W. H. Phelps, Jr. 1978. *Birds of Venezuela*. Princeton University Press, Princeton.
Miller, P. S. 1995. Selective breeding programs for rare alleles: examples from the Przewalski's horse and California condor pedigrees. *Conservation Biology* **9**: 1262–1273.
Myers, N. 1983. *A wealth of wild species*. Westview Press, Boulder, CO.
Myers, N. 1988. Threatened biotas: 'Hot spots' in tropical forests. *Environmentalist* **8**: 187–208.
Nabhan, G. P. 1996. The parable of the poppy and the bee. *Nature Conservancy* March/April:10–15
Nelson, G. 1989. Species and taxa: systematics and evolution. Pages 60–81 in D. Otte and J. A. Endler, editors. *Speciation and its consequences*. Sinauer Associates, Sunderland, Massachusetts.
Norton, B. G.1986. On the inherent danger of undervaluing species. Pages 110–137 in B. G. Norton, editor. *The preservation of species*. Princeton University Press, Princeton.
O'Brian, C. A., J. Brim-Box, and A. Daniels. 1995. Host fish attraction strategy and host fish identification for *Lampsilis subangulata*. Page 42 in Program and Abstracts. *The conservation and management of freshwater mussels II. Initiative for the future*. October 16–18, St. Louis.
O'Brien, S. J. 1994. When endangered species hybridize: the US hybrid policy. Pages 69–70 in G. K. Meffe and C. R. Carroll, editors. *Principles of conservation biology*. Sinauer Associates, Sunderland, Massachusetts.
O'Donnell, A. G., M. Goodfellow, and D. L. Hawksworth. 1995. Theoretical and practical aspects of the quantification of biodiversity among micro-organisms. Pages 65–73 in D. L. Hawksworth, editor. *Biodiversity: Measurement and estimation*. Chapman and Hall, London.
O'Hara, R. J. 1993. Systematic generalizations, historical fate, and the species problem. *Systematic Zoology* **42**: 231–246.
Parrish, J. D. 1995. Experimental evidence for intrinsic microhabitat preferences in the black-throated green warbler. *Condor* **97**: 935–943.
Paterson, H. E. H. 1985. The recognition concept of species. Pages 21–29 in E. S. Vrba, editor. *Species and speciation*. Transvaal Museum Monograph No. 4, Pretoria.
Polziehn, R. O., C. Strobeck, J. Sheraton, and R. Beech. 1995. Bovine mtDNA discovered in North American bison populations. *Conservation Biology* **9**: 1638–1643.
Ralls, K., and J. Ballou. 1983. Extinction: lessons from zoos. Pages 164–184 in C. M. Schonewald-Cox, S. M. Chambers, B. MacBryde, and L. Thomas, editors. *Genetics and conservation: A reference for managing wild animal and plant populations*. Benjamin/Cummings, Menlo Park, CA.
Raven, P., and E. O. Wilson. 1994. A 50-year plan for biodiversity surveys. Pages 83–84 in G. K. Meffe, and C. R. Carroll, editors. *Principles of conservation biology*. Sinauer Associates, Sunderland, Massachusetts.
Reisenbichler, R. R., J. D. McIntyre, M. F. Solazzi, and S. W. Landing. 1992. Genetic variation in steelhead of Oregon and northern California. *Transactions of the American Fisheries Society* **121**: 158–169.
Rhymer, J. M., M. J. Williams, and M. J. Braun. 1994. Mithochondrial analysis of gene flow between New Zealand mallards (*Anas platyrhynchos*) and grey ducks (*A. superciliosa*). *Auk* **111**: 970–978.

Rosenberg, K. V. 1996. Food for thought. *Birdscope* Spring:6.

Rothenbuhler, W. C. 1964. Behavior genetics of nest cleaning in honey bees. IV. Responses of F1 and backcross generations of disease-killed brood. *American Zoologist* **4**: 111–123.

Rowley, I., and G. Chapman. 1986. Cross-fostering, imprinting and learning in two sympatric species of cockatoo. *Behaviour* **96**: 1–16.

Ryder, O. A. 1986. Species conservation and systematics: the dilemma of subspecies. *Trends in Ecology and Evolution* **1**: 9–10.

Ryder, O. A. 1996. Saving endangered species. Page 272 in M. R. Cummings, editor. *Biology: Science and life*. West, Minneapolis, Minnesota.

Ryder, O. A., J. H. Shaw, and C. M. Wemmer. 1988. Species, subspecies and ex situ conservation. *International Zoo Yearbook* **27**: 134–140.

Savage, J. M. 1995. Systematics and the biodiversity crisis. *BioScience* **45**: 673–679.

Silva, J. and S. D. Strahl. 1991. Human impact on populations of chachalacas, guans and curassows (Galliformes: Cracidae) in Venezuela. Pages 36–52 in J. G. Robinson and K. H. Redford, editors. *Neotropical wildlife use and conservation*. University of Chicago Press, Chicago.

Simpson, B. B., and J. Cracraft. 1995. Systematics: the science of biodiversity. *BioScience* **45**: 670–672.

Soulé, M. 1985. What is conservation biology? *BioScience* **35**: 727–734.

Soulé, M. 1986. Conservation biology and the real world. Pages 1–12 in M. E. Soulé, editor. *Conservation biology: The science of scarcity and diversity*. Sinauer Associates, Sunderland, Massachusetts.

Stiassny, M. L. 1994. Systematics and conservation. Pages 64–66 in G. K. Meffe, and C. R. Carroll, editors. *Principles of conservation biology*. Sinauer Associates, Sunderland, Massachusetts.

Strahl, S. D., and J. Silva. 1989. Chickens of the trees. *Animal Kingdom* Nov/Dec: 56–65.

Taylor, E. B. 1995. Genetic variation at minisatellite DNA loci among North Pacific populations of Steelhead and Rainbow Trout (*Onchorhynchus mykiss*). *Journal of Heredity* **86**: 354–363.

Temple, A. R. 1989. The meaning of species and speciation: a genetic perspective. Pages 3–27 in D. Otte and J. A. Endler, editors, *Speciation and its consequences*. Sinauer Associates, Sunderland, Massachusetts.

Temple, S. A. 1978. Manipulating behavioral patterns of endangered birds: a potential management technique. Pages 435–443 in S. A. Temple, editor. *Endangered birds: Management techniques for preserving threatened species*. University of Wisconsin, Madison.

Temple, S. A. 1986. Recovery of the endangered Mauritian kestrel from an extreme population bottleneck. *Auk* **103**: 632–633.

Templeton, S., H. Hemmer, G. Mace, U. S. Seal, W. M. Shields, and D. S. Woodruff. 1986. Local adaptation, coadaptation, and population boundaries. *Zoo Biology* **5**: 115–125.

Terborgh, J. 1974. Preservation of natural diversity: the problem of extinction prone species. *BioScience* **24**: 715–722.

Terborgh, J. 1986. Keystone plant resources in the tropical forest. Pages 330–344 in M. E. Soulé, editor. *Conservation biology: The science of scarcity and diversity*. Sinauer Associates, Sunderland, Massachusetts.

Théry, M., and S. L. Vehrencamp. 1995. Light patterns as cues for mate choice in the lekking White-throated Manakin (*Corapipo gutturalis*). *Auk* **112**: 133–145.

Tinbergen, N. 1976. Ethology in a changing world. Pages 507–527 in P. P. G. Bateson and R. A. Hinde, editors. *Growing points in ethology*. Cambridge University Press, Cambridge.

US Congress, Office of Technical Assessment. 1987. *Technologies to maintain biological diversity, OTA-F-330*. US Government Printing Office, Washington.

Vane-Wright, R. I., C. J. Humphries, and P. H. Williams. 1991. What to protect? – Systematics and the agony of choice. *Biological Conservation* **55**: 235–254.

Wayne, R. K., and Jenks, S. M. 1991. Mitochondrial DNA analysis implying extensive hybridization of the endangered red wolf *Canis rufus*. *Nature* **351**: 565–568.

Wecker, S. C. 1963. The role of early experience in habitat selection by the prairie deer mouse, *Peromyscus maniculatus bairdi*. *Ecological Monographs* **33**: 307–325.

Wenzel, J. W. 1992. Behavioral homology and phylogeny. *Annual Review of Ecology and Systematics* **23**: 361–81.

West-Eberhard, M. J. 1983. Sexual selection, social competition, and speciation. *Quarterly Review of Biology* **58**: 155–183.

Whittaker, R. H. 1972. Evolution and measurement of species diversity. *Taxon* **21**: 213–251.

Williams, J. D., and M. Mulvey. 1994. Recognition of freshwater mussel taxa: a conservation challenge. Pages 57–58 in G. K. Meffe, and C. R. Carroll, editors. *Principles of conservation biology*. Sinauer Associates, Sunderland, Massachusetts.

Williams, P. H., C. J. Humphries, and R. Vane-Wright. 1991. Measuring bio-diversity: taxonomic relatedness for conservation priorities. *Australian Systematic Botany* **4**: 665–679.

Wilson, E. O. 1984. *Biophilia*. Harvard University Press, Cambridge, Massachusetts.

Woodruff, D. S. 1989. The problems of conserving genes and species. Pages 76–88 in D. Western and M. C. Pearl, editors. *Conservation for the twenty-first century*. Oxford University Press, New York.

W. P. A. 1985. Conservation strategy. UK, Reading: World Pheasant Association.

Part III:
Examples and case studies

The collection of examples and case studies in this last section demonstrates how behavioral studies can contribute uniquely to conservation in the way of theory, methods, and technology. Behavioral contributions to conservation flow from diverse avenues of research, as demonstrated by the numerous behavioral subdisciplines represented in this section, including behavioral ecology, behavioral toxicology, bioacoustics, behavioral genetics, sensory and neuroethology, or some mixture of these and other fields. The ways in which behavioral subdisciplines can contribute are only as limited as the imaginations and willingness of researchers to discover how their own fields of interest relate to conservation.

Whereas previous chapters have stressed the value of a theoretical approach, much of the usefulness of science to conservation will not come from the one percent inspiration rendered by theory, but rather in the day-to-day perspiration of censusing, inventorying, and monitoring populations. Chapter 9 (Baptista and Gaunt) shows how a great number of techniques and methods that were developed to study animal communication and their acoustic environments in the field can be (and have been) applied to these common conservation tasks. These authors also describe an astonishing number of additional applications of bioacoustics to conservation.

Assessing the integrity of populations, especially ones that are small or dwindling, also involves determining the amount of genetic variation necessary to prevent populations from going extinct and to allow the evolutionary process to continue. Chapter 10 (Parker and Waite) explores the role of two different tools, molecular techniques and behavioral observations and theories, for assessing the genetic stability of populations. It is readily apparent that neither genetics nor behavior by itself is sufficient, reinforcing the need for a multidisciplinary approach to conserving populations.

Behavior, and not simply the number of individuals in a population, is an important determinant of the amount and kind of genetic variation that is transferred from one generation to the next. Although social organization, such as mating or parental care systems, frequently places constraints on the transfer of genetic information across generations, it is commonly and erroneously assumed that social organization is a fixed characteristic of a species. Chapter 11 (Komdeur and Deerenberg) considers the conservation implications of variation in social systems resulting from environmental factors.

Earlier (Chapter 5) it was shown how hormone levels could be used to identify populations under stress. Chapter 12 (Smith and Logan) shows how behavior patterns themselves may be used as direct measures of stressful or potentially stressful environmental conditions. Behavioral toxicologists have long struggled with the practical and legal aspects of using behavior to demonstrate biological stress of chemically contaminated environments. These authors discuss the reciprocal benefits to be gained by a cross-disciplinary effort.

Ethologists have been both fascinated and challenged by their own perceptual biases to understand the perceptual worlds (umwelt) of animals, and these limitations confront conservationists as well. In some cases, the study of perceptual mechanisms of organisms is necessary to know which aspects of the environment to protect. For example, whereas we typically regard pollution in terms of chemically altered environments, we are also greatly altering the acoustic and visual environments that organisms rely upon to carry out basic biological functions, such as communication, predator avoidance, prey detection, thermoregulation, orientation, and habitat selection. Chapter 13 (Witherington) shows how research of the mechanisms of orientation by hatchling sea turtles offers management strategies that will allow humans and non-humans to share the light environment to the satisfaction of both. Our thinking about light worlds is further expanded in Chapter 14 (Endler). Traditionally, the effects of removing trees from a forest, for example, is understood in terms of community succession, edge effects, or disruption of corridors. This chapter explores even more subtle effects of habitat alteration on the reproductive and predator–prey activities of organisms.

For the various reasons touched upon throughout this volume, the transition into conservation will be easier for some disciplines than others. Disciplinary evolution as a whole normally occurs too slowly to be of help to conservation, thus individuals must set forth on their own. The final chapter (Ralls, Chapter 15) provides a personal account of how that transition can be made and the kinds of obstacles to be expected, both at the individual level (e.g., developing an expertise; asking questions that are relevant to conservation; learning conservation policy), and at the level of the discipline. The degree of success of any discipline making the transition to conservation will hinge on the willingness of specialists to reach consensus on management advice, especially in the face of conflicting or incomplete information, and the degree to which research related to conservation is encouraged and supported by colleagues. Ralls also sets the tone for ending this volume by describing her personal valuation and commitment to the future of biodiversity and the human experience with animals, capturing our own motivation for writing this volume.

Chapter 9

Bioacoustics as a tool in conservation studies

LUIS F. BAPTISTA AND SANDRA L. L. GAUNT

Monitoring and managing species are central issues to conservation and preservation of animal diversity. Bioacoustic research and tools provide insights into animal behavior that can aid these conservation efforts, sometimes in unexpected ways. Bioacoustics is concerned with sounds used by animals. Although this statement of purpose seems narrow, bioacoustics, in fact, crosses a broad spectrum of disciplines including taxonomy and systematics (Payne 1986; Fonseca 1991; Cocroft & Ryan 1995), communication theory (Hailman *et al.* 1985), vocal ontogeny (Nelson *et al.* 1995), behavioral genetics (Bentley & Hoy 1974; Marble & Bogart 1991; Baptista 1994, 1996) and behavioral ecology (Payne *et al.* 1988) to name a few. We will weave some of these topics in bioacoustics into the fabric of the conservation effort from the literature on groups of organisms that use the sound modality in communication, including insects, anurans, birds, and mammals.

The conservation mission includes the management and preservation of ecosystems or taxa within specific ecosystems. These endeavors often entail inventories of biological diversity or censusing and life history studies of one or more species within specific habitats. Biosurveys may necessitate capture of target species and monitoring by marking captured individuals with visual tags or radiomonitoring devices. Such activities may entail many hours of intensive work and may directly or indirectly affect survival or reproductive success of the study subjects (e.g., Gessaman & Nagy 1988; Sorenson 1995; Croll *et al.* 1996). Bioacoustic methodologies may be employed in conservation efforts to avoid such handling.

As conservation must work with organisms at many interactive levels from the ecosystem to the individual, this chapter is organized from that perspective. The bioacoustic interfaces with conservation are examined at the (1) environmental level for assessing reaction to and control with sound, (2) community level for censusing and biological diversity studies, (3) population level for the study of deme structure, changes in population structure, and in reintroduction efforts, (4) species and subspecies levels in identification and for the mapping of ranges of endangered taxa, and (5) the individual where identification is sometimes possible by voice printing.

The sound environment

Environmental quality

Animals have adapted to a wide range of environmental sounds from the roar of surf and torrents to the whisper of wind through the trees; these adaptations may be in the sounds animals produce to transmit effectively through the environment during communication (Wiley & Richards 1982; Martens & Geduldig 1989) and in the morphological structures involved in the acquisition of sound. The quality of the sound environment is increasingly impacted by the introduction of human-generated noise (from aircraft, traffic, heavy construction, etc.). In addition, certain human activities require the exclusion of animals in order to function. Thus, determining and monitoring the sound portion of the sensory environment, or umwelt, of animals is a concern for both the protection of animals and the maintenance of projects important to human activity. That the Acoustic Society of America devoted a symposium in 1993 and 1996 to setting sound standards to protect animals from man-made noise is an indication of the growing concern about human noise pollution.

If noise pollution alters the fitness of animals, e.g., interferes with food acquisition, mate attraction, or reproductive effort, then monitoring both the pollution and the animals' physiological and behavioral responses becomes an issue. Instrumentation is available that can be attached directly to an animal to monitor and store information about extraneous noises, such as aircraft overflights, over a long period (months), and/or monitor the stress on the animal using measures such as heart rate and body temperature (Bunch & Workman 1993; Kugler & Barber 1993; Larkin *et al.* 1996). Most of these instruments have been used with large mammals, but miniaturization will expand their application. Similarly, automated recording from acoustic arrays of microphones in the field (see below) or hydrophones in aquatic habitats can be used to monitor the presence or absence of animals under conditions of varying levels of environmental noise (Heyer *et al.* 1993).

Data from monitoring the environment and animals' responses to noise can be used to make recommendations for the protection of animals. Thus, large ungulates near a proposed extension of a US Air Force supersonic range were found to habituate readily to supersonic noise (Bunch & Workman 1993; Krausman *et al.* 1993), whereas increased noise pollution from a proposed highway through a small metropolitan park was shown to have potential deleterious effects on the breeding of remnant bird and anuran populations (A. Thompson & S. L. L. Gaunt, unpublished report).

Marine organisms are exposed to human-made noise both in air and water from aircraft, ships, icebreaking, dredging, seismic exploration, and sonar. Most available data on marine organisms have been on marine mammals, concerned with short-term behavioral reactions, and from observations without controlled tests or that are anecdotal (Richardson *et al.* 1995). A recent proposal, however, to study global scale ocean temperatures using the broadcast of low frequency 75 Hz acoustic signals over the ocean basins (Acoustic Thermometry of Ocean Climate, ATOC) by the Scripps Institute of Oceanography and associates has heightened interest on the effect of sound on marine life, especially dolphins and whales. Already baseline data on the effect of ATOC and ATOC-like systems are being gathered on indicator species in both Hawaii and California by the Marine Mammal Research Program (Clark 1995). Reliable physiological, long-term, and population consequences of sound on marine mammals should be forthcoming from these efforts.

Sound and animal control

Sound can also be used to repel animals when they interfere with human activity, in order to protect animals from human activity, or when animals are deemed noxious or nuisances. The need for sound recordings of animals for use in repelling animals is so great that together the US Fish and Wildlife Service and US Air Force have published a catalog of sources where animal sounds can be obtained (Schmidt & Johnson 1982) for use in such work. Birds are the primary target for dispersal with sound, but 'acoustic harassment devices' are available for marine animals (Richardson *et al.* 1995), and a sound repelling system for marine turtles is under investigation (Moein *et al.* 1993).

The effectiveness of acoustic harassment may decline over time, as some animals develop tolerance to noise. Behavioral habituation, waning of responsiveness, may occur when a repeated stimulus lacks any significant consequence. Starlings (*Sturnus vulgaris*) can become nuisances when they form large roosting aggregates in the autumn. They readily habituate to explosions used to disperse them (S. L. L. Gaunt personal observation). Similarly, polar bears habituate to pyrotechnic scaring devices and warning shots if not used with other reinforcement methods (Richardson *et al*. 1995). As will be repeated often, indiscriminate use of a methodology, without knowledge of the behavior of the animal of concern, can render its use ineffective or counterproductive, not to mention costly.

Community level

Surveying and censusing biological diversity

Monitoring the acoustic environment is a standard technique used in the study of diversity in insects, anurans, birds, and mammals. An experienced investigator is soon familiar with all the sounds of the taxa in a habitat being studied, and any sound out of the ordinary will not go unnoticed. Acoustic monitoring is especially useful in habitats, such as tropical rain forests, where animal diversity is great but visual survey and censusing is difficult, if not impossible, because of dense vegetation, low visibility, and secretive habits of the subjects. The late Theodore Parker III of the Museum of Natural History, Louisiana State University, was one of the greatest listeners of our time. He developed a keen knowledge of bird identification by vocalizations, and it was widely reported that he could recognize some 3000 neotropical bird species by voice.

Parker (1991) demonstrated the utility of audiocensusing in the Bolivian Amazon. There he sound recorded the dawn choruses of birds and found that within seven days he had captured on tape 85% of the avian species in the region. It took seven experienced ornithologists working with mist-nets 54 days to inventory the birds of the same region. Ten of the species missed in Parker's study were non-residents of the site. The remainder were species, such as forest raptors and canopy hummingbirds, that are often overlooked in brief surveys. Thus, sound recording may function as a much more efficient method to inventory birds in an area when the investigator has prior knowledge of the local fauna.

The latter point is a limitation, as learning to identify animals by vocalizations comes only with years of field experience. This process, however, can be greatly facilitated by working with experienced listeners. Parker availed himself to many biologists, and, most importantly, he actively trained local students to recognize the indigenous avifauna. Experience both in sound recording and sound identification can also be obtained by attending field courses such as are made available by the Cornell Laboratory of Ornithology's Nature Sound Recording Workshop, joining sound recording associations such as the Nature Sound Society sponsored by the Oakland Museum of Natural History, California, and contacting local colleges and universities for biology courses devoted in part or wholly to animal sound communication (such as the behavior and bioacoustic courses offered at The Ohio State University in the Department of Zoology).

Access to properly identified recorded sounds can also greatly facilitate identification of animals by vocalizations, and recorded sound archives have valuable reference resources for faunal surveys. Parker's collection of neotropical recordings is deposited at the Library of Natural Sounds (LNS), Cornell Laboratory of Ornithology, Ithaca, NY 14850, USA; other North American archives include the Borror Laboratory of Bioacoustics (BLB), Museum of Biological Diversity, The Ohio State University, Columbus, OH 43212, USA, and the Bioacoustics Laboratory and Archive (BLA), Florida Museum of Natural History, Gainesville, FL 32611, USA. Sound archives worldwide are documented by Kettle (1989), and their importance to research is described by Hardy (1984), Nikol'sky (1984), and Kroodsma *et al.* (1996).

Staff at these archives can also provide valuable information about the equipment and techniques for sound recording and analysis. The instruments of bioacoustics can readily be exploited by the conservationist. This is especially relevant today with many new advances in miniature, portable analog and digital recording machines, light-weight directional microphones, and the power of computer analysis, including the continuous, digital sound spectrograph analyzer. All of these make capturing, processing, manipulating, and analyzing ephemeral sound signals possible with greater facility and rapidity then ever before (Williams & Slater 1991; Baptista & Gaunt 1994). With the realization that the technology for sound recording is today changing more rapidly than since the introduction of the magnetic tape recorder after World War II, it is futile to make specific recommendations here, but an introduction to recording configurations, both analog and digital, can be obtained from other sources (Wickstrom 1982; Ewing 1989; Heyer *et al.* 1993; Kroodsma *et al.* 1996; G. F. Budney & R. W. Grotke, unpublished results).

Commercially available, synoptic records and tapes are excellent training tools in the identification of animal sounds. The BLA has lead the way in producing, under the ARA label, 22 such synopses including *New World Cuckoos and Trogons* (Hardy *et al.* 1987), *New World Pigeons and Doves* (Hardy *et al.* 1989), and *Vireos and Their Allies* (Barlow 1995). The last synoptic production by the recording artists D. J. Borror and W. W. H. Gunn, was of all the North American wood warblers (1985); it was produced and is available through the LNS, and a pamphlet is available that has sound spectrograms for each species. The Ohio Biological Survey is using a synoptic tape produced by the BLB to census and monitor amphibians. To catalyze statewide interest, high school teachers and students were trained and employed to identify all frogs and toads in the state. The project was begun in 1995, and compiling the results is an ongoing project.

Special systems for censusing

Automated recording systems have been used for determining the presence or absence of anuran species and to establish their temporal calling patterns (Heyer *et al.* 1993). These systems allow for 24-hour sampling at multiple sites without entailing investigator presence. More sophisticated systems can be voice activated, thus reducing and streamlining the replay and analysis time, and others will simultaneously monitor environmental conditions. Similar automated systems are being designed for birds, especially for censusing nocturnal migrating birds (Evans 1994, personal communication) and for monitoring aquatic organisms, especially fish (Fish & Mowbray 1970) and whales (Norris & Evans 1995). There are limitations to automated monitoring, not the least being protection of unattended equipment from changes in environmental conditions and from theft or vandalism. More critical are limitations that can result in errors in estimating population sizes (reviewed in Chapter 7 in Heyer *et al.* 1993).

Chiroptera uttering sounds in frequencies above the human hearing range (ultrasonic) may be detected with the use of commercially available bat detectors (Amabat II, Titley Electronic; Batbox III, Stage Electronics; Mini II, Ultra Sound Advice). Bat detectors lower ultrasonic sound to an audible frequency spectrum for censusing, and in some cases, species identification. The signals can also be recorded and sound spectrographed for comparative studies. As with any instrumentation, methodology for standardizing the use of bat detectors and an understanding of bat biology needs to be developed to avoid error in the estimation of animal numbers or species (Waters & Walsh 1994).

Other mammals use ultrasonics including a number of rodents (Warburton *et al.* 1989; Holman & Seale 1991; Blanchard *et al.* 1992), and possibly a reptile (Fenton & Lecht 1990). Elephants and possibly other large ungulates use sounds below human hearing (infrasonic) to communicate long distances (Garstang *et al.* 1995), and although we know of no censusing techniques exploiting these sounds to date, they too could be monitored with appropriate receiving devices (Langbauer *et al.* 1991).

The use of playback in censusing

There are limitations to relying solely on human auditory capacity in censusing populations, even for species using frequencies in the sound channel of normal human hearing. Human hearing sensitivity can vary between individuals, within individuals with age (Mayfield 1966), and because of distractions from background noise or inattentiveness. Animals also vary in their sound output, producing sounds in different parts of the day, breeding cycle, and season (Best 1981; Baskett 1993). Some birds, e.g., the California towhee (*Pipilo crisalis*) and Sedge warbler (*Acrocephalus schoenobaenus*) cease singing once they are paired (Marshall 1964; Catchpole 1973). Thus, a reliance on counts of singing males alone to estimate population size may be grossly biased towards unpaired males (Gibbs & Faaborg 1990).

The use of playback of recorded sounds or imitations of calls to attract animals can overcome some of these difficulties. Territorial species belonging to a number of taxa will respond to conspecific vocalizations with vocalizations, approach and aggressive displays, often at all stages of the breeding cycle, even in winter. Owls readily respond to imitation of calls (Foster 1965; Redpath 1994), and the response to such calls has been used as a censusing technique to ascertain the effect of forest fragmentation on woodland owls (Redpath 1995; see below).

Johnson *et al.* (1981) review the literature on the use of playback to census birds belonging to about 20 families. Although columbiforms are not on their list, it should be noted that various doves and pigeons also respond to the playback of conspecific voice with approach and/or calling (e.g., *Claravis pretiosa*, *Columba subvinacea*). At least *Claravis pretiosa* (L. F. Baptista personal observation) and the Pheasant pigeon (*Otidiphaps nobilis*, Delacour 1959) may be attracted by human imitations of calls in the same way that owl biologists call owls by imitation. Endangered Grenada doves (*Leptotila wellsi*) responded to playback of conspecific song in eight of 12 trials but ignored playback of allospecific vocalization (Blockstein & Hardy 1989).

Organisms other than birds have been censused using playback. Leonard & Fenton (1984) played conspecific calls to Spotted bats (*Euderma maculata*) and induced volant foraging bats to approach to within 50 m of the speaker. Frogs respond well to playback of conspecific calls and herpetologists have devised their own portable synthesizer enabling them to synthesize and play calls to frogs in the field (Pina & Channing 1981). Playback might be useful for species of orthopterans when sound elicits not just phonotaxic responses but also an acoustic response from the female or other males in a chorus (Ewing 1989).

Playback as a censusing technique can work well with (1) rare, threatened, or endangered species that are found in low densities, for example locally rare Yellow-billed cuckoos (*Coccyzus americanus*), (2) nocturnal species such as owls and caprimulgiforms (Kavanagh & Bamkin 1995), (3) secretive species such as various rails, and (4) species in areas of low visibility such as tropical forests or dense brushland (Johnson *et al.* 1981; Marion *et al.* 1981; Johnson & Dinsmore 1986).

Not all recorded vocalizations of a given species, however, will elicit strong responses equally. For example, some avian taxa exhibit regional song dialects, and the response will be more effective if the local dialect rather than alien dialects is used. Male Nuttall's white-crowned sparrows (*Zonotrichia leucophrys nuttalli*) respond more strongly to the local dialect than to an alien dialect, but even less so to the song of an alien subspecies (Petrinovich & Patterson 1981; Tomback *et al.* 1983) (Fig. 9.1).

Yet another caveat should be noted: not all organisms respond positively to vocalization playback. Some species of hummingbirds form male groups that sing from traditional territorial areas, leks. The song is produced almost constantly throughout the day and season, and the leks are easily located by the song emanating from them. A novel song, however, played back to a male Green violet-ear (*Colibri thalassinus*) will cause him to not only cease singing but to depart the lek usually in the opposite direction from the speaker (L. F. Baptista & S. L. L. Gaunt, personal observation). Once again, knowledge of the organism is required before assessing the appropriateness of implementing a method.

Portable computers are now available that can be used in the field for playback. Animal sounds can be digitized for immediate replay, for manipulation prior to playback, or synthesized *de novo* in the field for playback. Such systems are interactive and allow altering the approach used as conditions dictate. Several general systems are available, e.g., for a MAC-based system SINGIT! (Bradbury & Vehrencamp 1994), for

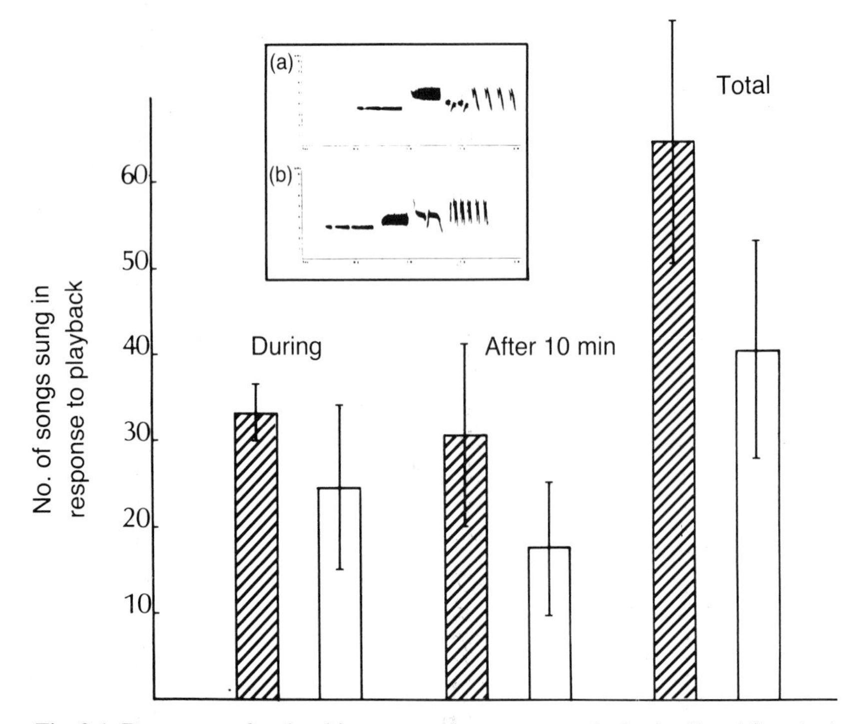

Fig. 9.1. Responses of male white-crowned sparrows to playback of local (hatched bars) versus alien (open bars) song dialect. Vertical bars represent ± 25 SE of songs sung either during five minutes of playback (first pair of bars), ten minutes following the five minute playback (second pair of bars), and cumulative number of songs sung in 15 minutes. Spectrograms are of the (a) local Tioga Pass song, and the (b) alien Virginia Lakes song. Males sang many more songs in each 15 minute experimental trial to the playback of local than to alien song (Wilcoxon $P<0.01$).

IBM systems, SIGNAL (Beeman 1994), and the aforementioned system used for anurans.

Population level

Deme structure

Population studies are often prerequisite to conservation programs and the study of deme structure may be important in management policies, for example in determining the sizes of patches or features of the habitat to be conserved. Deme structure analysis is dependent on genetically based phenotypic variation. Thus, vocal characters that are genetically determined are appropriate at this level, whereas caution must be used when vocalizations are acquired in part by learning. Indeed, in vocal learning, the relative degree of influence or interaction between genetic inheritance (genes) and inheritance via cultural transmission (memes) is unknown.

Bobwhite quail (*Colinus virginianus*) are widespread in distribution (e.g., Leopold 1959). Although their assembly 'hoy-poo' call sounds similar across populations, frequency and temporal patterns subjected to multivariate statistical analyses have revealed regional and covey differences (Goldstein 1978; Baker & Bailey 1987*a*). Cross-fostering studies have revealed that these calls develop independent of learning experience (Baker & Bailey 1987*b*). Theoretically, mapping of assembly call 'dialects' should allow investigators to study deme structure in these quail, a technique which may prove especially useful in studies of the endangered subspecies of Masked bobwhite (*C. v. ridgwayi*, Greenway 1967). Geographic variation of vocalizations has also been shown to be independent of learning in males of the endangered European quail (*Coturnix c. coturnix*, Guyomarc'h *et al.* 1984). Here too, vocal dialects may be useful in studying deme structure.

Various amazon parrots (*Amazona* spp.) produce flock specific calls (review in Baptista 1993) acquired by vocal tradition as in other parrots (e.g., Rowley & Chapman 1986; Kavanau 1987) and are not suitable in studying deme structure in psittacines. These same calls, however, may be useful in studying home ranges of particular flocks and how flocks partition the habitat. Once again, understanding the vocal communication system of the group in question is critical.

Killer whales (*Orcinus orca*) associate in social groups known as 'pods.' Several pods may sometimes associate together and share several calls recordable with hydrophones. It has been found that each pod has at least one call unique to the group (Ford & Fisher 1982), and pods are often recognized aurally by calls long before coming into visual recognition range. These unique calls have been used to study the home ranges of pods (Hoezel & Osborne 1986).

Advertising calls of Cricket frogs (*Acris crepitans*) vary in acoustic frequency, duration, and rate over their entire range in central and eastern North America (Nevo & Capranica 1985). A microgeographic study of this species in Texas, where the two subspecies of the Cricket frog (*A. c. crepitans* and *A. c. blanchardi*) are parapatric, demonstrated a cline in this variation from east to west (Ryan & Wilczynski 1991). Some of this variation is explained by acoustic adaptation of calls to the environment that enhance transmission in different habitats. Much of the clinal variation between habitats, however, was attributed to gene flow between populations.

Song dialects of birds may be used to trace the origins of colonizing populations, and, if they become pests, may require monitoring to protect crops or to manage the species in the new area. For example, Mundinger (1975) used song to study the spread of House finches (*Carpodacus mexicanus*) in New York, a species that can have adverse effects on fruit crops. Baptista (1975) traced origins of breeding populations of White-crowned sparrows on islands in the San Francisco Bay, California. Song of dialectal species may also be used to study the movements of specific populations on migration (DeWolfe & Baptista 1995).

Population declines and changes in structure

The trained ear not only detects species in an area but also notices absences. In Vincent Dethier's (1992) charming book *Crickets and Katydids*, he laments the loss over the years of various orthopteran sounds from the "concerts and solos" of the northeastern evening.

Songs of the Olive-sided flycatcher (*Contopus borealis*) and Swainson's thrush (*Catharus ustulatus*) present in a 1930 survey were conspicuously missing from a survey carried out 50 years later at Redwood Mountain, Tulare County, California (Marshall 1988). In a search for a cause for this decline, the wintering distribution was investigated; many song bird species breeding in North America winter in Central and South America and are faithful to their wintering areas, returning to the same spots year after year (e.g., Holmes & Sherry 1992). Birds often vocalize on their southern winter grounds, and these vocalizations can be used to study winter habitat occupancy. Olive-sided flycatchers, for example, sing while wintering in Mexico (Phillips *et al.* 1964). Miller (1963) located Swainson's thrushes by their call notes to find them in primary, humid, evergreen, upper tropical forest in the Andes of Colombia.

Some species hold feeding territories on their wintering grounds and often will respond to playback with approach and song. For example, flycatchers of the *traillii* complex (*Empidonax traillii, E. alnorum*) are distinguished by their 'fitz-bew' and 'fee-bee-o' songs. Playback of both song types along with the presentation of a stuffed model evoked 'fitz-bew' singers in Panama and 'fee-bee-o' birds in Peru (Gorski 1969, 1971). Playback thus enabled identification of the wintering grounds for these species with each song form. From this type of information, conservationists can monitor these habitats and plan conservation efforts prior to species decline.

Holmes *et al.* (1989) used the playback of song and chirp calls to census wintering American redstarts (*Setophaga ruticilla*) and Black-throated blue warblers (*Dendroica caerulescens*) in Jamaica. They found that redstarts were more aggressive during the early winter and that responses waned as the season progressed. Black-throated blue warblers, however, were equally aggressive throughout the winter. The effectiveness of this technique was tested (Sliwa & Sherry 1992) by comparing the results of ten minute point counts and subsequent five minute playback trials. They found that playback allowed them to detect 3.0 and 2.3 times more Redstarts and Black-throated blue warblers, respectively, than did point counts alone. More adult male than female and hatching year redstarts were attracted, so that this method of censusing redstarts is biased towards males. Both sexes of Black-throated blue warblers, however, were equally responsive to playback, so that overwintering sex ratios could also be studied using this method (see also identifying the sexes below).

Automated monitoring systems, discussed above, are being tested and will enable investigators to assess long-term changes in population sizes and compare different sites.

Sound and playback to assess, establish, or re-establish breeding populations

Fragmentation of natural habitats is increasing with human activity. The result can be a mosaic of interdependent patches that may or may not be capable of supporting metapopulations – contiguous populations of species now separated in space but not time. How habitat fragmentation affects metapopulations is of great concern, and metapopulation concepts (Harrison *et al.* 1988) work to model fragmented habitats in order to predict their dynamics and determine which should be maintained and how.

As an example of variables that confound predictions of success of the metapopulations, consider two closely related woodland-dependent species, the Tawny owl (*Strix aluco*) of Eurasia and the Northern spotted owl (*S. occidentalis*) of North America. Redpath (1995) tested the effect of woodland fragmentation on home range and reproductive success of the Tawny owl using vocalizations as a censusing tool, as discussed above, for both adults and fledglings. Somewhat surprisingly, this owl species had larger home ranges, higher population densities, and showed no decline in breeding success in the fragmented habitat compared with continuous woodland, there was no apparent cost to living in the more open landscape. On the other hand, the Northern spotted owl is highly susceptible to loss and fragmentation of old growth forest even within continuous forest landscapes (Thomas *et al.* 1990). It is, thus, difficult to make predictive generalities about the effects of fragmentation; the tawny owl appears to have a greater degree of behavioral flexibility such that it can take advantage of the patchy environment.

It becomes important, therefore, to know how different species assess habitat quality, especially when dispersing. There is evidence that many species of diverse taxa cue on the presence of conspecifics as a mechanism of habitat selection (reviews in Smith & Peacock 1990; Reed & Dobson 1993). Dispersing young of birds tend to wander and prospect for potential breeding areas. The quality of potential sites may be indexed by the vocalizations heard emanating from existing residents. Sounds coming from a locale may enable propagules to evaluate the safety of a site (Orians 1966).

Use of playback demonstrated this attractant quality in the Pied flycatcher (*Ficedula hypoleuca*) which nests in territorial clusters. By playing songs of established territory holders from nest boxes in vacant fragments, Alatalo *et al.* (1982) found that new arrivals settled preferentially in nest boxes near speakers. Thus, non-random settlement patterns result from conspecific attraction. Although it is not surprising that noisy colonial or semi-colonial species might use the sounds of their conspecifics as cues for habitat selection, the phenomenon has also been noted in non-colonial species (review by Stamps 1988).

The use of playback methodology linked to known biases for seabirds to select existing colonies as breeding sites has been directly applied to conservation efforts to re-establish breeding colonies on suitable new or historic breeding grounds (previously abandoned due to climatic events or human activity). By placing decoys at strategic sites and playing conspecific colony sounds through speakers, various seabird species have successfully been attracted to these sites, settled, and established breeding colonies (Kress & Nettleship 1988; Podolsky & Kress 1989; Podolsky 1990).

Conservation of a species often includes identification and preservation of specific parameters within the habitat essential to breeding biology. In the case of the Yellow-knobbed curassow (*Crax daubentoni*) the parameter is a display site. In this species each male calls from a traditional display tree utilized year after year. Buchholz (1995) noted the disappearance of a male following the loss of his display tree. Male

Yellow-knobbed curassows respond to playback of conspecific whistles by approaching or countersinging with the speaker. Thus, playbacks may be used to census males and also to locate display trees which may then be targeted for preservation in management programs.

Lekking areas are also traditional so that management of lekking species must include locating and preserving display sites. Many lekking species are loquacious. For example, we located the leks of *Phaethornis* hummingbirds and *Manacus* manakins in Costa Rica by listening for their display sounds – songs in the case of the hummingbirds, and mechanical snapping sounds in the case of the manakins.

Species and subspecies level

Vocalizations and taxonomy

In addition to identifying known species or detecting the loss of species by voice as discussed above, many new species have been first detected and characterized by their vocalizations. Thus, the conservationist trained to be attentive to animal vocalizations is prepared to detect unusual vocalizations that can be clues to the recognition of new taxa, adding to the inventory of biological diversity.

As early as 1789, Gilbert White noted that European warblers in his region sang three different song types which he identified as emanating from not one but three distinct species, namely the Willow warbler (*Phylloscopus trochilis*), Eurasian chifchaff (*P. collybita*), and Wood warbler (*P. sibilatrix*, Marler 1960). Song of the Short-toed treecreeper (*Certhia brachydactyla*) was the first cue that led C. L. Brehm (1820) to recognize it as a species distinct from the Treecreeper (*C. familiaris*). More recently yet another *Phylloscopus* spp. was determined to be a species complex; the Lemon-rumped warbler (*Phylloscopus proregulus*) of Sichuan Province, China is in fact two species, the second now named the Chinese leaf warbler (*P. sichuanensis*). It is most distinguishable by voice from the Lemon-rumped (Alström *et al.* 1992). The pygmy owls (*Glaucidium*) recorded between 1700 m and 1975 m along the eastern slope of the Andes in central Peru call differently from Least pygmy owls (*G. minutissimum*) recorded at 1660 m; the higher elevation population subsequently proved to be a new species (*G. parkeri*, Robbins & Howell 1995). Vocalizations were part of a suite of characters enabling investigators to raise the endangered Socorro dove (*Zenaida graysoni*) to full species status (Baptista *et al.* 1983).

Mammals too have been characterized on vocalizations. Based partly on studies of vocalizations, estimates of the existing number of bushbaby (*Galago*) species in Africa have been increased from six to 16 species (Zimmermann *et al.* 1988; Courtenay & Bearder 1989; Harcourt & Bearder 1989). The ultimate number may be as high as 40 species (Bearder 1995).

Sounds produced by insects are often used in their taxonomy as well. Closely related cicadas (Homoptera) are distinguishable by call (Fonseca 1991). A recent study of sibling grasshopper species in the genus *Chorthippus* of Europe demonstrated that, although these species exhibit few morphologic differences, they can be differentiated by their bio-acoustic behavior alone (Stumpner & von Helverson 1994).

Automated species recognition by voice

Computer algorithms for comparing digitized sounds make it theoretically possible to store samples of vocalizations of various taxa and later match them with samples of unknowns from continuous recordings at various times and sites. Spectrogram cross-correlation (Clark *et al.* 1987; Gaunt *et al.* 1994) has limited use in this area, but other algorithms such as dynamic programming (Williams 1993) and artificial neural networks (Potter *et al.* 1994) may make automated species recognition a valuable tool to conservation, especially for difficult to detect species such as nocturnal migrants (Evans 1994), tropical rain forest species (Robbins & Howell 1995), and aquatic species (Potter *et al.* 1994).

Automated species recognition by voice should be most feasible with organisms producing simple, stereotypic, pulsatile signals, for example many anurans, orthopterans, and chiropterans. The technique should also work for birds that produce simple stereotypic vocalizations, for example doves, tinamous, and caprimulgiforms. Songs of oscines may be less tractable in the early development of these techniques as their vocalizations are often quite variable, for example each Song sparrow (*Melospiza melodia*) in the San Francisco Bay Area may utter up to 20 different themes (Mulligan 1966). Finding unique common denominators between themes can be a formidable task. This is an area ripe for cooperation between behaviorists, conservationists, and computer scientists to interface in identifying areas of common needs and concerns that could be addressed with this technology.

Mapping territories and ranges

Huxley (1934) likened a bird's territory to an elastic disc. As numbers of individuals increase with the season or territory quality, territories compress, but only to a critical minimum (Zimmerman 1971). Thus, birds are fiercest at the center of the territory and least so towards the edge (Ickes & Ficken 1970; Melemis & Falls 1982).

One method of mapping territorial boundaries is to place a speaker in a bird's territory and attract it with playback. The speaker is then moved and the bird's responses noted until a point is reached where the bird ceases to react. This may be regarded as one point of its territorial boundary. The speaker may then be moved back to the middle of the territory and then moved in a different direction. By moving the speaker in different directions and plotting the points when the territory-owner ceases to respond, the investigator may soon plot the territorial boundaries. This technique has been used successfully in plotting territories of Savannah sparrows (*Passerculus sandwichensis*), White-throated sparrows (*Zonotrichia albicollis*), Ovenbirds (*Seiurus aurocapillus*), and Great tits (*Parus major*) (Falls 1981). It most probably can be used with any territorial species that responds to playback. As with using playback for controlling animals, however, care must be taken to avoid habituation to the broadcast signal.

Individual level

Recognition

The traditional method of distinguishing individuals in a population is to use observable individual intraspecific differences (external color or pattern) or capture and mark with visible bands or tags or radio tagging (Bibby *et al.* 1992). Individual pattern differences can change and thus be unreliable over time. As discussed earlier, marking requires capture and, if telemetry is used, surgical invasion of individuals with the possible risk of permanent injury or death. Observing marked individuals in a population over a long period enables the construction of population survival curves. Marking combined with radiotelemetry in secretive or nocturnal species enables the investigator to study the home ranges of individuals. Wildlife managers, however, often prohibit the capture and marking of sensitive or endangered species (e.g., Dahlquist *et al.* 1990). Radio-transmitters may affect survival directly or indirectly (Marks & Marks

1987; Paton *et al.* 1991). Radiotagged Spotted owls produce fewer young than untagged owls. One explanation for these results is that the added weight of the transmitter might reduce the amount of prey males are capable of carrying (Foster *et al.* 1992). Similar results were found in Chinstrap penguins (*Pygoscelis antarctica*) where chick mortality was a consequence of nest abandonment (Croll *et al.* 1996).

Recording and spectrographing sounds may allow long-term recognition of individuals by voice, thus avoiding the need for capture or recapture and disturbance. There is a growing body of experimental (playback) and descriptive literature indicating that individual birds of certain species may be 'fingerprinted' by unique characteristics in their vocalizations (e.g., Gilbert *et al.* 1994; Lambrechts & Dhondt 1995). This technique can be especially useful for endangered bird species, for example loons (*Gavia* spp., Gilbert *et al.* 1994), Wild turkeys (*Meleagris gallopavo*, Dahlquist *et al.* 1990), Spotted owls (Baptista 1993), Peregrine falcons (*Falco peregrinus*, Telford 1993), and various parrots (Baptista 1993).

Voice printing has also been used for individual long-term recognition of primates and other mammals. For example, individual Pygmy marmosets (*Cebuella pygmaea*) may be recognized by their trill vocalizations (Snowdon & Cleveland 1981; Snowdon 1987), Vervet monkeys (*Cercopithecus aethiops*) by their screams (Cheney & Seyfarth 1982), Spider monkeys (*Ateles geoffroyi*) by their whinny calls (Chapman & Weary 1990), and Chimpanzees (*Pan troglodytes*) by their long calls (Marler & Hobbett 1975). Among rodents, grasshopper mice (*Onychomys* spp.) produce a high-pitched, pure-toned whistle which identifies individuals (Hafner & Hafner 1979). Banner-tail kangaroo rats (*Dipodomys spectabilis*) communicate by foot-drumming against the substrate (Randall 1989), and playback experiments have shown that individuals may be distinguished by foot-drumming signatures (Randall 1994).

Sometimes individuals are difficult to recognize based on visual inspection of spectrograms alone as differences may be very subtle. Gaunt *et al.* (1994) used a computer spectrographic cross-correlation technique on hummingbirds in the genus *Colibri* and were able to distinguish individuals from nearest neighbors, and individuals from different populations based on correlation coefficients. This technique should serve well for taxa producing simple, pulsatile sounds (e.g., some anurans, orthopterans, and birds).

Sexing by voice

Studies of social behavior of individuals in a population often necessitates the identification of sex. In monomorphic species this may require capture and surgical laparotomy or taking blood samples. Alternatively, sexes may be determined by voice in species that display vocal sexual dimorphism.

Male Humpback whales may be distinguished from females by song (Tyack 1981). DNA-fingerprinting studies confirmed the fact that song was produced by only one sex (Medrano *et al.* 1994). Sexes may be distinguished in Kloss' gibbon (*Hylobates klossii*) by their long call (Haimoff & Tilson 1985).

Many sexually monochromatic birds may be sexed on voice. All ducks may be sexed by voice as only adult males have a resonating organ associated with the syrinx so that their calls tend to be higher-pitched and relatively pure-toned compared with those of females (Johnsgard 1961). Thus, voice is useful in sexing the endangered monochromatic Laysan duck (*Anas laysanensis*). Male and female spotted owls produce very different vocalizations (Forsman *et al.* 1984). Whooping cranes (*Grus americana*) and probably all cranes may be sexed by voice (Carlson & Trost 1992). At least two species of petrel can be sexed by voice (Taoka *et al.* 1989; Taoka & Okumura 1990). The Willow ptarmigan can be sexed from calls (Martin *et al.* 1995). The endangered Socorro mockingbird (*Mimodes graysoni*) can be sexed by calls (J. Martinez and L. F. Baptista, unpublished results).

Although song is usually a behavior expressed by males in birds, females of some species also sing. Female song has been described for one dove species, the Ring dove (*Streptopelia roseogrisea*, Cheng 1992). However, one of us (L.F.B.) has recorded songs in females of the following dove species: *Columba livia*, *C. squamosa, Zenaida graysoni*, *Tutur tympanistra*, *Treron vernans*, *Phapitreron leucotis*, and six species in the fruit-dove genus *Ptilinopus*. Female song is probably widespread among columbiforms and tends to be higher-pitched than that in males (e.g., Fig. 9.2). Sex of doves may thus be determined by call differences and may be especially useful when studying endangered taxa such as the Marianas fruit dove (*Ptilinopus roseicapillus*, King 1981).

When assessing minimum effective population size for maintaining endangered species, conservationists use the numbers of adult pairs in a population in their models (Lande 1995). In those species where males

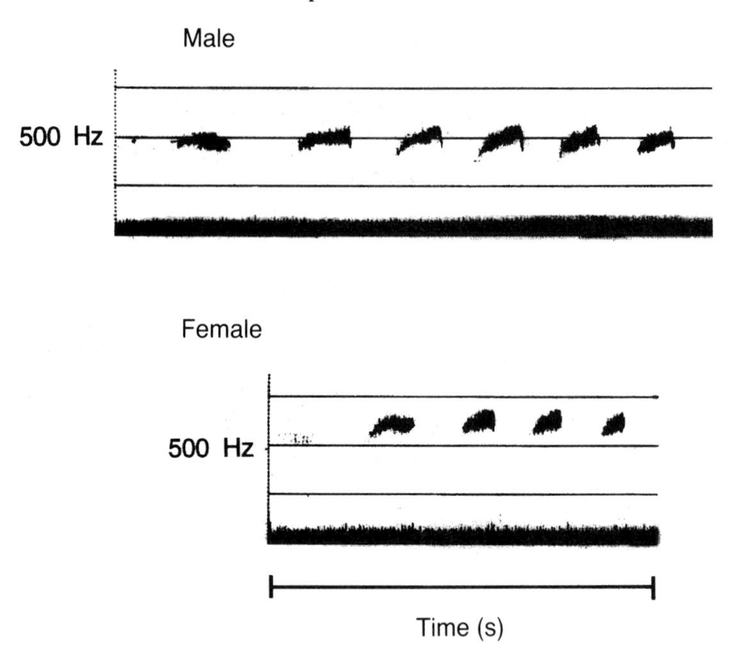

Fig. 9.2. Call of a male and a female Orange-billed fruit dove (*Ptilinopus iozonus*). Note that the female's call is higher in frequency and shorter in duration than the male's call. Both calls were recorded during an alternating duet performance between the male and female.

and females are equally responsive to playback (see above), countersing, or both, vocalizations may prove effective in assessing pair numbers.

Playback methods and individuals

Tape recorded conspecific sounds have been shown to stimulate gonadal development in several avian species including a parrot (Brockway 1969), a dove (Cheng 1992), a quail (Guyomarc'h & Guyomarc'h 1989), and an oscine sparrow (Morton *et al.* 1985). Theoretically, it would be possible to use taped vocalizations to stimulate breeding activity in captive breeding programs of endangered species.

The response of an animal to a vocal stimulus can be exploited to attract or capture individuals for marking, sexing, aging, and otherwise monitoring for conservation purposes. Recordings of distress calls of nestling White-crowned sparrows may be used to attract pairs to mist-nets to be banded (L. F. Baptista personal observation). Goshawks (*Accipiter gentilis*) respond best to alarm calls when they are attending nestlings but to wail and begging calls when attending fledglings (Kennedy & Stahlecker 1993). Ground or low shrub nesting warblers can be attracted to mist-nets by male song or female call notes; females are more elusive but can be captured by placing mist-nets near nests with or without playback of nestling calls (Dettmers 1993). Thus, knowing species' behavior allows assessment of which sound(s) to use and how to present them to effect the desired results.

An anecdotal account of attraction aiding a Humpback whale (*Megaptera noveangliae*) that wandered up the Sacramento river in California in 1985 is reported by Diana Reese (personal communication). She played recordings of Humpback whale feeding sounds to the lost cetacean and induced the whale to approach the speaker each time it heard conspecific sound. Over a period of eight hours she managed to lead the whale back to open ocean.

General comments on playback

Throughout this chapter the technique of replay of recorded sound stimuli to animals for various purposes has been discussed. There are two concerns for which investigators using playback should be cognizant, one ethical and the other theoretical.

Ethical considerations

Although playback may serve as an important tool in various aspects of conservation and biological diversity studies, playback used indiscriminately may cause disruption and even destruction of the subjects studied. For example, male Coppery-tailed trogons (*Trogon elegans*) share incubation responsibilities with their mates. Males are highly responsive to playback and may follow tape-recordings to great distances from the nest (Johnson *et al.* 1981). Eggs can thus become exposed to dangers from chilling or predation due to an incubating bird's prolonged absence. Similar deleterious effects can occur with playback at territorial boundaries. This situation may attract two territorial owners and thus precipitate fighting and possible injury of the combatants. One of us (L.F.B.) once used playback continued over a long period to attract timid male White-crowned sparrows in order to identify their color bands. It was noticed on the following day that nestlings of these target males had been pecked to death (L. F. Baptista, personal observation). Aggressive interactions may be followed by rise in androgen titers in the blood

(Wingfield 1984). It is possible that excessive playback precipitated elevated testosterone titers which in turn resulted in aggression redirected at the nestlings.

Thus, investigators need to know the species they study well enough to be aware of any consequences that might accrue to the animal when playback is used. Such self-monitoring can alleviate the potential for externally imposed regulation. Indeed, there are areas where local managers, especially where there is a large public presence, currently forbid the use of playback altogether.

Theoretical consideration

Conservationists would most likely use playback primarily to attract and identify individuals. We have already mentioned the need to avoid stimulus habituation. By using the technique, however, anyone may become intrigued with other experimental uses of playback (e.g., attractiveness of the stimulus to the responding individual). Thus any investigator using playback should be aware of how to select appropriate stimuli for playback and the need for adequate numbers of exemplars of the stimulus in order to avoid pseudoreplication in experimental design (Kroodsma 1990; McGregor 1992).

Conclusions

It should be evident that bioacoustics is less a discipline than it is a suite of approaches to and methods for the study of sound and sound communication. The situation is similar for conservation: both complement and depend upon many other disciplines. Bioacoustics and conservation intercept where animals are affected by and use sound. That nexus can be profound, in part, because the sound sensory modality, more than any other, is used so widely by animals. We as observers can exploit these sounds by monitoring and playback with appropriate training and instrumentation. The organisms then, not bioacoustics *per se*, will determine the usefulness of bioacoustic studies and methodologies to conservation.

Thus, a knowledge of animal behavior and communication precedes insights into the utility of using bioacoustic methodology. For example, there is much concern about noise pollution of the sound environment as it may impact animal fitness. This could occur if animals are sensitive to the sound frequency of the noise. Thus, acoustic monitoring of both sound pollution and the animals' responses to that noise is an area receiving much attention for both terrestrial and aquatic organisms. Oceanic noise pollution is of particular concern because of the proposed intentional introduction of low frequency noise to monitor global warming. If that noise is in the range of marine animal hearing sensitivity, it can mask and otherwise interfere with communication.

Possibly the area of greatest current use of bioacoustics in conservation is that of censusing communities for inventories of biological diversity, populations to monitor change, and individuals for identification. Censusing by listening can be augmented by broadcasting animal vocalizations (by imitation or recorded sound playback), thus attracting target species into view or stimulating a vocal response. Knowing how animals respond to vocalizations has led to the insight that these sounds can, in some situations, be used to attract animals with some surprising conservation consequences. Fragmented metapopulations can be monitored by sound, and playback has proven useful in establishing or re-establishing breeding populations into suitable remaining habitats.

The area of greatest opportunity for future interaction is that of automated recognition of sounds from continuous sound recordings. Even a well-trained and talented observer comes upon unknown vocalizations, and the novice can be at a loss to learn to identify the source of vocalizations in the time required for a project. Continuous sound recordings from unattended acoustic microphone arrays, although most useful in long-term censusing, produce prodigious amounts of data that take, at a minimum, an equal amount of time to analyze as to acquire. Automation will greatly facilitate that task and make the use of such arrays more feasible.

There have been and there will continue to be many applications of bioacoustic methodologies to conservation efforts. Sources for information and assistance are identified in the section on the community level 'surveying and censusing biological diversity.' Discussion can facilitate innovation and we encourage a continued dialogue between conservationists and bioacousticians.

Literature cited

Alatalo, R. V., A. Lundberg, and M. Björklund. 1982. Can the song of male birds attract other males: an experiment with the pied flycatcher *Ficedula hypoleuca*. *Bird Behaviour* **4**: 42–45.

Alström, P., U. Olsson, and P. R. Colston. 1992. A new species of *Phylloscopus* warbler from central China. *Ibis* **134**: 329–334.

Baker, J. A., and E. D. Bailey. 1987*a*. Sources of phenotypic variation in the separation call in northern bobwhite. *Canadian Journal of Zoology* **65**: 1010–1015.

Baker, J. A., and E. D. Bailey. 1987*b*. Ontogeny of the separation call in northern bobwhite. *Canadian Journal of Zoology* **65**: 1016–1020.

Baptista, L. F. 1975. Song dialects and demes in sedentary populations of the white-crowned sparrow (*Zonotrichia leucophrys nuttalli*). *University of California Publications in Zoology* **105**: 1–53.

Baptista, L. F. 1993. El estudio de la variacion geografico usando vocalizaciones y las bibliotecas de sonidos de aves neotropicales. Pages 15–30 in P. Escalante-Pliego, editor. *Curación moderna de colecciones ornitológicas*. American Ornithologists' Union, Washington, DC.

Baptista, L. F. 1994. The role of aviculture in science and conservation. *Avicultural Magazine* **100**: 183–188.

Baptista, L. F. 1996. Nature and its nurturing in avian vocal development. Pages 39–59 in D. E. Kroodsma and E. H. Miller, editors. *Ecology and evolution of acoustic communication in birds*. Cornell University Press, Ithaca, New York.

Baptista, L. F., W. I. Boarman, and P. Kandianidis. 1983. Behavior and taxonomic status of Grayson's dove. *Auk* **100**: 907–919.

Baptista, L. F., and S. L. L. Gaunt. 1994. Advances in studies of avian sound communication. *Condor* **96**: 817–830.

Barlow, J. C. 1995. *Songs of the vireos and their allies*. ARA Record no. 7, Gainesville, Florida.

Baskett, T. S. 1993. Biological evaluation of the call-count survey. Pages 233–268 in T. S. Baskett, M. W. Sayre, R. E. Tomlinson, and R. E. Mirachi, editors. *Ecology and management of the mourning dove*. Stackpole Books, Harrisburg, Pennsylvania.

Bearder, S. K. 1995. Calls of the wild. *Natural History* **104**: 48–57.

Beeman, K. 1994. *Signal users guide, version 2.2*. Engineering Design, Belmont, Massachusetts.

Bentley, D., and R. R. Hoy. 1974. The neurobiology of cricket song. *Scientific American* **231**: 34–44.

Best, L. B. 1981. Seasonal changes in detection of individual species. Pages 252–261 in C. J. Ralph and J. M. Scott, editors. *Estimating numbers of terrestrial birds*. Cooper Ornithological Society's Studies in Avian Biology, Volume 6. Allen Press, Lawrence, Kansas.

Bibby, C. J., N. D. Burgess, and D. A. Hall. 1992. *Bird census techniques*. Academic Press, New York.

Blanchard, R. J., R. Agullana, L. McGee, S. Weiss, and D. C. Blanchard. 1992. Sex differences in the incidence and sonographic characteristics of antipredator ultrasonic cries in the laboratory rat (*Rattus norvegicus*). *Journal of Comparative Psychology* **106**: 270–277.

Blockstein, D. E., and J. W. Hardy. 1989. The Grenada dove (*Leptotila wellsi*) is a distinct species. *Auk* **106**: 339–340.

Borror, D. J., and W. W. H. Gunn. 1985. *Songs of the warblers of North America*. Three 45 rpm monaural discs or two cassettes and booklet. The Library of Natural Sounds, Cornell Laboratory of Ornithology, Ithaca, New York.

Bradbury, J., and S. Vehrencamp. 1994. SINGIT! a program for interactive playback on the Macintosh. *Bioacoustics* **5**: 308–310.

Brehm, C. L. 1820. *Beiträge zur Vogelkunde in vollständigen Beschreibungen mehrerer neu entdeckter und vieler seltener, odor nicht gehörig beobachteter deutscher Vögel, 1*. J. K. G. Wagner, Neustadt in der Orla.

Brockway, B.F. 1969. Roles of budgerigar vocalization in the integration of breeding behavior. Pages 131–158 in R. A. Hinde, editor. *Bird vocalizations*. Cambridge University Press, London.

Buchholz, R. 1995. The descending whistle display and female visitation rates in the yellow-knobbed curassow, *Crax daubentoni*, in Venezuela. *Ornitologia Neotropical* **6**: 27–36.

Bunch, T. D., and G. W. Workman. 1993. Sonic boom, animal stress project report on elk, antelope and Rocky Mountain bighorn sheep. *Journal of the Acoustical Society of America* **93**: 2378.

Carlson, G., and C. G. Trost. 1992. Sex determination of the whooping crane by analysis of vocalizations. *Condor* **94**: 532–536.

Catchpole, C. K. 1973. The function of advertising song in the sedge warbler (*Acrocephalus schoenobaenus*) and the reed warbler (*A. scirpaceus*). *Behaviour* **46**: 300–320.

Chapman, C. A., and D. M. Weary. 1990. Variability in spider monkeys' vocalizations may provide basis for individual recognition. *American Journal of Primatology* **22**: 279–284.

Cheney, D. L., and R. M. Seyfarth. 1982. How vervet monkeys perceive their grunts: field playback experiments. *Animal Behaviour* **30**: 739–751.

Cheng, M. F. 1992. For whom does the female dove coo? A case for the role of vocal self-stimulation. *Animal Behaviour* **43**: 1035–1044.

Clark, C. W. 1995. An overview of the ATOC-Marine Mammal Research Program (ATOC-MMMR). *Conference on Biology of Marine Mammals* **11**: 23.

Clark, C. W., P. Marler, and K. Beeman. 1987. Quantitative analysis of animal vocal phonology: an application to swamp sparrow song. *Ethology* **76**: 101–115.

Cocroft, R. B., and M. J. Ryan. 1995. Patterns of advertisement call evolution in toads and chorus frogs. *Animal Behaviour* **49**: 283–303.

Courtenay, D. O., and S. K. Bearder. 1989. The taxonomic status and distribution of bushbabies in Malawi with emphasis on the significance of vocalization. *International Journal of Primatology* **10**: 17–34.

Croll, D. A., J. K. Nansen, M. E. Goebel, P. L. Boveng, and J. L. Bengtson. 1996. Foraging behavior and reproductive success in chinstrap penguins: the effects of transmitter attachment. *Journal of Field Ornithology* **67**: 1–9.

Dahlquist, F. C., S. D. Schemnitz, and B. K. Flachs. 1990. Distinguishing individual male wild turkeys by analyzing vocalizations using a personal computer. *Bioacoustics* **2**: 303–316.

Delacour, J. 1959. *Wild pigeons and doves*. All-Pets Books, Fond du Lac, Wisconsin.

Dethier, V. G. 1992. *Crickets and katydids, concerts and solos*. Harvard University Press, Cambridge, Massachusetts.

Dettmers, R. P. 1993. The role of male and territory characteristics in mate selection by female hooded warblers (*Wilsonia citrina*). M.S. Dissertation, Ohio State University, Columbus.

DeWolfe, B. B., and L. F. Baptista. 1995. Singing behavior, song types on their wintering ground and the question of leap-frog migration in Puget Sound white-crowned sparrows (*Zonotrichia leucophrys pugetensis*). *Condor* **97**: 376–389.

Evans, W. R. 1994. Nocturnal flight call of Bicknell's thrush. *Wilson Bulletin* **106**: 55–61.

Ewing, A. W. 1989. *Arthropod bioacoustics: Neurobiology and behavior*. Comstock Publishing Association, Ithaca. New York.

Falls, J. B. 1981. Mapping territories with playback: an accurate census method for songbirds. Pages 86–91 in C. J. Ralph, and J. M. Scott, editors. *Estimating numbers of terrestrial birds. Cooper Ornithological Society's Studies in Avian Biology 6*. Allen Press, Lawrence, Kansas.

Fenton, M. B., and L. E. Lecht. 1990. Why rattle snake? *Journal of Herpetology* **24**: 276–279.

Fish, M. P., and W. H. Mowbray. 1970. *Sounds of Western North Atlantic Fishes, A reference file of biological underwater sounds*. Johns Hopkins Press, Baltimore, Maryland.

Fonseca, P. J. 1991. Characteristics of the acoustic signals in nine species of cicadas (Homoptera, Cicadidae). *Bioacoustics* **3**: 173–192.

Ford, J. B., and H. D. Fisher. 1982. Killer whale (*Orcinus orca*) dialects as an indicator of stocks in British Columbia. *Report of the International Whale Commission* **32**: 671–680.

Forsman, E. D., E. C. Meslow, and H. M. Wight. 1984. Distribution and biology of the spotted owl in Oregon. *Wildlife Monograph* **87**: 1–64.

Foster, C. C., E. D. Forman, E. C. Meslow, G. S. Miller, J. A. Reid, F. F. Wagner, A. B. Carey, and J. B. Lint. 1992. Survival and reproduction of radio-marked adult spotted owls. *Journal of Wildlife Management* **56:** 91–95.

Foster, M. 1965. An early reference to the technique of owl calling. *Auk* **82**: 651–653.

Garstang, M., D. Larom, R. Raspet, and M. Lundque. 1995. Atmospheric controls on elephant communication. *Journal of Experimental Biology* **198**: 939–951.

Gaunt, S. L. L., L. F. Baptista, J. E. Sánchez, and D. Hernandez. 1994. Song learning as evidenced from song sharing in two hummingbird species (*Colibri coruscans and C. thalassinus*). *Auk* **111**: 87–103.

Gessaman, J. A., and K. A. Nagy. 1988. Transmitter loads affect the flight speed and metabolism of homing pigeons. *Condor* **90**: 622–668.

Gibbs, J. L., and J. Faaborg. 1990. Estimating the viability of ovenbird and Kentucky warbler populations in forest fragments. *Conservation Biology* **4**: 193–196.

Gilbert, G., P. K. McGregor, and G. Tyler. 1994. Vocal individuality as a census tool: practical considerations illustrated by a study of two rare species. *Journal of Field Ornithology* **65**: 335–348.

Goldstein, R. B. 1978. Geographic variation in the 'hoy' call of the bobwhite. *Auk* **95**: 85–94.

Gorski, L. J. 1969. Traill's flycatcher of the 'fitz-bew' song form wintering in Panama. *Auk* **86**: 745–747.

Gorski, L. J. 1971. Traill's flycatcher of the 'fee-bee-o' song forms wintering in Peru. *Auk* **88**: 429–431.

Greenway, J. C., Jr. 1967. *Extinct and vanishing birds of the world*. Dover Publications, New York.

Guyomarc'h, C., and J. C. Guyomarc'h. 1989. Stimulation of sexual development in female Japanese quail by male song: influence of ecoethological variables. *Biological Behavior* **14**: 52–65.

Guyomarc'h, J. C., Y. A. Hemon, C. Guyomarc'h, and R. Michel. 1984. Le mode de dispersion des males de Cailles des Bles *Coturnix c. coturnix* en phase de reproduction. *Comptes rendus de l'Academie des Sciences Paris (Series III)* **299**: 805–808.

Hafner, M. S., and D. J. Hafner. 1979. Vocalizations of grasshopper mice (genus *Onychomys*). *Journal of Mammalogy* **60**: 85–94.

Hailman, J. P., M. S. Ficken, and R. W. Ficken. 1985. The 'chick-a-dee' calls of *Parus atricapillus*: a recombinant system of animal communication compared with written English. *Semiotica* **56**: 191–224.

Haimoff, E. H., and R. L. Tilson 1985. Individuality in the female songs of wild Kloss' gibbons (*Hylobates klossii*) on Siberut Island, Indonesia. *Folia Primatology* **44**: 129–137.

Harcourt, C. S., and S. K. Bearder. 1989. A comparison of *Galago moholi* in South Africa with *Galago zanzibaricus* in Kenya. *International Journal of Primatology* **10**: 35–45.

Hardy, J. W. 1984. Depositing sound specimens. *Auk* **101**: 623–624.

Hardy, J. W., G. B. Reynard, and B. E. Coffee Jr. 1987. *Voices of the New World Cuckoos and Trogons*. ARA Record no.11, Gainesville, Florida.

Hardy, J. W., G. B. Reynard, and B. E. Coffee Jr. 1989. *Voices of the New World pigeons and doves*. ARA Record no.14, Gainesville, Florida.

Harrison, S., D. D. Murphy, and P. R. Ehrlich. 1988. Distribution of the bay checkerspot butterfly, *Euphydryas editha bayensis*: evidence for a metapopulation model. *American Naturalist* **132**: 360–382.

Heyer, W. R., M. A. Donnelly, R. W. McDiarmid, L.-A. C. Hayek, and M. S. Foster. 1993. *Measuring and monitoring biological diversity: Standard methods for amphibians*. Smithsonian Institution Press, Washington, DC.

Hoezel, A. R., and R. W. Osborne. 1986. Killer whale call characteristics: implications for cooperative foraging strategies. Pages 373–403 in B. C. Kirkevold, and J. S. Lockard, editors. *Behavioral biology of killer whales*. Alan R. Liss, New York.

Holman, S. D., and W. T. C. Seale. 1991. Ontogeny of sexually dimorphic ultrasonic vocalizations in Mongolian gerbils. *Developmental Psychobiology* **24**: 103–115.

Holmes, R. T., and T. W. Sherry. 1992. Site fidelity of migratory warblers in temperate breeding and neotropical wintering areas: implications of population dynamics, habitat selection, and conservation. Pages 563–575 in J. M. Hagan and W. Johnston, editors. *Ecology and conservation of neotropical migrant landbirds*. Smithsonian Institution Press, Washington, DC.

Holmes, R. T., T. W. Sherry, and L. Reitsma. 1989. Population structure, territoriality and overwinter survival of two migrant warbler species in Jamaica. *Condor* **91**: 545–561.

Huxley, J. 1934. A natural experiment on the territorial instinct. *British Birds* **27**: 270–272.

Ickes, R. A., and M. S. Ficken. 1970. An investigation of territorial behavior in the American redstart utilizing recorded songs. *Wilson Bulletin* **82**: 167–176.

Johnsgard, P. A. 1961. The tracheal anatomy of the Anatidae and its taxonomic significance. *Wildlife Trust* **12**: 58–69.

Johnson, R. R., B. T. Brown, L. T. Haight, and J. M. Simpson. 1981. Playback recordings as a special avian censusing technique. Pages 68–75 in C. J. Ralph, and J. M. Scott, editors. *Estimating numbers of terrestrial birds. Cooper Ornithological Society's Studies in Avian Biology* 6. Allen Press, Lawrence, Kansas.

Johnson, R. R., and J. J. Dinsmore. 1986. The use of tape-recorded calls to count Virginia rails and soras. *Wilson Bulletin* **98**: 303–306

Kavanagh, R. P., and K. L. Bamkin. 1995. Distribution of nocturnal forest birds and mammals in relation to the logging mosaic in south-eastern New South Wales, Australia. *Biological Conservation* **71**: 41–53.

Kavanau, J. L. 1987. *Lovebirds, cockatiels, budgerigars: Behavior and evolution*. Science Software Systems, Incorporated, Los Angeles, California.

Kennedy, P. L., and D. W. Stahlecker. 1993. Responsiveness of nesting northern goshawks to taped broadcasts of three conspecific calls. *Journal of Wildlife Management* **57**: 249–257.

Kettle, R. 1989. Major wildlife sound libraries. *Bioacoustics* **2**: 171–175.

King, W. B. 1981. *Endangered birds of the world*. The ICBP Red Data Book. Smithsonian Press, Washington, DC.

Krausman, P. R., M. C. Wallace, M. E. Weisenberger, D. W. DeYoung, and O. E. Maughan. 1993. Effects of simulated aircraft noise on heart rate and behavior of desert ungulates. *Journal of the Acoustical Society of America* **93**: 2377.

Kress, S. W., and D. N. Nettleship. 1988. Re-establishment of Atlantic puffins (*Fratercula arctica*) at a former breeding site in the Gulf of Maine. *Journal of Field Ornithology* **59**: 161–170.

Kroodsma, D. E. 1990. Using appropriate experimental designs for intended hypotheses in 'song' playbacks, with examples for testing effects of song repertoire sizes. *Animal Behaviour* **40**: 1138–1150.

Kroodsma, D. E., G. F. Budney, R. W. Grotke, J. Vielliard, R. Ranft, O. D. Veprintseva, and S. L. L. Gaunt. 1996. Natural sound archives. In D. E. Kroodsma and E. H Miller, editors. *Ecology and evolution of acoustic communication among birds*. Cornell University Press, Cornell, New York.

Kugler, A. B., and D. S. Barber. 1993. A method for measuring wildlife noise exposure in the field. *Journal of the Acoustical Society of America* **93**: 2878.

Lande, R. 1995. Mutation and conservation. *Conservation Biology* **9**: 782–791.

Lambrechts, M., and A. A. Dhondt. 1995. Individual voice discrimination in birds. Pages 115–139 in D. M. Power, editor. *Current Ornithology* **12**. Plenum Press, New York.

Langbauer, W. R. Jr., K. B. Payne, R. A. Charif, L. Rapaport, and F. Osborn. 1991. African elephants respond to distant playbacks of low-frequency conspecific calls. *Journal of Experimental Biology* **157**: 35–46.

Larkin, R. P., L. L. Pater, and D. J. Trzik. 1996. Effects of military noise on wildlife: A literature review. USACERL Technical Report 96/21.

Leonard, M. L., and M. B. Fenton. 1984. Echolocation calls of *Euderma maculatum* (Vespertilionidae): use in orientation and communication. *Journal of Mammalogy* **65**: 122–126.

Leopold, A. S. 1959. *Wildlife of Mexico*. Cambridge University Press, London, England.

Marble, B. K., and J. P. Bogart. 1991. Call analysis of triploid hybrids resulting from diploid-triploid species crosses of hylid tree frogs. *Bioacoustics* **3**: 111–119.

Marion, W. R., T. E. O'Meara, and D. S. Maehr. 1981. Use of playback recordings in sampling elusive or secretive birds. Pages 81–85 in C. J. Ralph, and J. M. Scott, editors. *Estimating numbers of terrestrial birds. Cooper*

Ornithological Society's Studies in Avian Biology 6. Allen Press, Lawrence, Kansas.

Marks, J. S., and V. S. Marks. 1987. Influence of radio collars on survival of sharp-tailed grouse. *Journal of Wildlife Management* **51**: 468–471.

Marler, P. 1960. Bird songs and mate selection. Pages 348–367 in W. E. Lanyon, and W. N. Tavolga, editors. *Animal sound communication*. Animal Institute of Biological Sciences, Washington, DC.

Marler, P., and L. Hobbett. 1975. Individuality in a long-range vocalization of wild chimpanzees. *Zeitschrift für Tierpsychologie* **38**: 97–109.

Marshall, J. T. 1964. Voice in communication and relationships among brown towhees. *Condor* **66**: 345–356.

Marshall, J. T. 1988. Birds lost from a giant sequoia forest during fifty years. *Condor* **90**: 359–372.

Martens, J., and G. Geduldig. 1989. Acoustic adaptations of birds living close to Himalayan torrents. Pages 123–131 in R. van den Elzen, K.-L. Schuchmann, and K. Schmidt-Koenig, editors. *Current topics in avian biology*. Proceeding of the 100th DO-G meeting. Deutschen Ornithologen-Gesellschaft, Bonn.

Martin, K., A. G. Horn, and S. J. Hannon. 1995. The calls and associated behavior of breeding willow ptarmigan in Canada. *Wilson Bulletin* **107**: 496–509.

Mayfield, H. 1966. Hearing loss and bird song. *Living Bird* **5**: 167–175.

McGregor, P. K. 1992. *Playback and studies of animal communication*. Plenum Press, New York.

Medrano, L., M. Salinas, I. Salas, P. Ladrón de Guevara, and A. Aguayo. 1994. Sex identification of humpback whales, *Megaptera novaeangeliae*, on the wintering grounds of the Mexican Pacific Ocean. *Canadian Journal of Zoology* **72**: 1771–1774.

Melemis, S. M., and J. B. Falls. 1982. The defense function: a measure of territorial behavior. *Canadian Journal of Zoology* **60**: 495–501.

Miller, A. H. 1963. Seasonal activity and ecology of the avifauna of an American equatorial cloud forest. *University of California Publications in Zoology* **66**: 1–78.

Moein, S., M. Lenhardt, D. Barnard, J. Keinath, and J. Musick. 1993. Marine turtle auditory behavior. *Journal of the Acoustical Society of America* **93**: 2378.

Morton, M., M. Peryra, and L. F. Baptista. 1985. Photoperiodically induced ovarian growth in the white-crowned sparrow (*Zonotrichia leucophrys gambeli*) and its augmentation by song. *Comparative Biochemistry and Physiology* **80A**: 93–97.

Mulligan, J. A. 1966. Singing behavior and its development in the song sparrow *Melospiza melodia*. *University of California Publications in Zoology* **81**: 1–76.

Mundinger, P. 1975. Song dialects and colonization in the house finch *Carpodacus mexicanus* on the east coast. *Condor* **77**: 407–422.

Nelson, D. A., P. Marler, and A. Palleroni. 1995. A comparative approach to vocal learning: intraspecific variation in the learning process. *Animal Behaviour* **50**: 83–97.

Nevo, E., and R. R. Capranica. 1985. Evolutionary origin of ethological reproductive isolation in cricket frogs, *Acris*. *Evolutionary Biology* **19**: 147–214.

Nikol'sky, I. D. 1984. Libraries of wildlife sounds – their use in scientific research and other applications. *Biology Nauki* **8**: 39–43.

Norris, J. C., and W. E. Evans. 1995. Sources of variance in acoustic censusing of cetaceans. *Conference on Biology of Marine Mammals* **11**: 83.

Orians, G. H. 1966. Social stimulation within black-bird colonies. *Condor* **63**: 330–337.

Parker, T. A. 1991. On the use of tape recorders in avifaunal surveys. *Auk* **108**: 443–444.

Paton, P. W. C., C. A. Zabel, D. L. Neal, G. N. Stegner, N. C. Tilghman, and B. R. Noon. 1991. Effects on radio tags on spotted owls. *Journal of Wildlife Management* **55**: 617–622.

Payne, R. B. 1986. Bird songs and avian systematics. Pages 87–126 in R. F. Johnston, editor. *Current ornithology 3*. Plenum Press, New York.

Payne, R. B., L. L. Payne, and S. M. Doehlert. 1988. Biological and cultural success of song memes in indigo buntings. *Ecology* **69**: 104–117.

Petrinovich, L., and T. L. Patterson. 1981. The responses of white-crowned sparrows to songs of different dialects and subspecies. *Zeitschrift für Tierpsychologie* **57**: 1–14.

Phillips, A., J. Marshall, and G. Monson. 1964. *The birds of Arizona*. University of Arizona Press, Tucson.

Pina, R. F., and A. Channing. 1981. A portable frog call synthesizer. *Moniore zoologico Italiano, Supplemento XV* **20**: 387–392.

Podolsky, R. H. 1990. Effectiveness of social stimuli in attracting Laysan albatross to new potential nesting sites. *Auk* **107**: 119–125.

Podolsky, R. H., and S. W. Kress. 1989. Factors affecting colony formation in Leach's storm-petrel. *Auk* **106**: 332–336.

Potter, J. R., D. K. Mellinger, and C. W. Clark. 1994. Marine mammal call discrimination using artificial neural networks. *Journal of the Acoustical Society of America* **96**: 1255–1262.

Randall, J. A. 1989. Individual footdrumming signatures in banner-tailed kangaroo rats *Dipodomys spectabilis*. *Animal Behaviour* **38**: 620–630.

Randall, J. A. 1994. Discrimination of footdrumming signatures by kangaroo rats, *Dipodomys spectabilis*. *Animal Behaviour* **47**: 45–54.

Redpath, S. M. 1994. Censusing tawny owls (*Strix aluco*) using imitation calls. *Bird Studies* **41**: 192–198.

Redpath, S. M. 1995. Habitat fragmentation and the individual: tawny owls *Strix aluco* in woodland patches. *Journal of Animal Ecology* **64**: 652–661.

Reed, J. M., and A. Dobson. 1993. Behavioral constraints and conservation biology: conspecific attraction and recruitment. *Trends in Ecology and Evolution* **8**: 253–256.

Richardson, W. J., C. R. Greene, Jr., C. I. Malme, and D. H. Thomson. 1995. *Marine mammals and noise*. Academic Press, New York.

Robbins, M. B., and N. G. Howell. 1995. A new species of pygmy-owl (Strigidae: *Glaucidium*) from the eastern Andes. *Wilson Bulletin* **107**: 1–6.

Rowley, I., and G. Chapman. 1986. Cross-fostering, imprinting and learning in two sympatric species of cockatoo. *Behaviour* **96**: 1–16.

Ryan, M. J., and W. Wilczynski. 1991. Evolution of intraspecific variation in the advertisement call of a cricket frog (*Acris crepitans*, Hylidae). *Biological Journal of the Linnaean Society* **44**: 249–271.

Schmidt, R. H., and R. J. Johnson. 1982. *Bird dispersal recordings: sources of supply*. University of Nebraska in cooperation with United States Fish and Wildlife Service and United States Air Force, Lincoln.

Sliwa, A., and T. W. Sherry. 1992. Surveying wintering warbler populations in Jamaica: point counts with and without broadcast vocalizations. *Condor* **94**: 924–936.
Smith, A. T., and M. M. Peacock. 1990. Conspecific attraction and the determination of metapopulation colonization rates. *Conservation Biology* **4**: 320–323.
Snowdon, C. T. 1987. A naturalistic view of categorical perception. Pages 332–354 in S. Harnad, editor. *Categorical perception*. Cambridge University Press, New York.
Snowdon, C. T., and J. Cleveland. 1981. Individual recognition of contact calls in pygmy marmosets. *Animal Behaviour* **28**: 717–727.
Sorenson, M. D. 1995. Effects of neck collar radios on female redheads. *Journal of Field Ornithology* **60**: 523–528.
Stamps, J. A. 1988. Conspecifics as cues to territory quality: a preference of juvenile lizards (*Anolis aeneus*) for previously used territories. *American Naturalist* **131**: 329–347.
Stumpner, A., and O. von Helverson. 1994. Song production and song recognition in a group of sibling grasshopper species (*Chorthippus dorsatus, Ch. dichrous* and *Ch. loratus*: Orthoptera, Acrididae). *Bioacoustics* **6**: 1–23.
Taoka, M., T. Sato, T. Kamada, and H. Okumura. 1989. Sexual dimorphism of chatter calls and vocal sex recognition in Leach's storm petrels (*Oceanodroma leucorhoa*). *Auk* **106**: 498–501.
Taoka, M., and H. Okumura. 1990. Sexual differences in flight calls and the cue for vocal sex recognition of Swinhoe's storm petrels. *Condor* **92**: 571–575.
Telford, E. A. 1993. The use of sonographic analysis in identifying individual peregrine falcons *Falco peregrinus*. *Journal of Raptor Research* **27**: 82.
Thomas, J. W., E. D. Forsman, J. B. Lint, E. C. Meslow, B. R. Noon, and J. Verner. 1990. *A conservation strategy for the Northern spotted owl*. US Department of Agriculture and US Department of the Interior, Portland, Oregan.
Tomback, D. F., D. B. Thompson, and M. C. Baker. 1983. Dialect discrimination by white-crowned sparrows: reactions to near and distant dialects. *Auk* **100**: 452–460.
Tyack, P. 1981. Interactions between singing Hawaiian humpback whales and conspecifics nearby. *Behavioral Ecology and Sociobiology* **8**: 105–116.
Warburton, V. L., G. D. Sales, and S. R. Milligan. 1989. The emission and elicitation of mouse ultrasonic vocalizations: the effects of age, sex and gonadal status. *Physiological Behavior* **45**: 41–47.
Waters, D. A., and A. L. Walsh. 1994. The influence of bat detector brand on the quantitative estimation of bat activity. *Bioacoustics* **5**: 205–221.
White, G. 1789. *The natural history and antiquities of Selbourne*. Benjamine White and Son, London.
Wickstrom, D. C. 1982. Factors to consider in recording avian sound. Pages 2–52 in D. E. Kroodsma and E. H. Miller, editors. *Acoustic communication in birds*. Academic Press, New York.
Wiley, R. H., and D. G. Richards. 1982. Adaptations for acoustic communication in birds: sound transmission and signal detection. Pages 131–181 in D. E. Kroodsma and E. H. Miller, editors. *Acoustic communication in birds*. Academic Press, New York.

Williams, J. M. 1993. Objective comparison of song syllables: a dynamic programming approach. *Journal of theoretical Biology* **161**: 317–328.

Williams, J. M., and P. J. B. Slater. 1991. Computer analysis of bird sound: a guide to current methods. *Bioacoustics* **3**: 121–128.

Wingfield, J. 1984. Environmental and endocrine control of reproduction in the song sparrow *Melospiza melodia* II. Agonistic interaction as environmental information stimulating secretion of testosterone. *General and Comparative Endocrinology* **56**: 417–424.

Zimmerman, J. L. 1971. The territory and its density dependent effect in *Spiza americana*. *Auk* **88**: 591–612.

Zimmermann, E., S. K. Bearder, G. A. Doyle, and A. B. Anderson. 1988. Variations in vocal patterns of Senegal and South African lesser bushbabies and their implications for taxonomic relationships. *Folia Primatology* **51**: 87–105.

Chapter 10

Mating systems, effective population size, and conservation of natural populations

PATRICIA G. PARKER AND THOMAS A. WAITE

Large populations may become small for many reasons. Population declines are chiefly attributed to extrinsic factors, such as climate change or anthropogenic effects like habitat fragmentation or hunting pressure. Once small, populations may be at risk of extinction through a variety of intrinsic processes as well. By chance fluctuations, for example, small populations may have insufficient numbers of one sex to sustain productivity, or to find mates. In addition to these demographic concerns, small populations may experience genetically based problems. The number of successfully reproducing individuals in a population is the main determinant of the genetic diversity that it maintains. Populations low in genetic diversity may lack the variation necessary to respond to environmental change and so may be at risk over evolutionary time.

More immediately, individuals that are homozygous, as a result of having parents that were themselves closely related, are likely to have reduced viability and the population may suffer from inbreeding depression (Frankham 1995 and references therein). For a given number of sexually mature, potentially reproductive adults, the greatest genetic diversity is maintained when all individuals reproduce and all are equally successful. Of course, this never happens, and the exact pattern of reproduction in any population depends upon behavioral interactions among individuals. These behavioral interactions usually result in recognized patterns of reproduction by members of each sex and age class that we call the *mating system* of the population. In mating systems in which competition over mates results in skewed reproductive success among members of either sex, heterozygosity (the proportion of individuals heterozygous at a particular locus, a general index of genetic variation) should be lower than in the same population with less variance in reproductive success. In this chapter, we explore the relationship between mating system and genetic diversity, and consider the costs and benefits of behavioral and genetic study in attempts to conserve populations.

Is genetic variation really important?

Recent empirical evidence has supported the concern that the viability of small, dwindling populations may be reduced by the loss of genetic variation as a result of inbreeding (Jimenez *et al.* 1994; Frankham 1995). Even with mounting evidence, however, for the fitness consequences of loss of genetic variation in captive and natural populations (Ralls & Ballou 1983; Vrijenhoek *et al.* 1985; Quattro & Vrijenhoek 1989), the potential consequences of inbreeding for population viability have often been dismissed by conservation biologists. (This tendency to be dismissive of genetic consequences can be partly traced to an influential paper that argued that demographic factors are likely to be more immediate threats than genetic factors to the viability of small populations (Lande 1988).) Population decline in cheetahs, for example, has been correlated with their extremely low levels of genetic polymorphism and resultant susceptibility to disease (O'Brien *et al.* 1983; O'Brien & Everman 1988). This interpretation was recently rejected when it was discovered that newborn cheetahs in nature are frequent victims of predation (Caro & Laurenson 1994). It seems as if the revelation of proximate ecological problems somehow nullifies the genetic concerns of small populations. Nevertheless, even the author of the paper arguing the demographic priority (Lande 1988) now emphasizes the potential importance of genetic dynamics of declining populations in reducing viability (Lande 1994, 1995; Lynch *et al.* 1995). We are optimistic that we are approaching a mature synthesis in which genetic, demographic, and life historic parameters of populations all are recognized for their influence on population viability (Avise 1994).

With the objective of fostering more effective conservation through greater collaboration and more efficient targeting of financial resources, we consider the theoretical and practical tradeoffs of using behavioral indicators of effective population size versus genetic ones. Behavioral studies cannot always tell us what we need to know about patterns of genetic recombination (Gowaty & Mock 1985), nor can genetic analysis always provide the least expensive or least ambiguous indication of the mating system of all populations. In fact, especially in the case of cryptic or otherwise intractable species, both behavioral and genetic information must be used to generate an adequate estimate of the mating parameters of a population. Our exploration of this question has five parts.

First, we review the processes that create differences between actual population size (N) and the parameter most important to long-term genetic variation, effective population size (N_e). Although many scholars will already be familiar with this concept, in this section we review the essentials of this technical concept before we discuss the application of N_e to threatened populations. Second, we describe a simple relationship between recognized categories of mating systems and the seasonal effective population size.

Third, we apply a demographic estimator of N_e to evaluate the value of genetic study of animals whose matings are deceptively cryptic (e.g., extra-pair fertilizations). In avian and mammalian mating systems, actual patterns of recombination of gametes between males and females may be inconsistent with observed patterns of mating behavior (e.g., Birkhead & Møller 1992; Amos *et al.* 1995). We call these cryptic matings. Colonies of purple martins, for example, consist of dozens of pairs of birds that nest as socially monogamous pairs in close proximity. Wagner *et al.* (1996) have shown that approximately 21% of young are not sired by their apparent fathers, the male nest attendants, but instead are sired by a few older males within the colony. Dozens of recent studies report comparisons of observed (social) mating systems and their underlying genetic mating systems as determined by the application of molecular markers (e.g., Birkhead & Møller 1995). While Nunney (1993) has made a thorough evaluation of the theoretical influence of mating system on N_e, we focus on the extent to which the difference between observed and actual mating system influences the effective population size by the application of recent demographic models that incorporate the effects of overlapping generations (Nunney 1993; Nunney & Campbell 1993; Caballero 1994; Nunney & Elam 1994). We then use published data describing the differences between apparent and actual patterns of reproductive success (reflecting the social and genetic mating systems, respectively) in natural populations to examine the effect of these differences on estimates of N_e/N.

Fourth, we evaluate the effectiveness of recent attempts in our laboratory to apply molecular tools to describe the mating system of a demographically intractable species in which behavioral observations of mating relationships cannot be made. Sea turtles mate at sea before coming to shore, and thus their patterns of recombination are not detectable through behavioral observations. We consider whether genetic analysis of females and their offspring provides a valuable indication of the mating behavior of these endangered species. Fifth, and finally, we propose a new, system-sensitive approach to assessing N_e that maximizes the conservation gains that can be achieved through behavioral and genetic study.

A review of effective population size

The effective size of a population (N_e) is the best predictor of its ability to maintain genetic variation. This quantity is the size of an ideal population (i.e., one that experiences no mutation, selection or migration and in which mating is random) that would undergo genetic drift at the same rate as the focal population of size N (Wright 1931), where N is the number of sexually mature individuals capable of reproduction. In actual populations, fluctuations in N through time, skewed sex ratios, and variation in reproductive success that exceed Poisson expectation, frequently cause N_e to fall well below N (Wright 1969; Crow & Kimura 1970; Lande & Barrowclough 1987; Frankham 1996). Of these parameters, the combination of breeding sex ratios and patterns of reproductive success across individuals within populations describe behavioral aspects of populations.

The patterns of reproductive success across individuals characterize the mating system of the population. Recent work suggests that the influence of mating system on N_e may often be overwhelmed by other factors such as fluctuating population size (Frankham 1996; see also Nunney 1996). A population that has been through a recent bottleneck loses much variation in the process; once genetic variation is gone, the states of otherwise important influences are irrelevant. Similarly, some combinations of life history characters (such as small maturation time/adult lifespan ratios; Waite & Parker, 1996) may overwhelm the effects of mating system. Not all populations, however, have been through recent fluctuations in size, and many taxa have life history characteristics that allow an important influence of mating system; its influence deserves attention independent of the effects of fluctuation in N and life history characteristics. Behavioral studies of mating systems can provide valuable data for this purpose.

Why study mating systems?

The mating system of a population describes the breeding sex ratio and the number of mates attained by members of each sex. Numerical examples of the influence of various mating systems (Fig. 10.1) illustrate the relationship between effective population size and extent of reproductive failure in a hypothetical population. This model represents a population of organisms with non-overlapping generations, or a simple *within-season* calculation of the effective population size for more long-lived organisms or those with overlapping generations, using Wright's (1931) equation $N_e = (4n_m \times n_f)/(n_m + n_f)$, where n_m and n_f are the number of breeding males and females, to correct for the effect of different breeding sex ratios under different mating systems. These relationships are likely to be quite different for organisms with long generation times and overlapping generations; we choose to illustrate with organisms with non-overlapping generations because it is a simple place to start and the contribution of mating systems is clear and uncluttered by other effects. We will see later that this does not mean that mating system has little potential effect in more complex systems.

We start with a hypothetical population of size $N = 100$, comprising 50 adult females and 50 adult males. All females are assumed to produce one clutch or litter per season (reproductive success follows a Poisson

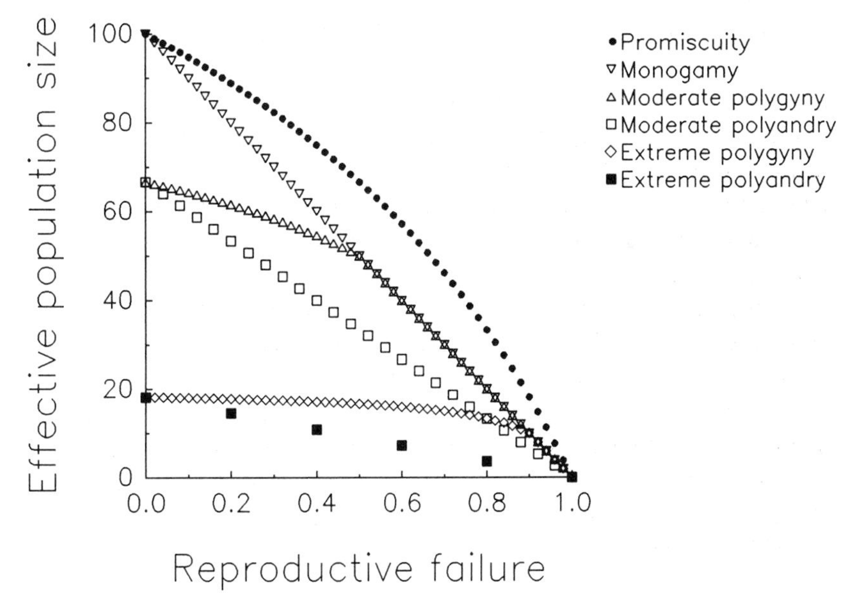

Fig. 10.1. Influence of mating systems on the relationship between extent of reproductive failure (proportion of breeding attempts unsuccessful) and annual effective population size, $N_e = 4(N_f N_m)/(N_f + N_m)$, for a population beginning with 50 adult males and 50 adult females. N_f = the number of successfully breeding females, and N_m = the number of successfully breeding males. See text.

distribution) and failure means a complete failure to reproduce. Promiscuity describes the extreme situation in which every female mates with every male. Therefore, a nesting failure of one female would reduce N_f by one but would have no effect on N_m, as each male's genotype would be retained within the brood of other successful females. Monogamy is defined here as exclusive mating between one female and one male, so that a failure removes both from the effective population. Two categories of polygyny are shown. In moderate polygyny, 25 (50%) males mate with two females each, to the exclusion of other males; in extreme polygyny, only five (10%) males mate with ten females each.

Failures are assumed to occur evenly across harems, so that in moderate polygyny, all males lose one female before any lose both; hence, the curve for polygyny converges with that for monogamy. Polyandry is defined in parallel fashion: in moderate polyandry, 25 (50%) females mate with two males each, to the exclusion of other females; in extreme polyandry, five (10%) females mate with ten males each.

When some adults are excluded from breeding, as in polygynous or polyandrous systems, N_e drops sharply below that for monogamous or promiscuous systems, especially when all breeding attempts are successful (Fig. 10.1). As the proportion of breeding attempts that fail increases, a common feature of declining populations, these differences in N_e persist. For example, if 50% of breeding attempts were to fail, N_e would be 50 assuming monogamy and 66.7 assuming promiscuity. By contrast, N_e would be only 9.1 assuming extreme polyandry (e.g., Faaborg *et al.* 1995). It is obvious that the mating system has a large effect on the effective size of populations, at least within breeding seasons.

Are genetic studies necessary when matings can be observed?

'Cryptic' reproduction in vertebrate populations is commonplace; exchange of genetic material outside of social bonds has been revealed in birds (for review, see Birkhead & Møller 1992, 1995) and mammals (e.g., Keane *et al.* 1994). In fact, the number of socially monogamous bird species that have been examined to date is perhaps the largest of any taxon (>40 species), and all but a few of them have demonstrated some deviation from genetic monogamy, with as many as >80% of young sired by males other than their putative fathers in some species; they are sired instead by extra-pair fertilization. If it is generally true that birds, the taxon in which social monogamy prevails (Lack 1968), are not genetically monogamous as a rule, then the variance in male reproductive success may be larger than calculations based on social relationships would suggest.

Monogamous and truly promiscuous (random) mating systems are characterized by relatively low variance in reproductive success in both sexes. In polygynous mating systems, variance in reproductive success among males is relatively high, while that among females remains relatively low; the converse is true in polyandrous mating systems. If we agree that patterns of variance in reproductive success correlate strongly with our recognized categories of mating systems, then this becomes a useful mathematical 'handle' by which we can evaluate the effect of mating systems on N_e. We turn our attention now to a model in which basic demographic features of populations are used to estimate N_e/N. The model incorporates effects of overlapping generations, and we emphasize N_e/N rather than simply N_e. N_e/N is a useful switch, as it reminds us that the effective population is different from the standing crop of potentially reproductive adults (N), and may often be only a small portion of N.

A model of N_e/N

A recent model allows estimation of N_e/N using minimal demographic information (Nunney 1993; Nunney & Campbell 1993; Nunney & Elam 1994) as a simplification of a previous model (Hill 1972, 1979). Specifically, the model requires estimates for time to sexual maturation (M), average adult lifespan (A), generation time T (measured as $M + A-1$), breeding sex ratio (r), standardized variances (variance/mean^2) for female and male reproductive success (I_{bf} and I_{bm}), and standardized variances for female and male adult lifespan (I_{Af} and I_{Am}):

$$N_e/N = \frac{4r(1-r)T}{r[A_f(1 + I_{Af}) + I_{bf}] + (1-r)[A_m(1 + I_{Am}) + I_{bm}]} \tag{10.1}$$

As these values are reasonably easily estimated for long-term population studies of birds, we will assess the importance of cryptic reproduction in estimating effective population size by comparing estimates based on observational patterns of reproductive success from behavioral associations with those using molecular determination of parentage. We will specifically address two possibilities: (1) that some males perform far better and others far worse than behavioral observations would suggest (this should increase the variance in male reproductive success and N_e/N should go down); and (2) that some males thought unproductive actually did reproduce (this should act to decrease variance in male reproductive success, and N_e/N should go up).

Redistribution of fertilizations that increase I_{bm}

To address the first possibility, we first illustrated the behavior of Nunney's model in a hypothetical population of birds, using demographic parameter values that are realistic for many species of passerine birds (Fig. 10.2). We then allowed I_{bm} to increase by taking nestlings formerly attributed to relatively unsuccessful males and attributing them instead to the most successful male (Fig. 10.2). By increasing this skew in stepwise fashion, and producing N_e/N estimates for each step, we illustrated the hypothetical decline in N_e/N as I_{bm} increased (Fig. 10.3). N_e/N fell from 0.52 to 0.13 as male reproductive success became increasingly skewed. Thus, the effect of mating system on N_e in organisms with overlapping generations and long generation time is still potentially large, comparable in magnitude to the difference between monogamy and extreme polyandry in species without overlapping generations and 50% reproductive failure in Fig. 10.1.

Next, we illustrated the magnitude of the decline in actual data sets from natural populations, starting with two parentage studies of populations of socially monogamous birds in which it seemed that some males were at a distinct advantage in achieving extra-pair fertilizations (Kempenaers *et al.* 1992; Wagner *et al.*, 1996), putting others at a distinct disadvantage. In breeding colonies of Purple martins (*Progne subis*), older males are typically successful at siring the young of their social mates as well as those of females mated to younger males (Wagner *et al.*, 1996). In this two-year study, 21% and 8% of young were sired by extra-pair fertilizations, mostly by a few older males. Despite this skew in reproductive success, the resulting decline in N_e/N was small: N_e/N fell from 0.60 to 0.52, producing a decline in N_e from 35 to 30 in a population of purple martins breeding in 1992; N_e/N fell from 0.64 to 0.62 in the same population in 1993, producing a decline in N_e from 29 to 28.

In Blue tits (*Parus caeruleus*), larger males are significantly better represented as extra-pair fertilization sires than (random) expectation based on their frequency in the population (Kempenaers *et al.* 1992). For this study population, in which 11% of young were sired by extra-pair fertilizations, N_e/N fell from 0.67 to 0.66 after calculating the actual variances in male reproductive success based on molecular determination of parentage, producing a decline in N_e from 48 to 47. This work suggests

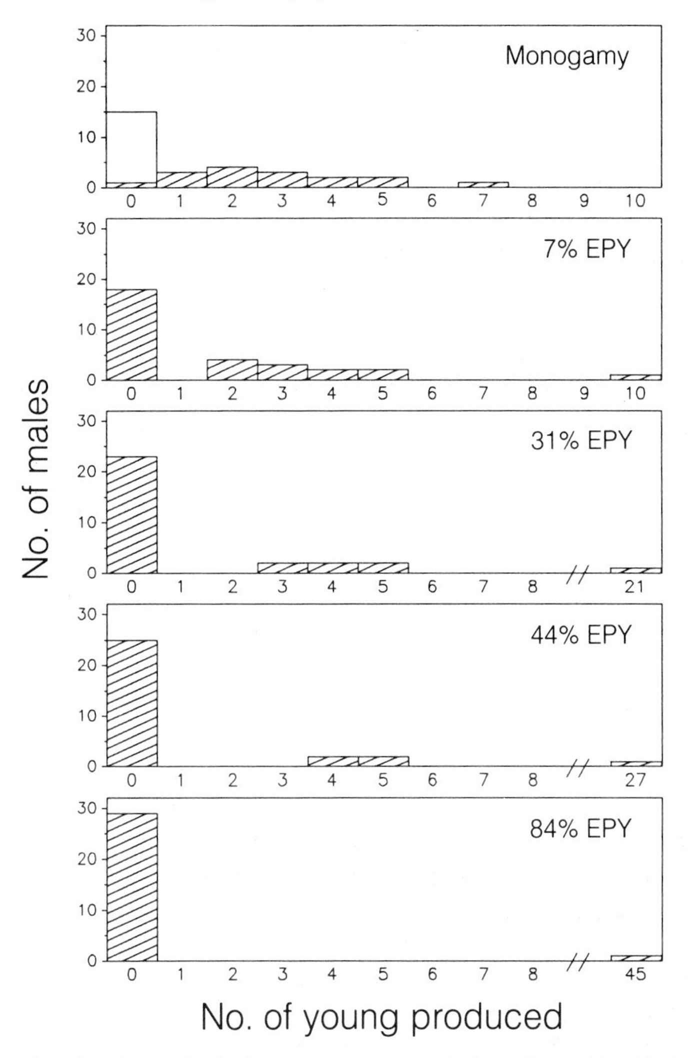

Fig. 10.2. Hypothetical monogamous population of passerine birds, erected with parameter values approximating those of passerines. Top panel: of 30 pairs in a neighborhood, 15 reproduce successfully in a particular season, and those produce three young per successful pair, with reproductive success following a Poisson distribution among successful breeders. Subsequent panels show the progressive skewing of 'actual' reproductive success, as nestlings formerly attributed to territory-holding males are taken away from the least successful males and attributed to the males already most successful. In the final panel, one of the 15 apparently breeding males has actually sired all the young in the neighborhood, 84% as extra-pair young (EPY), and 16% 'legitimately' in the nest(s) on his own territory.

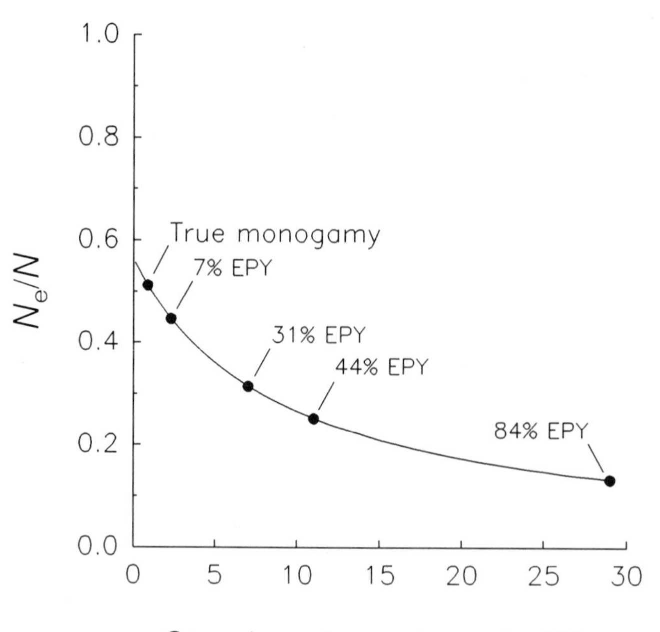

Fig. 10.3. Effects of the stages of hypothetical skew in reproductive success (RS) set up in Fig. 10.2 on standardized variance in male RS and thenceforth on N_e/N, using Nunney's minimum demographic model. In moving from monogamy to the extreme point at which all young are sired by a single male (84% extra-pair young, or EPY, in the neighborhood), N_e/N falls from 0.52 to 0.17. Other parameter values used were characteristic of passerine birds: maturation time (M) = 1 year; annual adult survival (v) = 0.60; expected adult lifespan = 1/(1–v); generation time (T) = $M + A$–1.

that, at least for these populations of Purple martins and Blue tits, the redistribution of fertilizations among males does not increase the magnitude of the standardized variance in male reproductive success sufficiently to have a large effect on N_e/N (T. A. Waite & P. G. Parker, unpublished data).

More reproducing individuals than observations would suggest

Cooperatively breeding birds are often considered socially monogamous birds with delayed dispersal; physiologically mature individuals serve as helpers, often of their parents' subsequent breeding attempts, before dispersing to attempt breeding themselves. Under this interpretation, only dominant individuals in the 'breeding' positions within social groups

Table 10.1 Estimates of N_e before and after application of molecular markers to determine parentage in two species of cooperatively breeding birds

	Stripe-backed wrens		Arabian babblers	
	Apparent	Real	Apparent	Real
No. of fathers	79	93	33	37
I_{bm}	0.20	0.53	0.75	1.14
N_e/N	0.83	0.69	0.74	0.61
N	160	202	66	90
N_e	133	139	48	55

In both, some subordinate males previously thought to be non-reproductive actually reproduced. We therefore add the number of subordinate males to the number of potentially reproductive adults (N), and recalculate N_e/N based on revised I_{bm} (Figs. 10.4 and 10.5).

should be counted when tallying N, the number of potentially reproductive individuals in a population. Data from studies of molecular determination of parentage in cooperatively breeding Stripe-backed wrens (*Campylorhynchus nuchalis*) and Arabian babblers (*Turdoides squamiceps*) can be used to test this interpretation. These data also allow a preliminary comparison of N_e/N assuming monogamy of dominants with that under the genetic mating system.

In both species, subordinate males sometimes reproduce (wrens: Rabenold *et al.* 1990; babblers: K. Lundy, P. Parker & A. Zahari, unpublished data). In Stripe-backed wrens, recalculation of I_{bm} including the category of subordinate males as potential breeders raises the standardized variance of male reproductive success from 0.20 to 0.53 (Fig. 10.4); likewise, including physiologically mature and occasionally breeding subordinate male Arabian babblers in N causes I_{bm} to increase from 0.75 to 1.14 (Fig. 10.5 and Table 10.1). In these cases, increasing the number of breeders increases the standardized variance substantially because it means that a large category of individuals, previously thought to be entirely non-reproductive, must be added to N, the number of potentially breeding individuals. As expected, as I_{bm} grows, N_e/N falls from 0.83 to 0.69 in Stripe-backed wrens, and from 0.74 to 0.61 in Arabian babblers.

The net effect on N_e, however, produced by multiplying the reduced N_e/N by the now-larger N (augmented by occasionally successful subordinates previously considered to be non-breeders), is to raise it from 133 to 139 in Stripe-backed wrens, and from 48 to 55 in Arabian babblers (Table 10.1). Although the magnitude of this effect is not large, perhaps that is

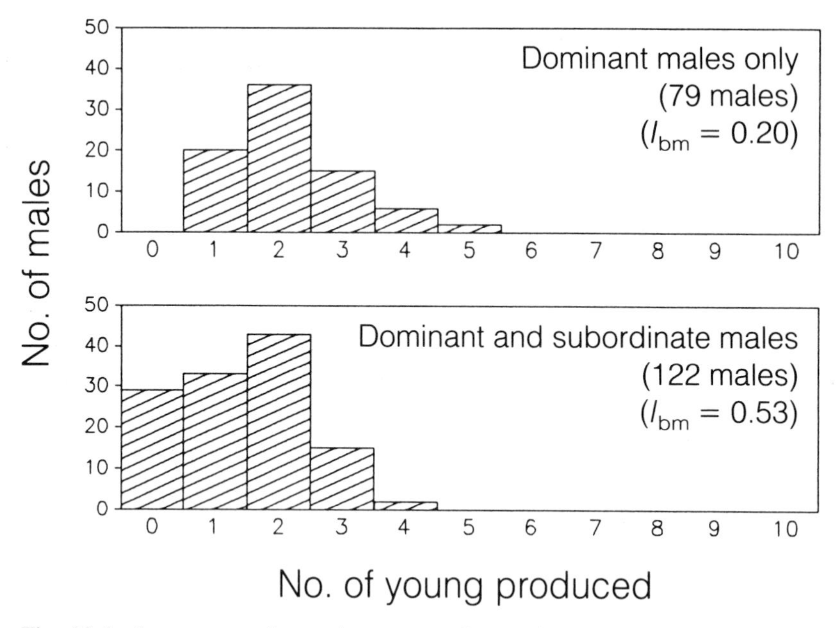

Fig. 10.4. Apparent and actual patterns of reproduction among male Stripe-backed wrens. When it was discovered that subordinate males sometimes reproduced, we added them to N as potential reproductives, and changed the average age at reproduction accordingly. This adds a significant number of males to N, and elevates I_{bm} considerably because few of the subordinates potentially capable of reproduction actually achieve any genetic parentage.

not surprising, given that about 10% of the young Stripe-backed wrens (and <10% of the young Arabian babblers) are sired by males other than the dominant male in any social group. It will be interesting to compare these results with comparable data from cooperatively breeding birds in which a larger proportion of young are sired by extra-pair fertilizations (e.g., Fairy wrens, in which almost 80% of young are sired by males other than dominants; Mulder *et al.* 1994).

What if mating behavior cannot be observed?

In some species, the prevalent mating system is unknown and thus information regarding the numbers of breeding individuals may be misleading. Adult Green turtles (*Chelonia mydas*) and Leatherback turtles (*Dermochelys coriacea*) are among the seven extant species of sea

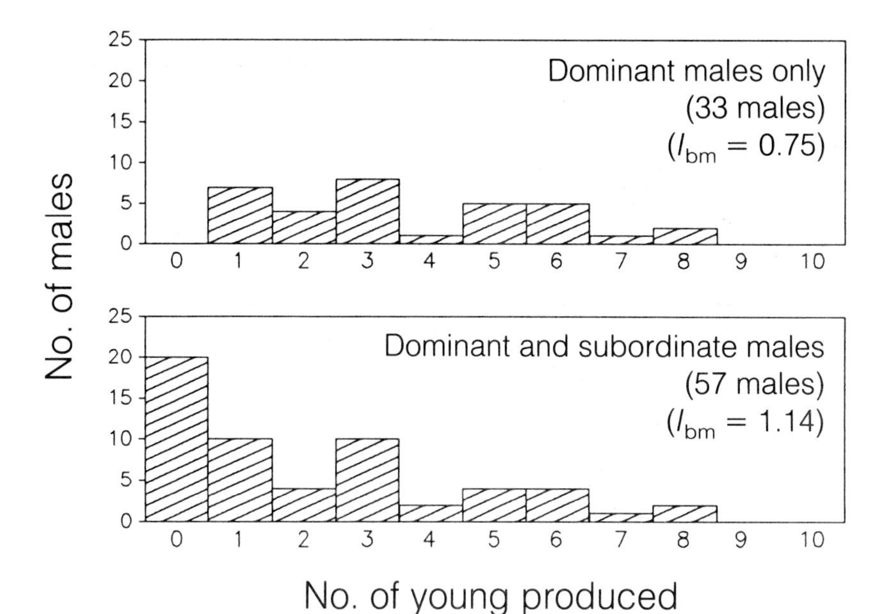

Fig. 10.5. Apparent and actual patterns of reproduction among male Arabian babblers. See legend, Fig. 10.4.

turtles; all are endangered or threatened. As an example of their population declines, Leatherback turtles nesting at Las Baulas, Costa Rica, have declined from 1500 in 1988 to 1989 to only 193 females in 1993 to 1994 (J. Spotila, unpublished data). Adult sea turtles migrate hundreds or even thousands of kilometers between nesting and feeding grounds (Meylan 1982; Eckert *et al.* 1989), and immature individuals move widely during their 25 to 30 year development to sexual maturity (Carr 1980). Individuals from different nesting populations converge on feeding grounds (Pritchard 1976; Meylan 1982), and so females probably encounter multiple potential mates who probably come from several different populations. This convergence of adults on communal feeding grounds and the likelihood that females store sperm (Solomon & Baird 1979) make multiple paternity within clutches highly plausible, provided that population density and, hence, the rate of encountering potential copulation partners, is sufficiently high.

The mating system of sea turtles remains a conspicuous gap in our knowledge of the life histories of these endangered species because mating occurs at sea. If the mating system of Green turtles is promiscuous, the effective size of nesting populations may be substantially larger than censuses of females at nesting beaches would suggest if each female mated with only a single male; the discrepancy would be even greater if the sex ratio favored females and multiple females mated exclusively with the same male. If multiple paternity is found with little sharing of males between females, conservation biologists could then place less emphasis on genetic aspects of population management.

We are using multilocus minisatellite DNA fingerprinting (Jeffreys *et al.* 1985) and single-locus microsatellite DNA technology (Tautz 1989) to investigate whether multiple paternity occurs within clutches of Green turtles and Leatherback turtles. In most other studies using DNA fingerprinting to analyze parentage, the DNA of both putative parents has been available for analysis (e.g., Jeffreys *et al.* 1985; Burke & Bruford 1987; Rabenold *et al.* 1990; Decker *et al.* 1993). Male sea turtles remain at sea while females go ashore to lay their eggs, and so nesting females and emerging hatchlings are the only easily accessible members of sea turtle populations.

We can distinguish the number of fathers contributing to each clutch, even when samples from only mothers and offspring are available, through the application of minisatellite and microsatellite DNA technology (FitzSimmons *et al.* 1995). For Green turtles, results to date suggest that females often use sperm from multiple males to fertilize their clutches (Peare *et al.* 1994; Parker *et al.* 1996), but for leatherbacks, only a single male is represented per clutch (J. P. Rieder, unpublished data). In order to distinguish a promiscuous from a polyandrous system, in the case of the green turtle, and a monogamous from a polygynous system in the leatherbacks, the next step is to determine the extent to which particular males are represented in clutches of multiple females. This can be performed by reconstruction of paternal genotypes by summing paternal alleles within full-sib clusters identified within nests. Although our study of sea turtle mating systems will always be limited by the inaccessibility of males, knowing something about the number of males mating with each female and the representation of particular males across females will help to estimate the breeding sex ratio and how the available genetic variability is distributed within a breeding population. In cases like these, behavioral study alone cannot provide accurate information on mating patterns for species that are difficult to observe; neither does genetic analysis provide the full picture in some of these cases. Rather than placing demographic approaches, which include behavioral methods, at odds with conservation genetics, often it will be more expedient to use these disciplines in a complementary manner, recognizing the tradeoffs of each. Only then can we achieve the synthesis we need to fully understand the dynamics of populations and their conservation implications.

Mating systems and effective population size: A summary to date

The breeding sex ratio and distribution of reproductive success among individuals within populations, or their mating systems, may have large effects on effective population size (Fig. 10.1). The effective size of populations, in turn, is the most sensitive predictor of the ability of any population to maintain genetic variability and persist through time; loss of variability is associated with both short-term fitness consequences (O'Brien & Everman 1988; Jimenez *et al.* 1994) and the longer-term consequences of having little variation upon which selection can act (Vrijenhoek *et al.* 1985).

Using a recent demographic estimator of N_e/N that incorporates effects of overlapping generations (Nunney & Elam 1994), the effect of skew in standardized variance in reproductive success for either sex is potentially substantial (Figs. 10.2 and 10.3). Using data from natural populations exhibiting such a skew induced by extra-pair fertilizations, the estimates of N_e/N changed little from those based on the more traditional assumption that nest attendance indicates parentage.

Taken together, these findings indicate just two of the many ways in which behavioral aspects of populations can have important implications for their conservation, and that behavioral observations themselves, when possible, may often be sufficient to characterize the parameters of population models. Understanding the distribution of reproductive success within a population is a key part of our understanding of population dynamics. The behavioral observations made during field studies of natural populations may be sufficient to characterize the mating system, from the perspective of conservation models, despite the fact that the underlying genetic mating system may differ as a result of cryptic matings.

We base this conclusion on the results of four studies of bird populations in which two analyses were performed: (1) the assumption of monogamy was tested by asking whether the male and female nest attendants, or dominants in the case of cooperative breeders, were the parents of young in the nest (an exclusion analysis); and (2) when that was not the case, the actual parents were found within the population for a large portion of the young for which a nest attendant, usually the male, was excluded. The second phase, assignment of young to particular adults with which they have no social relationship, is technically demanding and often impractical in very large populations or colonies. Both phases of analysis, however, are necessary to calculate the actual variance in reproductive success among individuals in populations. These data accumulate across species, so we will have a much clearer idea of the generality of our conclusions. We already know from the large number of exclusion analyses that have been performed on bird populations (reviewed in Birkhead & Møller 1992, 1995), that the proportions of young sired by extra-pair fertlizations in the four studies on which we concentrated (8 to 21%) are not atypical for passerine birds, although many populations show higher levels. We must wait, however, until a larger sample exists for assignment studies, before we know how those exclusions are redistributed among males in the population, the magnitude of their effect on I_{bm}, and the likelihood that such redistributions would significantly change the effective size of populations.

Basic research on mating behavior in natural populations has much to contribute to their conservation. Students of mating systems should be advised to devote careful attention to the collection of basic population data as they make their behavioral observations; the selective environment, including its social and demographic elements, in which behaviors are expressed, is critical to understanding their evolution. The consequences of competition for reproductive opportunities are the very patterns of distribution of reproductive success across individuals that we call the mating system. Combined with demographic data on survivorship, these data are essential components of population models. Behaviorists studying natural populations should keep on doing what they do, keeping in the front of their minds that the data they collect today may be more useful tomorrow if they reflect the behavioral and demographic dynamics of their population as completely as possible.

In conclusion, it is obvious that there will be times when neither field studies of behavior nor genetic studies (modeling or molecular) alone will be sufficient to provide accurate estimations of effective population size. It is also clear that for certain taxa, when behavioral study is possible, data from behavioral observation may be sufficient, and probably less expensive, to characterize mating patterns for estimation of N_e. Additional genetic analyses of species, however, chosen to represent different combinations of life history traits, such as those with long maturation times relative to adult life span like many fish, will help us develop predictive models that combine genetics and behavior for conservation planning. Carefully targeted studies can be used to generate genetic patterns for inferring the mating system, and thus N_e, of behaviorally intractable species. Finally, there is growing evidence that mating systems often vary within species or even change over short periods within a population (Komdeur and Deerenberg, Chapter 11). We need to develop reliable models that predict changes in mating systems under specific social and ecological conditions, particularly in environments that are being rapidly altered by human activity. Variation, whether measured as genetic heterozygosity or behavioral strategies, is linked to population viability. Genetic and behavioral conservationists should be partners in the fight to preserve wild populations.

Literature cited

Amos, B., S. Twiss, P. Pomeroy, and S. Anderson. 1995. Evidence for mate fidelity in the gray seal. *Science* **268**: 1897–1899.

Avise, J. C. 1994. *Molecular markers, natural history, and evolution*. Chapman and Hall, New York. 511 pp.

Birkhead, T. R., and A. P. Møller. 1992. *Sperm competition in birds: Evolutionary causes and consequences*. Academic Press, London.

Birkhead, T. R., and A. P. Møller. 1995. Extra-pair copulation and extra-pair paternity in birds. *Animal Behaviour* **49**: 843–848.

Burke, T., and M. W. Bruford. 1987. DNA fingerprinting in birds. *Nature* **327**: 149–152.

Caballero, A. 1994. Developments in the prediction of effective population size. *Heredity* **73**: 657–679.

Caro, T. M., and M. K. Laurenson. 1994. Ecological and genetic factors in conservation: a cautionary tale. *Science* **263**: 485–486.

Carr, A. 1980. Some problems of sea turtle ecology. *American Zoologist* **20**: 489–498.

Crow, J. F., and M. Kimura. 1970. *An introduction to population genetics theory*. Harper and Row, New York.

Decker, M. D., P. G. Parker, D. J. Minchella, and K. N. Rabenold. 1993. Monogamy in black vultures: Genetic evidence from DNA fingerprinting. *Behavioral Ecology* **4**: 29–35.

Eckert, K. L., S. A. Eckert, T. W. Adams, and A. D. Tucker. 1989. Internesting migrations by leatherback sea turtles (*Dermochelys coriacea*) in the West Indies. *Herpetologica* **45**: 190–194.

Faaborg, J., P. G. Parker, L. DeLay, T. J. de Vries, J. C. Bednarz, S. Maria Paz, J. Naranjo, and T. A. Waite. 1995. Confirmation of cooperative polyandry in the Galapagos hawk (*Buteo galapagoensis*) using DNA fingerprinting. *Behavioral Ecology and Sociobiology* **36**: 83–90.

FitzSimmons, N. N., C. Moritz, and S. S. Moore. 1995. Conservation and dynamics of microsatellite loci over 300 million years of marine turtle evolution. *Molecular Biology and Evolution* **12**: 432–440.

Frankham, R. 1995. Inbreeding and extinction: a threshold effect. *Conservation Biology* **9**: 792–799.

Frankham, R. 1996. Effective population size/adult population size ratios in wildlife: a review. *Genetical Research* **66**: 95–107.

Gowaty, P. A., and D. W. Mock, editors. 1985. *Avian monogamy*. AOU Ornithological Monographs, no. 37. American Ornithologists' Union Washington, DC.

Hill, W. G. 1972. Effective size of populations with overlapping generations. *Theoretical Population Biology* **3**: 278–289.

Hill, W. G. 1979. A note on effective population size with overlapping generations. *Genetics* **92**: 317–322.

Jeffreys, A. J., W. Wilson, and S. L. Thein. 1985. Individual-specific fingerprints of human DNA. *Nature* **316**: 75–79.

Jimenez, J. A., K. A. Hughes, G. Alaks, L. Graham, and R. C. Lacy. 1994. An experimental study of inbreeding depression in a natural habitat. *Science* **266**: 271–273.

Keane, B., P. M. Waser, S. R. Creel, N. M. Creel, L. F. Elliott, and D. J. Minchella. 1994. Subordinate reproduction in dwarf mongooses. *Animal Behaviour* **47**: 65–75.

Kempenaers, B., G. R. Verheyen, M. Van Den Broeck, T. Burke, C. Van Broeckhoven, and A. A. Dhondt. 1992. Extra-pair paternity results from female preference for high quality males in the blue tit. *Nature* **357**: 494–496.

Lack, D. 1968. *Ecological adaptations for breeding in birds*. Methuen, London.

Lande, R. 1988. Genetics and demography in biological conservation. *Science* **241**: 1455–1459.

Lande, R. 1994. Risk of population extinction from fixation of new deleterious mutations. *Evolution* **48**: 1460–1469.

Lande, R. 1995. Mutation and conservation. *Conservation Biology* **9**: 782–791.

Lande, R., and G. F. Barrowclough. 1987. Effective population size, genetic variation, and their use in population management. Pages 87–123 in M. E. Soulé, editor. *Viable populations for conservation*. Cambridge University Press, New York.

Lynch, M., J. Conery, and R. Bürger. 1995. Mutation accumulation and the extinction of small populations. *American Naturalist* **146**: 489–499.

Meylan, A. B. 1982. Sea turtle migration – evidence from tag returns. Pages 91–100 in K. A. Bjorndal, editor. *Biology and conservation of sea turtles*. Smithsonian Institution Press, Washington, DC.

Mulder, R. A., P. O. Dunn, A. Cockburn, K. A. Lazenby-Cohen, and M. J. Howell. 1994. Helpers liberate female fairy-wrens from constraints on extra-pair mate choice. *Proceedings of the Royal Society of London, Series B* **255**: 223–229.

Nunney, L. 1993. The influence of mating system and overlapping generations on effective population size. *Evolution* **47**: 1329–1341.

Nunney, L. 1996. The influence of variation in female fecundity on effective population size. *Biological Journal of the Linnaean Society*, in press.

Nunney, L., and K. A. Campbell. 1993. Assessing minimum viable population size: demography meets populations genetics. *Trends in Ecology and Evolution* **8**: 234–239.

Nunney, L., and D. R. Elam. 1994. Estimating the effective population size of conserved populations. *Conservation Biology* **8**: 175–184.

O'Brien, S. J., and J. F. Everman. 1988. Interactive influence of infectious disease and genetic diversity in natural populations. *Trends in Ecology and Evolution* **3**: 254–259.

O'Brien, S. J., D. E. Wildt, D. Goldman, C. R. Merril, and M. Bush. 1983. The cheetah is depauperate of genetic variation. *Science* **221**: 459–462.

Parker, P. G., T. A. Waite, and T. Peare. 1996. Paternity studies in animal populations. Pages 413–423 in T. B. Smith and R. K. Wayne, editors. *Molecular genetics in conservation*. Oxford University Press, Oxford.

Peare, T., P. G. Parker, and T. A. Waite. 1994. Multiple paternity in green turtles *(Chelonia mydas)*: conservation implications. Pages 115–118 in *Proceedings of the fourteenth annual symposium on sea turtle biology and conservation*. US Department of Commerce, NOAA, National Marine Fisheries Service, Miami, FL.

Pritchard, P. C. H. 1976. Post-nesting movements of marine turtles (Cheloniidae and Dermochelyidae) tagged in the Guianas. *Copeia* **1976**: 749–754.

Quattro, J. M., and R. C. Vrijenhoek. 1989. Fitness differences among remnant populations of the endangered Sonoran topminnow. *Science* **245**: 976–978.

Rabenold, P. P., K. N. Rabenold, W. H. Piper, J. Haydock, and S. W. Zack. 1990. Shared paternity revealed by genetic analysis in cooperatively breeding tropical wrens. *Nature* **328**: 538–540.

Ralls, K., and J. Ballou. 1983. Extinction: lessons from zoos. Pages 164–184 in C. M. Schonewald-Cox, S. M. Chambers, B. MacBryde, and L. Thomas, editors. *Genetics and conservation: A reference for managing wild animal and plant populations*. Benjamin-Cummings, New York.

Solomon, S. W., and T. Baird. 1979. Aspects of the biology of *Chelonia mydas* L. *Oceanographic and Marine Biology Annual Review* **17**: 347–361.

Tautz, D. 1989. Hypervariability of simple sequences as a general source for polymorphic DNA markers. *Nucleic Acids Research* **17**: 6463–6471.

Vrijenhoek, R. C., M. E. Douglas, and G. K. Meffe. 1985. Conservation genetics of endangered fish populations in Arizona. *Science* **229**: 400–402.

Wagner, R. H., M. D. Schug, and E. S. Morton. Confidence of paternity, actual paternity and parental effort by purple martins. *Animal Behaviour*, in press.

Waite, T. A., & P. G. Parker. 1996. Dimensionless life histories and effective population size. *Conservation Biology*, in press.

Wright, S. 1931. Evolution in Mendelian populations. *Genetics* **16**: 97–139.

Wright, S. 1969. *Genetics and the evolution of populations*, Volume 2. *The theory of gene frequencies*. University of Chicago Press, Chicago.

Chapter 11

The importance of social behavior studies for conservation

JAN KOMDEUR AND CHARLOTTE DEERENBERG

A social system can be defined as the outcome of a set of consistent social relationships (Hinde 1976, 1983), seemingly having a biological function (Lott 1991). The type of mating relationship or offspring care, for example, is described as the mating or parental care system. Contrary to the typological view, social systems are not fixed features of species but often show intraspecific variation. Hence, mating systems can alternate between monogamy and polygamy, parental care can vary between two-parent care and cooperative nesting, and nesting strategies can range from colonial to territorial.

The variation seen in the social systems of many species is probably explained by individual attempts to maximize fitness under local conditions. Thus, variations in social behavior have been viewed as adaptions to variable environmental conditions (Bekoff *et al.* 1984; Lott 1984; Moehlman 1989; West-Eberhard 1989). Changes in the quantity and distribution of limiting resources such as food, territories, or nesting sites can have profound effects on the structure of a social system. Variation in social behavior may sometimes occur in response to changes in the demographic characteristics of a population. The prevailing adult sex ratio, for example, may limit the number of mates available in one sex and result in populations switching from monogamy to polygamy. The age structure of a population may also influence the social system; as animals gain experience with age, for example, they may switch from monogamy to polygamy. Intraspecific variation in social systems has conservation relevance if by studying the factors shaping social systems, social interactions could be managed or engineered to minimize the risk of extinction or reduce human conflicts with animals.

Studies of social organization also relate directly to problems associated with the management of small populations, which has become a major activity for conservation biology (Soulé 1985, 1986, 1987). To prevent extinction there are several features of an endangered population that must be kept under surveillance. Population size must be adequate to ensure the reproductive success of breeding units. In the long term, reproductive success depends on genetic variation (Nunney & Campbell 1993), a stable age structure, and an adult sex ratio that is appropriate to the social system of the species concerned. Effective management also requires the ability to predict a population's or community's response to changes imposed by the management activities.

In view of the fact that social systems frequently place constraints on which individuals can breed, and thereby determine the reproductive success of individuals and critical population sizes, understanding how environmental variation affects social structure should be a major concern in conservation management. In this chapter we first describe how social organization may vary intraspecifically and how this variation affects effective population size and small population demography. Subsequently, we discuss environmental conditions that have been linked to variation in social organization, with the focus on mating and parental care systems. Finally, we sketch how conservation biologists can use this information to design and implement more effective management plans.

Small population demography: Implications for variation in social behavior

All environments are variable in the conditions and resources that they provide, and these environmental uncertainties contribute to population fluctuations, or even cause small populations to go extinct. Catastrophes, such as fires and volcanic eruptions, may kill all or most members of local populations. Habitat fragmentation is another important factor pushing populations toward extinction. In subpopulations produced by fragmentation, individuals breed more with one another than with members of other sub populations (Gilpin 1987). In a heavily fragmented landscape, social structure may be disrupted by overlap among the pairs, or on the contrary, by separation of pairs. In the Northern spotted owl (*Strix occidentalis caurina*), for example, fragmentation resulted in an increased proportion of adult–subadult pairs and instances of adult nomadism. The inability to find suitable mates led to reduced breeding output (Carey *et al.* 1992). Extinction in small populations may also occur 'by chance' because of demographic uncertainties, such as random increases in death rates or decreases in birth rates (Gilpin & Soulé 1986; Goodman

1987). Thus, an isolated subpopulation, being smaller, holds a higher risk of extinction (review in Mills & Smouse 1994), and once extinct, may not recover by an influx from other sub populations. Thus, it is important to know what aspects of a small population determine its vulnerability.

Effective population size

Ideally, population size is not estimated by the actual number of individuals present (N), but rather by the so-called 'effective population size' (N_e) (Begon *et al.* 1990), which is the size of a 'genetically idealized' population equivalent to the actual population in genetic terms. N_e is usually less (and often much less) than N, and depends on the variables determining the social system of a species. Without knowing the variations in the social system, one could over- or underestimate the effective breeding population. In breeding systems with more males than females, polyandry as an alternative to monogamy could theoretically increase the effective population size by increasing the genetic representation of individuals in the subsequent generation. There are some species, for example, that breed monogamously under some conditions and at other times in cooperative polyandry, where several males attend a single female and clutch at a single nest (Lott 1991). In relation to the prevalent social system, there are two aspects of effective population size that require consideration. On a short-term time scale we first need to consider the number of breeding units; in the long term we may also need to safeguard genetic variation.

The effective breeding population is more a function of the breeding system than of the number of animals. Especially in facultatively polygamous or cooperatively breeding species, the number of breeding units may be much smaller than expected based on monogamous breeding. In a saturated environment, for example, the average breeding group of the Seychelles warbler (*Acrocephalus sechellensis*) comprises the breeding pair and two non-breeding helpers (Komdeur 1992). In this case the effective breeding population is only half of the total population. In an unsaturated environment with equal numbers of males and females, however, one expects that the birds will breed monogamously, because fitness benefits of immediate breeding would be larger than helping benefits. Indeed, after translocations of warblers to unoccupied islands, only breeding pairs without helpers were formed and the effective breeding population increased twofold (Komdeur 1992; Komdeur *et al.* 1995).

In small populations, simple stochastic variation may lead to biased sex ratios that originate in response to specific environmental conditions (Wibbels *et al.* 1991; Yanega 1993; Komdeur 1996*a*). Biased sex ratios may also result from sex-specific mortality rates (Weatherhead *et al.* 1995), sometimes as a result of risk-prone behavior of usually the males, or male–male competition (Sukumar 1991; Berger & Cunningham 1995).

If one sex becomes relatively more abundant, the mating system may move from monogamy towards polygamy (Balfour & Cadbury 1979; Picozzi 1984; Struhsaker & Pope 1991; Taborsky & Taborsky 1991), although other factors may interfere to determine the mating system, such as mate-searching efficiency, the maximal number of females a male can include within his home range, length of the mating season, and the female's estrous period (Sandell & Liberg 1992). Polygamy as opposed to monogamy, however, does not necessarily affect reproductive success negatively. For example, in the fish *Cichlasoma nigrofasciatum* with facultative biparental care (Keenleyside *et al.* 1990), the young survive equally well with only maternal care one week after spawning. In the Argentine ant (*Iridomyrmex humilis*) with a highly male-biased sex ratio, male interference may decrease the amount of sperm queens store in the field. Most queens, however, use only a small proportion of their sperm supply during their lifetime (Keller & Passera 1992). By contrast, a biased sex ratio may decrease reproductive success, e.g., through increased aggressive behavior (Starfield *et al.* 1995) or through increased hybridization with other, closely related species (Schlyter *et al.* 1991). In addition, a biased sex ratio and, correspondingly, increased polygamy may reduce population growth by reducing the effective population size and overall reproductive success (Macedo 1992), or that of secondary mates (Smith *et al.* 1994).

The social structure of small populations may also change dramatically when the age structure has been disrupted, thus affecting reproductive success and population growth. In cooperatively breeding species, the same bird may be a helper at one stage in its life and a breeder at another (Brown 1987; Koenig & Mumme 1987; Stacey & Koenig 1990; Komdeur 1992, 1996*b*,*c*; Russell & Rowley 1993; Emlen 1994, 1995). Animals usually gain experience with age, thereby increasing dominance, access to mates, and reproductive success (Davies 1992). In conclusion, the effective population size of a population not only depends on the number of adult males and females, but also largely on the social behavior of the species.

Genetic variation

Preservation of genetic diversity for maintaining viable populations is often a major goal in any species conservation program. The degree of polygamy in a breeding system will often determine the genetic diversity transmitted to the next generation (see Parker & Waite, Chapter 10). In the California quail, *Callipepla californica*, for example, males outnumber females because of higher female mortality, and so serial polyandry is favored over monogamy (Francis 1965). Polygamy may be favored as a result of other behavioral tendencies. In the Harris hawk, *Parabutea unicinctus* (Mader 1975), for example, there is intense competition for breeding territories, and breeding females control all suitable territories, thereby excluding some females from breeding altogether. In either species, serial or cooperative polyandry increases the genetic diversity of the next population because the genes of males that would be excluded under a monogamous system are transmitted to the next generation's gene pool.

The mean fitness of a population is generally assumed to be positively associated with heterozygosity (Gilpin & Soulé 1986; Lande & Barrowclough 1987). Hence, inbreeding, which leads to high levels of homozygosity and the expression of deleterious recessive alleles, is typically thought to reduce the viability of a population. Both the operative sex ratio and the growth rate of the population can greatly affect the effective population size, thereby leading to inbreeding and enhancing the extinction rate (Briton *et al.* 1994; Mills & Smouse 1994). Small, inbreeding populations are also more susceptible to the effects of genetic drift (i.e., the frequency of genes in a population determined by chance rather than evolutionary advantage). In addition, the lack of genetic variation may limit a population's ability to respond evolutionarily to induced changes in environmental conditions (Gilpin & Soulé 1986; Lande & Barrowclough 1987).

An assessment of the level of genetic variation and the occurrence of inbreeding in natural populations, therefore, is informative for its management. Such an assessment, however, will be possible only from genealogic data based on known genetic parentage integrated with information on effective population size derived from studies of demography and social organization. DNA fingerprinting techniques are now being used for the detection and regular monitoring of population genetic variability (see Dhondt 1996). Contrary to expectation, small isolated populations can still show a high level of genetic variation, as was shown for a colony of 39 Waldrapp ibises (*Geronticus eremita*). Despite the fact that it was founded by only six birds and that some of the founders were related, DNA fingerprinting revealed highly polymorphic banding patterns, indicating a high level of genetic variation (Signer *et al.* 1994).

It is also not clear how much genetic variation is required to ensure a population's persistence in the long term. Merola (1994) argued that genetic uniformity as present in the Cheetah (*Acinonyx jubatus*) does not necessarily appear to compromise the survival of the species, at least in the short term. For other species, the evidence suggests that the level of inbreeding leads to reduced fitness. Lions (*Panthera leo*) in the Ngorongoro Crater, Tanzania, form a small and naturally isolated population (Packer *et al.* 1991). All members of the current Crater population are descended from only 15 founders present in 1964 to 1965. By 1975, the population had recovered to its current level of 75 to 125 animals. Over the past 25 years there has been considerable variance in the reproductive success of both sexes. The Crater lion population shows a significant lack of genetic diversity compared to the much larger Serengeti populations. High levels of inbreeding are correlated with increased levels of sperm abnormality in lions, and there is evidence that the reproductive performance of the Crater lions has decreased as a result of decreasing heterozygosity. For some species, however, even if inbreeding occurs frequently, it does not appear to significantly decrease either the survival of progeny or the level of individual fitness. Neither in the Splendid fairy wren, *Malurus splendens* (Rowley *et al.* 1993), nor in the Seychelles warbler (Kappe & Komdeur 1996) does the level of inbreeding lead to reduced fitness in the short term; the production of fledglings by related pairs was no lower than that of unrelated pairs.

An important topic of social behavior in need of greater study is the degree to which inbreeding avoidance mechanisms actually exist in wild animals, or whether so-called inbreeding avoidance is merely a consequence of other behavioral or demographic constraints. If these constraints cease to operate, as populations become smaller and more isolated as a result of environmental change, then the loss of heterozygosity and inbreeding depression may begin to affect species or populations not otherwise considered endangered. On the other hand, true inbreeding avoidance mechanisms operating in small populations may further reduce the effective population size.

' influences on intraspecific variation in social

..nd distribution of food

..ne factor that has been found to have significant influence on social organization is the availability and distribution of food. There is in general both temporal and geographic variation in food abundance. Apart from year-to-year differences, food abundance is likely to vary with changing seasons. The observation that Townsend's voles (*Microtus townsendii*) were monogamous in spring and polygynous in summer (Lambin & Krebs 1991) may be related to this seasonal variation in food abundance together with increasing numbers inducing smaller and more overlapping home ranges. Also Red foxes (*Vulpes vulpes*) are not typically monogamous but could become polygamous when more food is present in the male territory (Zabel & Taggeart 1889). Especially in species where uniparental care has little effect on reproductive success, such as in Northern jacanas, *Jacana spinosa* (Betts & Jenni 1991), or Tree swallows, *Tachycineta bicolor* (Dunn & Hannon 1992), sequential polygyny or polyandry is recorded in situations of higher food availability.

The effect of food abundance on the breeding system is most obvious in cooperatively breeding species, in which the dominant pair is assisted in raising young by several helpers, often offspring from previous years. Depending on food availability in the natal territory, a young bird may become a helper or leave the territory to become a breeder. Increased food abundance in the natal territories is associated with the appearance of helpers at the nest in species that otherwise show biparental care, such as the Northwestern crow, *Corvus kaurinus* (Zerbek & Butler 1981), and the Seychelles magpie robin, *Copsichus sechellarum* (Komdeur & McCulloch 1996). In Burying beetles (*Nicrophorus*) larger food supply in the form of carcasses not only supported larger number of larvae, but also resulted in communal breeding (Trumbo 1992).

The pattern of feeding sites has been reported to induce changes in the breeding system. In the Musk-ox (*Ovibos moschatus*) group size increased when feeding sites became less abundant in winter. Group size in turn determined the breeding system (Heard 1992). Both the abundance and the spatial distribution of the food may affect the spacing of male and female territories or home ranges. Mating success of territorial males or the degree of polygyny usually increases with the degree of overlap with female home ranges. In the Iguanian lizard (*Uta palmeri*) females shifted their home ranges into food supplemented areas (Hews 1993), thereby promoting polygyny.

Quantity and quality of territories or nest sites

Densities of animals may vary inversely with the area of suitable habitat for breeding, thereby inducing changes in the social organization. Apart from human-induced habitat fragmentation or removal, a breeding habitat may increase in wet years and decrease in years of drought. Patch size, for example, was associated with the degree of polygamy in Gunnison's prairie dog (*Cynomys gunnisoni*). Uniformly distributed resources led to smaller groups with single females, while abundant and patchily distributed resources held groups including several females (Travis & Slobodchikoff 1993). Similarly, higher densities of Tree lizards (*Urosaurus ornatus*) in one habitat type were related to a greater proportion of polygyny compared with a habitat type with fewer animals and a lower degree of polygyny. Food abundance did not differ between those habitats (M'Closkey *et al.* 1990*a*,*b*). Populations in a saturated environment may breed cooperatively, such as the Seychelles warbler. This species was entirely confined to Cousin Island (29 ha), which was completely covered with warbler territories. Young birds remain on their natal territories as helpers until breeding vacancies arise (Komdeur 1992). When the warblers were introduced onto unoccupied islands, locally produced young set up breeding territories in the abundant unoccupied habitat, and there was no helping. As the new population increased, however, and all the good quality habitats became occupied, the young began to remain as helpers rather than breed in poor territories (Komdeur 1992). Thus an understanding of how territory quality determines animal breeding and movements is crucial to conservation management.

Management tools: Inducing changes in social systems

Given the flexibility of social systems, management strategies could be developed to manipulate these systems by varying the environmental determinants of the variation. There are several management options available for managing social systems to obtain the highest effective population size and genetic variation given the number of animals of either gender. In this section we discuss the applicability of translocations or the creation of travel corridors, supplementary feeding, and the removal of individuals.

Translocations and travel corridors

A translocation is the intentional release of animals in an attempt to establish, re-establish, or augment a population and may consist of more than one release. Translocations have varied goals that include bolstering genetic heterogeneity of small populations, reducing the risk of species loss due to catastrophes, and speeding recovery of species after their habitats have been restored and subsequently recovered from the negative effects of environmental toxicants or other limiting factors. In the face of increasing species extinction rates (Ehrlich & Ehrlich 1981; Ehrlich 1988; Wilson 1988) and impending reduction in overall biological diversity (Wilson 1988), translocation of rare species may become an increasingly important conservation technique. Increased rates of population extinction may be expected in small fragmented habitats or on small islands, and translocations or the creation of travel corridors may be required to maintain community compositions, especially for species with limited dispersal abilities. The effects of habitat fragmentation or removal can be further minimized by retaining travel corridors to connect the relict areas.

Social studies may indicate the abundance of superfluous non-breeding animals in the parental population. These superfluous animals can be regarded as a 'surplus' and used for translocations. Translocations are more successful when the founders have breeding experience. For example, translocations of experienced Seychelles warblers, which had fledged young in a previous year, and of birds of the same age with no breeding experience, showed that on the new islands experienced birds produced significantly more young than inexperienced birds (Komdeur 1996*b*). In addition, translocations of exclusively wild-caught animals are more likely to succeed than are those of exclusively captive-reared animals (Griffith *et al.* 1989; Bright & Morris 1994). The low success rate of translocations of small numbers of endangered, threatened, or sensitive species, even in excellent habitat quality (Griffith *et al.* 1989), means that translocation must be considered long before it becomes a last resort for these species.

Current knowledge on the association between numbers of released animals and translocation success, and of the genetic factors that influence the new population's persistence is weak (Olivieri *et al.* 1990; Nunney & Campbell 1993). It was often thought that a high genetic diversity, obtained by translocating many founders, may enhance persistence (Geist 1987). This theme currently enjoys considerable scientific interest. It is often not realized that a small number of individuals may carry most of the genetic variation that is present in the species (Joseph *et al.* 1996). Whether this degree of genetic variation can be maintained depends on the rate of population growth after the translocation. For example, the pool of genetic variation may become reduced through inbreeding when the population grows slowly following a transfer. A viable management solution to populations recovering slowly will be the repeated addition of individuals from the parent population. A study of the social system of the receiving population is important to prevent new incoming individuals disrupting the social system and thereby failing to accomplish the objectives of the translocation. For example, the social system may indicate saturation of the habitat. The addition of animals may then decrease the reproductive success of the resident population through disturbance.

Supplementary feeding

It is easy to imagine a scenario in which attempts to sustain a small population could include supplementation of food; however, one has to bear in mind that supplementation could have either a negative or positive effect on the species' future. For example, Red foxes express polygyny when the food resources within a territory are superabundant (Zabel & Taggeart 1989). In this species supplementation could take a form that leads to an increase in defendability of food by a single male and thus increases the likelihood of a polygynous system with non-breeding males, thus reducing effective population size. In such a situation, food supplementation would be a short-term management step that would be contradictory to long-term management goals. In a saturated environment, supplementary feeding may allow more individuals to live there than would otherwise be possible, but this is not sustainable in the long term. In such a case, the 'surplus' might be used for the establishment of new populations, as has been done in the Seychelles magpie robin (Komdeur 1996*d*). Through habitat loss, the total magpie robin population declined from 41 birds in 13 territories in 1977 to 17 birds in eight territories in 1988, and was entirely confined to one island. Supplementary feeding resulted in more young produced than was necessary to compensate for adult mortality. In the absence of suitable areas for establishing new territories, many young magpie robins become 'floaters' and negatively affect the breeding success of other robins.

Removal of individuals

The males of polygynous large mammals such as elephant, rhino, and tiger have a greater variance in reproductive success than females, leading to selection pressures favoring a 'high risk-high gain' strategy for promoting reproductive success. This strategy usually brings males into greater conflict with people. In polygynous species, the removal of a certain proportion of 'surplus' adult males is not likely to affect the fertility and growth rate of the population (Sukumar 1991); therefore, removal of individuals could be a management tool that would effectively reduce human–animal conflict, and simultaneously maintain the viability of the population. One has to be careful, however, because selective removal of males may result in a skewed sex ratio or may increase the rate of genetic drift and lead to inbreeding depression. For example, Fitzsimmons *et al.* (1995) hypothesized that selective removal of large-horned rams in Bighorn sheep (*Ovis canadensis*) contributed to losses of fitness by reducing genetic variability. Plans for managing mammals through the culling of males should ensure that the appropriate effective population size is maintained.

Importance of social behavior studies for conservation: Conclusion

The selection of any strategy for conservation is not without risk. For each strategy it is important to identify and quantify the risks to population viability, which requires the integration of not only ecological theory, empirical data, and professional judgment, but also the incorporation of studies of the species' social systems. Intraspecific differences in social organization may vary in time or geographically as populations adapt to local conditions. Impacts of social behavior on population dynamics seem greatest in species that exist in relatively small populations, such as those on islands or those that are dwindling because of human encroachment. Hence, a knowledge of the factors that produce variability in social systems and affect population viability has valuable conservation application as managers strive toward the long-term persistence of endangered populations.

Literature cited

Balfour, E., and C. J. Cadbury. 1979. Polygyny, spacing and sex ratio among Hen Harriers *Circus cyaneus* in Orkney, Scotland. *Ornis Scandinavica* **10**: 133–141.

Begon, M., J. L. Harper, and C. R. Townsend. 1990. *Ecology: Individuals, populations, and communities*. Blackwell Scientific Publications, Boston.

Bekoff, M., T. J. Daniels, and J. L. Gittleman. 1984. Life history patterns and the comparative social ecology of carnivores. *Annual Review of Ecology and Systematics* **15**: 191–232.

Berger, J., and C. Cunningham. 1995. Predation, sensitivity and sex: why female Black Rhinoceroses outlive males. *Behavioral Ecology* **6**: 57–64.

Betts, B. J., and D. A. Jenni. 1991. Time budgets and the adaptiveness of polyandry in Northern Jacanas. *Wilson Bulletin* **103**: 578–597.

Bright, P. W., and P. A. Morris. 1994. Animal translocations for conservation: performance of dormice in relation to release methods, origin and season. *Journal of Applied Ecology* **31**: 699–708.

Briton, J., R. K. Nurthen, D. A. Briscoe, and R. Frankham. 1994. Modelling problems in conservation genetics using *Drosophila*: consequences of harems. *Biological Conservation* **69**: 267–275.

Brown, J. L. 1987. *Helping and communal breeding in birds*. Princeton University Press, Princeton, New Jersey.

Carey, A. B., S. P. Horton, and B. L. Biswell. 1992. Northern spotted owls: influence of prey base and landscape character. *Ecological Monographs* **62**: 223–250.

Davies, N. B. 1992. *Dunnock behaviour and social evolution*. Oxford University Press, Oxford.

Dhondt, A. A. 1996. Molecular techniques in conservation and evolutionary biology: a quantum leap? *Trends in Ecology and Evolution* **11**: 147–148.

Dunn, P. O., and S. J. Hannon. 1992. Effects of food abundance and male parental care on reproductive success and monogamy in tree swallows. *Auk* **109**: 488–499.

Ehrlich, P. R. 1988. Loss of diversity. Pages 21–27 in E. O. Wilson, editor. *Biodiversity*. National Academy Press, Washington.

Ehrlich, P. R., and A. H. Ehrlich. 1981. *Extinction: The causes and consequences of the disappearance of species*. Random House, New York.

Emlen, S. T. 1994. Benefits, constraints and the evolution of the family. *Trends in Ecology and Evolution* **9**: 282–285.

Emlen, S. T. 1995. An evolutionary theory of the family. *Proceedings of the National Academy Sciences of the USA* **92**: 8092–8099.

Fitzsimmons, N. N., S. W. Buskirk, and M. H. Smith. 1995. Population history, genetic variability, and horn growth in Bighorn Sheep. *Conservation Biology* **9**: 314–323.

Francis, W. J. 1965. Double broods in California quail. *Condor* **67**: 541–542.

Geist, V. 1987. On speciation in Ice Age mammals, with special reference to cervids and caprids. *Canadian Journal of Zoology* **65**: 1067–1084.

Gilpin, M. E. 1987. Spatial structure and population vulnerability. Pages 125–139 in M. E. Soulé, editor. *Viable populations for conservation*. Cambridge University Press, Cambridge.

Gilpin, M. E., and M. E. Soulé. 1986. Minimal viable populations: the processes of species extinctions. Pages 13–34 in M. E. Soulé, editor.

Conservation biology: The science of scarcity and diversity. Sinauer Associates, Sunderland, Massachusetts.

Goodman, D. 1987. The demography of chance extinction. Pages 11–34 in M. E. Soulé, editor. *Viable populations for conservation*. Cambridge University Press, Cambridge.

Griffith, B., J. M. Scott, J. W. Carpenter, and C. Reed. 1989. Translocation as a species conservation tool: status and strategy. *Science* **245**: 477–480.

Heard, D. C. 1992. The effect of Wolf predation and snow cover on Musk-ox group size. *American Naturalist* **139**: 190–204.

Hews, D. K. 1993. Food resources affect female distribution and male mating opportunities in the Iguanian Lizard *Uta palmeri*. *Animal Behaviour* **46**: 279–291.

Hinde, R. A. 1976. Interactions, relationships and social structure. *Man* **11**: 1–17.

Hinde, R. A. 1983. A conceptual framework. Pages 1–7 in R. A. Hinde, editor. *Primate social relationships*. Blackwell Scientific Publications, Oxford.

Joseph, L., S. Degnan, P. R. Baverstock, D. Parkin. 1996. Genetics and the conservation of birds. *Bird Conservation International*, in press.

Kappe, A., and J. Komdeur. 1996. Population bottlenecks and genetic variability in the Seychelles Warbler: a field experiment. *Evolution*, submitted.

Keenleyside, M. H. A., R. C. Bailey, and V. H. Young. 1990. Variation in the mating system and associated parental behavior of captive and free-living *Cichlasoma nigrofasciatum* (Pisces, Cichlidae). *Behaviour* **112**: 202–221.

Keller, L., and L. Passera. 1992. Mating system, optimal number of mating, and sperm transfer in the Argentine Ant *Iridomyrmex humilis*. *Behavioural Ecology and Sociobiology* **31**: 359–366.

Koenig, W. D., and R. L. Mumme. 1987. *Population ecology of the cooperatively breeding Acorn Woodpecker*. Princeton University Press, Princeton, New Jersey.

Komdeur, J. 1992. Importance of habitat saturation and territory quality for evolution of cooperative breeding in the Seychelles warbler. *Nature* **358**: 493–495.

Komdeur, J. 1996*a*. Facultative sex ratio bias in the offspring of Seychelles Warblers. *Proceedings of the Royal Society London, B*, **263**: 661–6.

Komdeur, J. 1996*b*. Influence of helping and breeding experience on reproductive performance in the Seychelles Warbler: a translocation experiment. *Behavioral Ecology*, **7** 310: 130–7.

Komdeur, J. 1996*c*. Influence of age on reproductive performance in the Seychelles Warbler. *Behavioral Ecology*, in press.

Komdeur, J. 1996*d*. Breeding of the Seychelles Magpie Robin *Copsychus sechellarum* and its implications for conservation. *Ibis*, in press.

Komdeur, J., G. Castle, A. Huffstad, R. Mileto, W. Prast, and J. Wattel. 1995. Transfer experiments of Seychelles warblers to new islands: changes in dispersal and helping behaviour. *Animal Behaviour* **49**: 695–708.

Komdeur, J., and N. McCulloch. 1996. Cooperative breeding in the Seychelles Magpie Robin: test of the habitat-variance model. *Behavioral Ecology*, submitted.

Lambin, X., and C. J. Krebs. 1991. Spatial organization and mating system of *Microtus townsendii*. *Behavioral Ecology and Sociobiology* **28**: 353–364.

Lande, R., and G. F. Barrowclough. 1987. Effective population size, genetic variation, and their use in population management. Pages 87–123 in M. E. Soulé, editor. *Viable populations for conservation*. Cambridge University Press, Cambridge.

Lott, D. F. 1984. Intraspecific variation in the social systems of wild vertebrates. *Behaviour* **88**: 266–325.

Lott, D. F. 1991. *Intraspecific variation in the social systems of wild vertebrates*. Cambridge University Press, Cambridge.

Macedo, R. H. 1992. Reproductive patterns and social organization of the Communal Guira Cuckoo (*Guira guira*) in central Brazil. *Auk* **109**: 786–799.

Mader, W. J. 1975. Extra adults at Harris' Hawk nests. *Condor* **77**: 482–485.

M'Closkey, R. T., R. J. Deslippe, C. P. Szpak, and K. A. Baia. 1990*a*. Tree Lizard distribution and mating system: the influence of habitat and food resources. *Canadian Journal of Zoology* **68**: 2083–2089.

M'Closkey, R. T., R. J. Deslippe, C. P. Szpak, and K. A. Baia. 1990*b*. Ecological correlates of the variable mating system of an iguanid lizard. *Oikos* **59**: 63–69.

Merola, M. 1994. A reassessment of homozygosity and the case for inbreeding depression in the Cheetah, *Acinonyx jubatus*: implications for conservation. *Conservation Biology* **8**: 961–971.

Mills, L. S., and P. E. Smouse. 1994. Demographic consequences of inbreeding in remnant populations. *American Naturalist* **144**: 412–431.

Moehlman, P. D. 1989. Intraspecific variation in canid social system. Pages 143–163 in J. L. Gittleman, editor. *Carnivore behavior, ecology and evolution*. Cornell University Press, Ithaca.

Nunney, L., and K. A. Campbell. 1993. Assessing minimum viable population size: demography meets population genetics. *Trends in Ecology and Evolution* **8**: 234–239.

Olivieri, I., D. Couver, and P. H. Gouyon. 1990. The genetics of transient populations: research at the metapopulation level. *Trends in Ecology and Evolution* **5**: 207–210.

Packer, C., A. E. Pusey, H. Rowley, D. A. Gilbert, J. Martenson, and S. J. O'Brien. 1991. Case study of a population bottleneck: lions of the Ngorongoro crater (Tanzania). *Conservation Biology* **5**: 219–230.

Picozzi, N. 1984. Breeding biology of polygynous Hen Harriers *Circus circus cyaneus* in Orkney. *Ornis Scandinavica* **15**: 1–10.

Rowley, I., E. Russell, and M. Brooker. 1993. Inbreeding in birds. Pages 304–328 in N. W. Thornhill, editor. *The natural history of inbreeding and outbreeding*. The University of Chicago Press, Chicago and London.

Russell, E. M., and I. Rowley. 1993. Philopatry or dispersal: competition for territory vacancies in the Splendid Fairy-wren, *Malurus splendens*. *Animal Behaviour* **45**: 519–539.

Sandell, M., and O. Liberg. 1992. Roamers and stayers: a model on male mating tactics and mating systems. *American Naturalist* **139**: 177–189.

Schlyter, F., J. Hoeglund, and G. Stromberg. 1991. Hybridization and low numbers in isolated populations of the Natterjack, *Bufo calamita*, and the Green Toad, *Bufo viridis*, in southern Sweden: possible conservation problems. *Amphibia-Reptilia* **12**: 267–282.

Signer, E. N., C. R. Schmidt, and J. Jeffreys. 1994. DNA variability and parentage testing in captive Waldrapp Ibises. *Molecular Ecology* **3**: 291–300.

Smith, H. G., U. Ottosson, and M. Sandell. 1994. Intrasexual competition among polygynously mated female Starlings (*Sturnus vulgaris*). *Behavioral Ecology* **5**: 57–63.

Soulé, M. E. 1985. What is conservation biology? *BioScience* **35**: 727–734.

Soulé, M. E. 1986. *Conservation biology: The science of scarcity and diversity*. Sinauer Associates, Sunderland, Massachusetts.

Soulé, M. E. 1987. Introduction. Pages 1–10 in M. E. Soulé, editor. *Viable populations for conservation*. Cambridge University Press, Cambridge.

Stacey, P. B., and W. D. Koenig, editors. 1990. *Cooperative breeding in birds: Long-term studies of ecology and behaviour*. Cambridge University Press, Cambridge.

Starfield, A. M., J. D. Roth, and K. Ralls. 1995. 'Mobbing' in Hawaiian Monk *Seals (Monachus schauinslani)*: the value of simulation modeling in the absence of apparently crucial data. *Conservation Biology* **9**: 166–174.

Struhsaker, T. T., and T. R. Pope. 1991. Mating system and reproductive success: a comparison of two African forest monkeys (*Colobus badius* and *Cercopithecus ascanius*). *Behaviour* **117**: 182–205.

Sukumar, R. 1991. The management of large mammals in relation to male strategies and conflict with people. *Biological Conservation* **55**: 93–102.

Taborsky, B., and M. Taborsky. 1991. Social organization of North Island Brown Kiwi: long-term pairs and three types of male spacing behaviour. *Ethology* **89**: 47–62.

Travis, S. E., and C. N. Slobodchikoff. 1993. Effects of food resource distribution on the social system of Gunnison's Prairie Dog (*Cynomys gunnisoni*). *Canadian Journal of Zoology* **71**: 1186–1192.

Trumbo, S. T. 1992. Monogamy to communal breeding: exploitation of a broad resource base by burying beetles (*Nicrophorus*). *Ecological Entomology* **17**: 289–298.

Weatherhead, P. J., F. E. Barry, G. P. Brown, and M. R. L. Forbes. 1995. Sex ratios, mating behavior and sexual size dimorphism of the Northern Water Snake, *Nerodia sipedon*. *Behavioral Ecology and Sociobiology* **36**: 301–311.

West-Eberhard, M. J. 1989. Phenotypic plasticity and the origins of diversity. *Annual Review of Ecology and Systematics* **20**: 249–278.

Wibbels, T., F. C. Killebrew, and D. Crews. 1991. Sex determination in Cagle's Map Turtle: implications for evolution, development, and conservation. *Canadian Journal of Zoology* **69**: 2693–2696.

Wilson, E. D. 1988. *Biodiversity*. National Academy Press, Washington.

Yanega, D. 1993. Environmental influences on male production and social structure in *Halictus rubicundus* (Hymenoptera: Halictidae). 1993. *Insectes Sociaux* **40**: 169–180.

Zabel, C. J., and S. J. Taggeart. 1989. Shift in Red Fox, *Vulpes vulpes*, mating system associated with El Niño in the Bering Sea. *Animal Behaviour* **38**: 830–838.

Zerbek, N. A., and R. W. Butler. 1981. Cooperative breeding of the Northwestern Crow, *Corvus kaurinus*, in British Columbia. *Ibis* **123**: 183–189.

Chapter 12

Linking environmental toxicology, ethology, and conservation

EDMUND H. SMITH AND DENNIS T. LOGAN

Some level of realized or proposed anthropogenic stress threatens most, if not all, ecosystems today. Human societal and economic forces affect even those sites that have been preserved specifically to protect biodiversity and natural habitats. The anthropogenic stresses that threaten ecosystems include climate change; habitat destruction; direct mortality from hunting, fishing, and withdrawing water for industry and water supply; changes in non-toxic chemical levels from discharges to the atmosphere (e.g., nitrogen and acidic deposition), land (e.g., farming), and surface waters (e.g., sewage); and the introduction of toxic chemicals such as pesticides, metals, and synthetic organic compounds. As human societies have increasingly realized their dependence on and integral part in nature, they have sought to preserve and restore natural ecosystems. Conservation biology and restoration ecology grew in response.

Pollutants are often a source of ecological stress, and environmental toxicology is the discipline that identifies the types and levels of pollutants that potentially stress ecosystems. Early ecotoxicity tests frequently exposed laboratory fish or invertebrates to chemicals for one or several life cycles. Acute responses such as mortality and chronic responses such as decreased reproduction provided bases for inferring ecological relevance. Modern toxicity tests are shorter (typically 96 hours) because the labor and costs of such tests are high and the number of pollutants and effluents to be tested is large. Responses may also include processes such as fertilization rate that are further removed from ecological processes, and environmental toxicologists are continually pressed to demonstrate ecological relevance in their testing.

Use of behavioral endpoints, particularly in combination with electronic monitoring and computerized analysis, now provides greater sensitivity – behavioral responses usually occur at concentrations below those affecting survival or reproduction – and shorter response times (typically minutes to hours) than conventional testing. Where pollutants threaten ecosystems, behavioral toxicology holds promise as a powerful tool for restoration and conservation if tests of the simple behavioral responses required in the laboratory can be linked to ecological processes in nature. Fulfilling this promise requires the multidisciplinary efforts of environmental toxicologists, ethologists, restoration ecologists, and conservation biologists.

Despite similar roots, goals and problems, environmental toxicology, restoration ecology, and conservation biology seem to be evolving separately. The increasing isolation of these fields is cause for alarm as we come to realize that answers to many environmental and ecological questions transcend the boundaries of traditional disciplines (Brown 1995). The purpose of this chapter is to provide ethologists, restoration ecologists and conservation biologists with a perspective on how one aspect of environmental toxicology – behavioral toxicology – may be used successfully in restoration and conservation projects. In doing so, we hope to stimulate thought about how the gaps among the disciplines can be bridged. We review the applications, limitations and methods of behavioral toxicology, particularly in regard to one of the most highly studied areas, aquatic toxicology. We discuss some issues behavioral toxicologists have faced to gain acceptance by regulatory agencies, site managers, and the courts for standardizing methods and assuring data quality. The lessons learned from these experiences may be useful to those in other disciplines who use or hope to use behavior in their studies and management plans. Lastly, we present a strong advocacy for increasing and improving the communication between restoration ecology, conservation biology, ethology and behavioral toxicology. Collaboration and exchange of both ideas and methods are necessary if we are to solve the complex environmental problems now facing us.

The role of behavioral toxicology at restoration and conservation sites

Ecology and its subdisciplines of restoration biology and conservation biology have populations and ecosystems as their domain. In contrast, environmental toxicology is concerned with responses to chemicals at the organism or suborganism level when organisms are exposed in the field or, more commonly, in controlled experimental conditions. This focus on the organism or even suborganism level may appear to separate environmental toxicology from the ecological disciplines. Yet it is the potential ramifications of anthropogenically-induced behavioral changes on natural intra- and interspecific interactions that connect behavioral toxicology and ecology. Changes in feeding behavior, for example, may translate to changes in energy budgets and energy flow through a food web or in the accumulation of energy reserves that affect survival of a population during starvation. Likewise, the impairment of behavioral function can have immediate consequences for fish populations, such as range limitation or extirpation from reproductive failure or mortality (Lipton *et al.* 1996).

The challenge in considering ecotoxicological methods for restoration and conservation is to discern how to apply those methods to predict or reveal population or higher level impacts. The link must be made that understanding behavior through small-scale studies of individuals or groups of individuals can reveal not only stressful conditions but also biotic interactions that may affect population structure and dynamics. When that link is made, protocols, standards, and operational procedures developed from toxicological studies can be applied to issues in conservation biology. Behavioral toxicology can play an important role in conservation biology in many ways. For example, behavioral toxicology can: (1) provide part of the multidisciplinary approach to problem solving that is so necessary in understanding complex systems; (2) help identify stressors that may cause a loss in biodiversity; (3) promote conservation compromises; (4) provide insight into habitat management issues; and (5) provide data on the actual and potential effects of human exploitation of unexploited systems. The following section discusses factors that ethologists interested in toxicological research need to contemplate when considering behavioral tests for use in restoration and conservation.

General considerations for behavioral testing

Behavioral responses are important to survival because they perform essential life functions. These include habitat selection, competition, predator avoidance, prey selection, and reproduction (Little *et al.* 1985). Behavioral toxicity testing is based on the premise that exposure to chemical or other stressors can induce behavioral responses that exceed the normal range of variability (Marcucella & Abramson 1978). Toxicity may elicit many forms of behavioral change, and knowing only that a species responds in a certain way to a certain chemical concentration is insufficient for restoration and conservation. Extrapolation of behavioral test results to questions of survival of a population or to the effects on ecosystems is also necessary, although difficult (Miller 1980).

Olla *et al.* (1980) proposed that any significant change from the norm induced by a perturbation is deleterious. They reasoned that if the behavioral response removes or lessens the effect of the perturbation, the probability of death or the energetic cost incurred by other adaptive responses related to maintaining homeostasis may be lowered or eliminated (Olla *et al.* 1980). Whether that proposition proves to be true, the link between test results and ecological effects is required in conservation and restoration. To that end, the design of the behavioral test protocol should incorporate behavioral data for single species, relate the test approach to survival of the species in its environment, and relate the results of the tests to populations or communities (Smith & Logan 1993). We discuss each of these issues in the following sections.

Design of behavioral tests

Species selection

The selection of the best species to represent the habitat that is or will be exposed to a stressor is a critical first step in the design of a behavioral test. There is no 'standard' behavioral test species. Selection depends on behavioral traits that can be easily measured (e.g., locomotion or change in position), local conditions in the study area, and the availability of representative species.

Representative species should fit the following criteria (Rand 1985; Smith & Logan 1993): (1) the species should represent the habitat under stress; (2) the species should be easy to acquire and maintain; (3) the species should be sensitive to a wide range of stressors; (4) the behavioral data base on the species should be extensive enough to determine consistency in response to known stress and to understand particular behavioral patterns (e.g., locomotion, feeding) under normal and stressed conditions; (5) the individuals should be small (e.g., juveniles) to reduce the exposure chamber size, to allow adequate replication, and perhaps to increase sensitivity, as the young of a species are often more sensitive than adults; (6) the species should be active and in motion all the time (especially in locomotor studies where stop and start movements increase response variance and decrease the chance of detecting a response). In addition, for use in conservation and restoration, the species should (7) have habits or niches that potentially expose them to the stressor. Where multiple species are to be tested, the species should represent different taxonomic groups, trophic levels, or ecological assemblages to provide the greatest potential diversity in sensitivity and response.

The selection of test species may entail some level of behavioral research and observation to establish the critical behavioral database (item 4) before the actual testing begins, although control response can be observed during testing. In addition to the items above, selection of a behavioral test species depends on the objectives of the study. For testing avoidance or preference to a stressor, indigenous species may be appropriate for generating site-specific data (Rand 1985). But if the objective is to detect whether the stressor affects the general behavioral characteristics of the aquatic organisms (e.g., the stressor may be a narcotic that generally suppresses the behavioral response), then non-endemic species may be used if endemic species cannot be maintained easily in the laboratory. In such cases, surrogate species should be as taxonomically and ecologically close to endemic species as possible, and the uncertainty introduced by using surrogate species should be discussed along with the test results.

Acclimation time and conditions

Whether test organisms are collected from the field, cultured in the laboratory, or purchased from a reliable dealer, it is very important that they are hardy and viable (Smith & Logan 1993). The transfer from field to laboratory holding conditions or long-term maintenance in holding conditions can result in poor condition of the test organisms, which may cause an unusual response. Careful observation of the condition of test organisms before testing can help assure reproducible and representative test results. To help ensure that laboratory responses of the organism tested are representative of responses for the species, test organisms should be maintained in the laboratory under test conditions that are as close to their natural habitat as possible.

Acclimation before behavioral testing is very important, and the criteria used to establish that acclimation has occurred should be reported. Acclimation periods may vary considerably and range from a few minutes to many days, depending on the species. Some fish have been acclimated for several weeks to several months, while some crabs have been acclimated for one to two hours (Kittredge *et al.* 1974) and daphnids for up to five days (Smith & Bailey 1989). The observed differences between fish and invertebrate acclimation times may be due more to differences in life cycle and physiology of the species observed than to any phylogenetic differences between fish and invertebrates.

Smith & Bailey (1988, 1989) provide an example of using behavioral criteria to determine acclimation to test conditions. They established the acclimation periods for invertebrates and fish from one-month observations of test chambers using a continuous video-computer monitoring system. The movement functions measured included swimming speed, rate of change of swimming direction, and changes in the position of the fish within the test chamber. The resulting long-term behavioral data base was compared with responses of naive invertebrate or fish species placed in the test chamber. The criterion for acclimation was that the behavior of the test organisms did not vary significantly from control data.

Response time and sensitivity

Behavioral testing can provide rapid and sensitive responses to many contaminants compared with the endpoints used in conventional toxicity testing. In studies of the behavioral response time for four aquatic species, Smith & Bailey (1988) found the detection time for phenol (0.7 mg l^{-1}) by Rainbow trout (*Oncorhynchus mykiss*) to be about 11 minutes, while Steelhead (anadromous *O. mykiss*) and Salmon responded after seven minutes of exposure. *Daphnia* spp. detected the presence of 0.43 mg l^{-1} phenol within 1.5 minutes of exposure. Phenol is a well-studied chemical in conventional toxicity tests, although it does not usually elicit avoidance even at acutely lethal concentrations (Giattina & Garton 1983). For comparison, a typical toxicologic test procedure requires between 48 and 96 hours to complete (sensitive chronic tests may be longer), and the mean acute and chronic values for Fathead minnow (*Pimephales promelas*) are 36.0 and 2.56 mg l^{-1}, respectively (USEPA 1980). The use of behavior as an early warning system for the protection of drinking water removed from surface waters has gained much attention in recent years (Gruber & Diamond 1988) because of the early response time to stressors.

Besides being quick, behavioral response can be sensitive, and detection of low contaminant concentrations has been reported often in the literature. For example in salmonids, olfactory attractants were detected at levels of 10^{-4} M for amino acids and of 1:400 dilution (test material:diluent) for krill extract (Smith & Bailey 1989). Striped bass (*Morone saxatilis*) showed strong attraction to oil refinery effluent as low as 1:100.

Rhythms in behavior

Rhythms occur commonly in animal behavior and must be considered in toxicological testing. For example, Fingerman *et al.* (1979) used the periodic increases in undirected locomotive behavior in fiddler crabs as an indicator of rhythmic patterns. The Common goldfish (*Carassius auratus*) has been shown (Spencer 1939) to be markedly diurnal, and the rhythm persists in constant conditions for longer than rhythms in most fish species. Some rhythms, such as those involved in animal locomotion, are readily observed, while others, such as daily or annual rhythms, require long-term observations before they become apparent. Biological rhythms may be driven by a biochemical clock, which is endogenous in the sense that the rhythmic mechanism is independent of external events, or they may be exogenous, in the sense that they are cued by and entrained to rhythmic external stimuli, such as fluctuations in light intensity. Endogenous rhythms may arise because of the physiological or mechanical organization of the organism.

The importance of measuring the type and magnitude of invertebrate rhythms becomes evident when tests are developed to detect lowest observable effect levels (LOELs) of response. When not taken into account, rhythms add to the unexplained variation, which may or may not be affected by the introduction of a stressor. Unexplained variation decreases the probability of detecting the real effects of the stressor. Cyclic behavioral and physiologic rhythms may also affect the response level to environmental stimuli and toxic perturbations of the test organisms (Olla *et al.* 1980). For example, rhythms in the closing of mollusk valves (Davenport & Manley 1979) or the daily activity rhythms of the Limpet (*Patella vulgata*) (Dicks 1976) may cause differential exposure and susceptibility to stressors during the rhythmic cycle. Limpets are nocturnal feeders. During the day, they pull their shells tight against the substratum, but during the night, they raise their shells (Dicks 1976), which can increase their exposure to some stressors. Rhythms themselves may be affected by stressors. Although most invertebrates and many fish exhibit some form of behavioral rhythms (e.g., circadian, tidal, lunar, daily, seasonal), many investigators conducting behavioral toxicologic tests have not determined whether rhythms occur and, if so, the type and the magnitude (Smith & Logan 1993).

Endpoint selection

The presence of a strong stressor may elicit a variety of responses. There are three basic steps in response to potential toxicant exposure (Olla *et al.* 1980; Rand 1985): (1) the stressor is sensed; (2) it is recognized as harmful or beneficial; and (3) some form of response is taken (e.g., avoidance or attraction). Organisms may not sense a chemical that is either new to the organism or not similar to other substances found in the environment (Slobodkin & Rapoport 1974). Some stressors, such as carbon monoxide, may not be sensed at all. If the chemical is sensed and resembles a beneficial substance that is familiar to the organism or acts as an attractant, the organism may not take evasive action even though the chemical ultimately may be toxic (Smith & Bailey 1990). In fish, the first forms of response may be locomotor changes. Because fish may be irritated by a stressor or may attempt to escape exposure, initial reactions may include changes in orientation to water flow, speed of movement, and preferred location (Smith & Bailey 1988). As these sensory reactions subside, more severe aberrations in locomotor behavior may occur as the toxicant saturates the detoxifying enzymes or alters neurochemical or metabolic processes (Drummond *et al.* 1986; McKim *et al.* 1987).

It is important to bear response levels in mind when selecting endpoints for the behavioral test (Smith & Logan 1993). If the goal of the test is to determine the threshold of recognition of a stressor by the organism, then endpoints such as antennular flicking rate or increased shell closure rates (Dunning & Major 1974) may be selected for invertebrates, whereas altered swimming movements (surfacing, resting on the bottom or increase in speed) might be selected for fish (Little & Finger 1990). If the test is to decide the species' ability to mitigate exposure, then the endpoint may include avoidance/preference testing, strong locomotor response, feeding alterations or interference of the reproductive process.

Methods for behavioral tests

Movement functions

Swimming capacity and swimming activity are two general measures of swimming behavior commonly used to assess contaminant-related changes in locomotion (Little & Finger 1990). Swimming capacity is a measure of orientation to water flow (Dodson & Mayfield 1979) and the physical capacity to swim against it (Howard 1975). Swimming activity includes such variables as frequency and duration of movements (Cleveland *et al.* 1989), speed and distances traveled during movement (Miller 1980), frequency and angle of turns (Rand 1977*a*,*b*), position in the water column, and form and patterns in swimming. With the arrival of computer-assisted analysis of video images, movement functions can be recorded, analyzed statistically, and plotted. Batschelet (1965) and Smith & Logan (1993) summarize and define some movement functions used in current computer applications. Linear velocity or speed is a fundamental variable for the analysis of translocation movement, while angular velocity (or rate of change of direction) is the first derivative of continuous direction of travel with respect to direction. Linear acceleration is the first derivative of linear velocity of an organism's path with respect to time. The index of path characterization is a ratio between the distance in a straight line from the first point in the organism's path to a given point and the actual distance traveled by the organism from the first point of the path to the given point. Cartesian positional coordinates include x and y (typically horizontal) and z (vertical) in three-dimensional applications. No numerical value or range of values has been defined as the norm for swimming activity of any aquatic species; therefore, detection of abnormal activity is based on comparisons of responses of exposed fish with activity measured during either a baseline or re-exposure period or in a control treatment (Little & Finger 1990). The uses of movement functions for both invertebrates and fish play important roles in following behavioral changes in toxicological studies, and these are referred to throughout this chapter.

Undirected and directed behavioral responses

The movement of aquatic organisms may involve spontaneous 'undirected' (free-running) locomotor activity characterized by circadian and seasonal rhythmicity (treated earlier in this chapter) and not related to the direction of a stimulus. In this type of behavior, the direction of movement at any instant may be completely random. Movement may also involve orientation or 'directed' locomotor responses to gradients of natural stimuli such as light, temperature, salinity, substratum, chemicals such as pheromones or food odor, water currents, and man-made stimuli such as chemical toxicants (Rand 1985). A wide range of test methods, endpoints, and species have been used for testing the behavioral response to toxicants in water. For example, Davenport & Manley (1979) selected shell closure in mussels as an endpoint for copper exposure, although they found it difficult to relate the avoidance (shell closure) to survival in the field. On the other hand, migratory behavior of *Daphnia* spp. to a light source was disrupted by lindane, an insecticide that affected the physiological mechanism responsible for directing migration (Goodrich & Lech 1990). Lindane is used primarily on phytophagous and soil-inhabiting insects, public health pests and animal ectoparasites and has been applied in seed treatment, on livestock and on hardwood lumber (USEPA 1992). Strong avoidance, an increase in swimming activity, and a change in swimming direction by *Daphnia magna* when exposed to phenol, copper sulfate, and xylene – chemicals commonly used to standardize toxicity tests – were recorded using the protocols and methods described by Smith & Bailey (1989). Rainbow trout that were exposed to dioxin (29 pg l^{-1}), a very strong carcinogen, for 28 days exhibited a reduced frequency of activity, a change in swimming posture, and a change in position in the water column (Mehrle *et al.* 1988). For comparison, a 96-hour exposure for rainbow trout fry to 10 pg l^{-1} dioxin caused both histopathological and teratogenic effects at day 72 post exposure, and 6-hour exposures of immature individuals to 107 pg l^{-1} caused some deaths beginning at day 78 post exposure (Eisler 1986). These studies show that both undirected and directed behavior can be used to measure response to low levels of toxicants in water.

Avoidance/preference tests

Several experimental approaches have been used to determine the reaction of aquatic organisms to contaminant gradients (Larrick *et al.* 1978; Smith & Bailey 1990). Generally, the organism is given a choice between contaminated and uncontaminated areas, and the endpoint is the observed choice. These tests evaluate the ability of the organism to sense the presence of a stressor and respond by moving either toward or away from the stressor or taking no action. If the organism does not detect the presence of a toxicant or does not sense danger, no response of avoidance or preference may occur even though the stressor may affect survival. Many chambers and test protocols have been used for avoidance/preference studies including Y-tube or maze systems (Giattina & Garton 1983). Some authors (Smith & Bailey 1988, 1990; Benfield & Aldrich 1991) have developed laminar-flow dichotomous choice chambers to avoid turbulent mixing and concentration fluctuations inherent in most Y-maze designs. A large selection chamber, unencumbered with maze walls and monitored by a video camera simplifies the automated recording of long-term data. These data can be either analyzed in real time or evaluated from video tape by computer (Smith & Bailey 1989). Often the avoidance/preference chambers are used to test conditioned learning – a sensitive indicator of chemically induced stress in aquatic organisms. Learning is a critical element in many behavioral responses (e.g., territoriality, defense, or migration), and it has potential ecologic significance for the general behavioral activities of organisms in the natural environment (Rand 1985).

Avoidance/preference testing was used to investigate the potential of industrial waste discharges to interfere with the migration of anadromous fish in the San Francisco Bay and Delta system. Striped bass and Chinook salmon, *Oncorhynchus tshawytscha* (now an endangered species), were used to assess the potential effect of the effluents from two oil refineries discharging into the northern section of San Francisco Bay (Smith & Bailey 1990). Salmon are very sensitive to the oil refinery waste, and the results clearly indicated a strong preference for the effluent at all levels of dilutions tested. At the lowest level (1:1000 dilution), the response appeared to occur before the effluent had reached the designed dilution level within the test chambers, the initial time from which response is measured. Although adult migrating fish were not tested, some field data on adults were collected from each discharge area. At one refinery, adult salmon were found in the discharge channel, an area that does not connect to any present or historic spawning area. This suggests that the salmon were attracted to the discharge rather than the natal stream where they would normally spawn. At the other refinery, striped bass found near the site of the discharge channel had abnormal fin loss. There is strong indication that the loss or damage to fins found on juvenile striped bass in the field and duplicated in the laboratory may have been caused by contact with the effluent. In these cases, where the attraction to the effluent may have resulted in inappropriate selection of a spawning area or increased exposure to deleterious contaminant concentrations, attraction may impact on the populations. The results of these behavioral tests together with long-term bioassay data (21-day flow-through tests) and field studies were used to set discharge limits at the refineries. Now that winter chinook salmon are an endangered species and striped bass populations in the San Francisco Bay are declining, conservation efforts must consider not only over-harvesting but also cumulative impacts of many bay and delta discharges as major causes of the declines of fish populations.

Acidity is an example of an environmental change other than pollutants that can be measured and tested using behavior. Changes in the pH of natural waters can occur under many sets of conditions encountered by site managers. Low pH can result from acidic precipitation, discharges, acid mine waste, excessive secondary productivity, and other causes. Avoidance/preference testing was used to elucidate potential impacts on fish migration caused by exposure to changes in pH and salinity. Buckler *et al.* (1995) demonstrated the effect on behavior of low pH on Atlantic salmon (*Salmo salar*). After a 60-day exposure to pH 6.5 and lower, spontaneous swimming of test fish was significantly less than that of controls. Prey strike frequencies of fish exposed to pH 5.0, 5.5, and 6.5 were also significantly reduced and ranged from 57 to 60% of control fish strike frequencies. The strike frequencies of fish exposed to pH 4.5 were inhibited most severely: 45% of control fish response. The authors suggested that the impairment of normal behavioral functions of fish exposed to acidic conditions compromised their ability to adapt and survive in natural systems. Proposed changes in land management in adjacent watersheds or upstream of the protected site might be tested using the very sensitive avoidance/preference protocols prior to implementing any management alterations.

Sediment behavioral tests

The most common behavior tests used to assess the potential toxicity to sediments are avoidance and burrowing (Swartz 1987). The selection of important behavioral endpoints for sediment testing is often difficult because of the limitations of direct observation, knowledge of the test species' behavior in the test sediment type (which may differ from the preferred sediment type), and the difficulty of evaluating the results in relationship to the survival of the organism in nature. Amphipods have often been used to test contaminated sediments from sewage outfalls and toxic sediment collected from large industrial areas. Test organisms typically burrow into natural sediments. Reburial of exposed organisms following introduction to clean sediments and the emergence from toxic sediments during laboratory tests have been used to characterize sediments from selected sites (Swartz *et al.* 1985). Field abundance records of the amphipod *Rhepoxynius abronius* around test sites correspond closely with the behavioral test results in that amphipod densities are low where behavioral tests indicate that the amphipods do not remain buried in sediments. Preliminary progress has been made in relating behavioral response data to field conditions and the survival of the exposed organisms, although much more work needs to be carried out. Swartz (1987) pointed out the importance of relating the functional significance of sediment sources and predators.

Predator prey interactions

Many morphological traits and behavioral mechanisms have evolved among aquatic organisms to reduce the probability of detection and lessen the chances of capture (O'Brien 1979; Little *et al.* 1985). Exposure to stressors may increase an organism's vulnerability to predation through disruption of these defenses. Disruption of growth is an early indicator of chronic stress monitored in many conventional toxicological tests. Growth is important to survival, particularly in early life stages, as survival improves as an organism increases in size (Ware 1975; Logan 1985; Wahl 1992). Stresses that affect growth may therefore affect survival (Logan 1986), and this size-dependent mortality has been employed in modeling the effects of anthropogenic stresses on fish populations (Cowan *et al.* 1993; Rose & Cowan 1993).

Changes in feeding behavior have often been measured in toxicity tests, and such changes can affect growth and thereby affect survival. Stressors may impair the mechanisms predators employ for detecting prey. Visual recognition of form, size, movement, and contrast between organism and background are important in the predatory responses of visual feeders. Stressors can also alter the chemosensory responses to food or mask natural odors detected by the taste buds. For example, Bluegills (*Lepomis macrochirus*) were unresponsive when the food odors were presented simultaneously with parathion at 10 mg ml^{-1} (Rand 1977*b*). Acute toxicity levels for parathion, an organophosphorus pesticide, range from about 50 mg ml^{-1} for sensitive species such as bluegill to about 2.5 mg ml^{-1} for resistant minnows. One of the most sensitive chronic toxicity levels is the development of deformities in bluegills after a 23-month exposure at 0.34 mg ml^{-1} (USEPA 1986). Loss of the ability to detect food odor, lack of search behavior for food, and food rejection are all indications that there may be an alteration in chemoreceptors or some part of the central nervous system. Contaminants can cause the taste buds themselves to suffer structural degeneration (Bardach *et al.* 1965).

Prey selection depends on the predator's morphology, prior experience with the prey, the abundance and diversity of prey, and the competition among species. The influence of stressors on prey selection has received little study (Little *et al.* 1985). Fish exposed to a contaminant are often unresponsive to familiar prey. Predators must have sufficient swimming capacity to pursue and overtake their prey, and swimming speed is one of the movement functions that can be monitored in behavioral testing. Prey exposed to stressors also loose the ability to avoid capture by unaffected predators (Little *et al.* 1993*a*). More research is needed on the effects of stressors on predator–prey and other species interactions and particularly on field validation of conclusions reached from laboratory studies.

Extrapolation of laboratory results to ecosystem effects

Although the behavioral toxicological literature clearly shows that different types of behavior are altered by exposure to stressors in the laboratory, questions remain as to whether these behavioral effects occur and are ecologically meaningful in the natural environment and whether the behavioral changes affect intra- and interspecific species dynamics and interactions. One of the most thorough studies to address such questions includes both laboratory and field observations (Krebs & Burns 1977). In 1969, a spill of number 2 fuel oil provided an opportunity for a seven-year study of the effects of oil contamination of salt-marsh sediments on the population dynamics and behavior of a fiddler crab (*Uca pugnax*). The crabs were exposed to oil in both the sediment and the water. Behavioral observations were made in the field directly after the spill. Effects of the oil on the crabs were also studied in the laboratory by timing the escape responses from a standard stimulus. Field observations after the oil spill showed dead and moribund *Uca*. Many migrated from the oiled creek areas and burrowed into the sandy sediments above the mean tide line. Surviving crabs that remained in the oiled areas displayed aberrant locomotor and burrowing behavior. Adults also showed physiological disorders, such as increased molting and display of mating colors at the wrong time of the year. The normally rapid escape response was slow, and crabs appeared lethargic and showed loss of equilibrium.

Laboratory experiments supported the field observations. The escape response times of crabs ingesting oil sediments more than doubled. Impaired activity, loss of equilibrium, immobilization, and death were observed with exposure to increasing concentrations of toxicant. Slowed escape responses may have also left the crabs more vulnerable to predation (Ward *et al.* 1976).

Population data revealed that *Uca* densities were reduced for at least seven years. The impairment of locomotor ability and other behavior accounted for the persistent reduction in fiddler crab populations. The ecological significance of reduction of fiddler crab populations by oil contamination is that it affects the energy flow within salt marshes and between marshes and coastal waters.

In another study, Little *et al.* (1993*b*) found that the impact of pulsed exposures of an insecticide (esfenvalerate) on bluegill survival in the laboratory was in close agreement with mortality estimates measured in realistic field exposures (Fairchild *et al.* 1992). The use of behavioral end points, together with standard measurements of impacts on growth and survival, can provide estimates of toxicity that are protective of fish populations in the field (Little *et al.* 1993*b*).

As Rand (1985) pointed out there are two important aspects in the extrapolation of laboratory data to effects in the field: (1) laboratory studies of behavioral toxicity with aquatic organisms may provide more meaningful results when field studies are also conducted; and (2) physical, chemical or biological characteristics in the natural environment may mitigate or enhance the potential behavioral toxicity of stressors. There is no certainty that at any field site, an observed impact (or lack of one) can or must always result from effects, phenomena or mechanisms observed in the laboratory. Matches between laboratory and field responses do occur (as pointed out above) but these may not necessarily 'validate' either approach. Most behavioral toxicology testing uses single species, so much research is needed on development and field validation of multispecies tests. Extrapolation to the population level has met with more success than extrapolation to community or ecosystems levels. The following example illustrates how behavioral toxicity testing has provided population-level information used in a restoration and management plan.

Example: A marsh restoration project

Along the northern shore of San Francisco Bay, over 2000 acres of altered marsh have been used as salt evaporation ponds for 70 years. The State of California recently purchased the land with the intention of restoring it as wetlands. The goals of the site management plan include maintaining part of the established biodiversity supported by highly saline ponds that will not be drained, while discharging large volumes of salt from other ponds designated for restoration without adversely affecting the ecology of receiving waters – the bay and Napa River. The proposed restoration plan includes draining designated salt ponds to allow them to return to salt water and estuarine marsh.

Salt was produced by evaporating bay water in a series of large open-air ponds. Salinity in the ponds is typically high, and three levels of concentration are called 'crystal solids,' 'bittern,' and 'pickle.' Although the salinities of these evaporates may be well above the levels that most estuarine and bay organisms can tolerate, such species normally inhabit salinity gradients and would be expected to sense and avoid potentially threatening salinities. The potential discharge sites of the saline water to the river or bay support species eaten by juvenile Striped bass (*Morone saxatilis*), and these forage populations would be reduced if discharges caused toxicity or elicited avoidance. Both the Napa River and the bay are migration corridors for anadromous fish such as Striped bass, Steelhead, and salmon, including the endangered Winter chinook salmon. Avoidance of high salinity discharges could hinder normal migrations and adversely affect these populations.

Before the State purchased the salt ponds, a behavioral toxicity study was conducted to determine whether the toxicity of the saline discharges was high, in which case the transfer of the property from private ownership to the state as a marsh restoration project would be in danger, or low, in which case the State could allow the discharge and the sale. In the latter case, the levels of discharge that might cause stress to the species in the receiving waters had to be determined. The major objectives of the resulting avoidance/preference study were to assess the potential for the discharge of various concentrations and mixtures of salt to impair the upstream and downstream migration of Striped bass and to evaluate the possible acute and chronic effects of the discharge components on salmon and a resident species of shrimp (*Crangon* spp.).

Avoidance/preference and locomotor responses were selected as endpoints because they represent aspects of behavior that may affect the movement or feeding of the populations in the river or bay. Only juvenile fish were tested, and test exposures lasted about an hour. Laboratory behavioral tests on Striped bass showed that exposure to 1% and 5% dilutions of crystal resulted in a statistically significant increase in swimming speed. The fish were attracted to a 1% dilution of crystal, although at 5% dilution, no attraction or avoidance was evident. The fish avoided a 1% dilution of bittern fraction. Using standard bioassay procedures, this dilution of bittern caused a reduction in growth and some mortality after five days of exposure. The difference in response time for behavioral and conventional tests was evident. The information from behavioral and conventional toxicity testing were used to develop a restoration plan and management program. The cooperative work between the behavioral toxicologist and conservation site manager allowed the State to modify the restoration plan to protect biodiversity both on-site and in nearby receiving waters by diluting the discharge and spreading the release of saline waters over high winter flow periods during which dilution would be most effective and concentrations would not reach levels that would alter behavior.

There is always concern in both conventional and behavioral toxicology about possible disparity between field and laboratory results and between adult and juvenile response. These concerns apply here because these studies were conducted on only juvenile fish in the laboratory. Although the laboratory studies were not validated by field work in this project, Smith & Bailey (1989, 1990) tested the response of juveniles to other effluents in San Francisco Bay and observed that adult fish in the field show similar avoidance or attraction to the discharges. Their results helped show that laboratory behavioral toxicity test results in this study could be extrapolated to help protect wild populations.

Lessons from aquatic toxicology

In spite of the previous example, the use of behavior as a tool to understand environmental stress is often met with doubt by biologists unfamiliar with the application, methods and reliability of data analysis used in behavioral testing. This reluctance to use the sensitive behavioral tests is even more pronounced in multidisciplinary studies where ecologists, engineers, site managers, and regulatory agencies are involved. The acceptance of behaviorists, therefore, on restoration or conservation teams to assist in site assessment, to identify and predict possible impacts from stressors, and to participate in restoration and management planning is problematic. Behavioral toxicologists have begun the following two-stage effort to change the image of the discipline. They (1) educate scientists, environmental managers, and the general public about behavioral testing; and (2) review the processes within the discipline, attempt to bolster areas of insufficient research, and change many of the ways behavioral tests are conducted. Many of these same problems are also faced by more traditional behaviorists, and perhaps through cooperative efforts, both disciplines can share in the solutions.

One problem with behavioral results in aquatic toxicology is that for many stressors results are difficult to compare and evaluate because so many different techniques and test species have been used (Rand 1985). As a result, many regulators and site managers and some biologists believe that behavioral testing cannot be repeated and that the same data can be interpreted in many ways. These objections result in part from the lack of standard test protocols. Little *et al.* (1993*b*) called for more and better behavioral testing procedures, especially for using sensitive early life stage fishes. These new methods must be developed and refined into effective tools that can be readily and clearly applied in stress analysis. Based upon this need and the urging of behavioral toxicologists, the American Society of Testing and Materials (ASTM) has established a subdivision of the E−47 section of the society dealing with environmental standards. Several guidance documents have been presented and approved by the Main Committee of the Society (e.g., ASTM 1994). A great deal of work remains in writing, reviewing and seeking approval of the various standards subcommittees within ASTM for behavioral protocols and testing procedures.

The acceptance of behavioral tests by regulatory agencies is becoming more common. In fact, under various USA federal regulations (e.g., the Comprehensive Environmental Response, Compensation and Liability Act, CERCLA ; the Oil Pollution Act, OPA) behavioral avoidance is recognized as a valid test procedure. The behavioral tests have been used under CERCLA (Section 43 CFR Part 11) to establish damage to natural resources. Behavioral test results have been applied to Natural Resource Damage Assessments (DeLonay *et al.* 1996), most of which can find their way to the court room. Natural resource damage assessments performed under CERCLA and OPA have the restoration of injured natural resources as the ultimate objective (Lipton *et al.* 1996), and the testing should have relevance to the development of restoration goals, objectives, and approaches.

Along with the development of standard test protocols and the growing acceptance of behavior as a valid test procedure by regulators, quality assurance/quality control (QA/QC) protocols must be developed. Standard operating procedures are very important to ensure that data are collected systematically, that all equipment is calibrated, that staff are properly trained, and that each step in the laboratory procedures is recorded and checked. These steps are particularly important when litigation is involved. Equally important is repeatability established through intra- and interlaboratory testing. Very few intralaboratory behavior tests have been conducted, most likely because no standard procedures of the testing protocol have been developed. Lipton *et al.* (1996) is one of the few studies to have reported intra- and interlaboratory calibration of the test behavioral protocol used in a natural resource damage assessment. These authors found very good reproducibility between laboratories that applied the same test procedures and organisms and followed well established QA/QC protocols.

While behavioral toxicology has been used successfully under the regulatory mandate of various federal and state acts, the lessons learned by aquatic toxicology could well be adapted by conservation biologists who face the determination of a species under the USA Endangered Species Act. This act states that the determination of a species' status as threatened or endangered is to be based solely on the best scientific data. In addition, recovery plans are to provide objective and measurable criteria. All of the standardization of test methods, development of repeatable results through intralaboratory testing, and QA/QC procedures lead to not only good science but the acceptance of behavior as a respectable and necessary component of the conservation plan.

In short, with the development of new methods and increased interest in behavioral testing, the need for standardized methods has become increasingly important. Methods and procedures must be developed to: (1) produce repeatable results; (2) meet standardization requirements of regulatory agencies or review by a court of law; (3) embrace a wide variety of behavioral test species and test objectives; and (4) expand understanding of normal behavior for aquatic organisms – especially those used in conventional acute and chronic toxicity testing.

Advantages and disadvantages of using behavior in stress analysis

There are several advantages of using behavior to assess the response of organisms to stressors. These include quick response time, sensitivity to low levels of contaminants, demonstrable relationships between stressors and behavioral response, insight into relationships between stressors and population-level responses, information complementary to other disciplines and approaches to provide understanding about the ecological response to a stressor, and early indications of more serious complications from exposure to a stressor. There are also disadvantages of using behavior, although some issues listed below will most likely be resolved as more information and testing is conducted. The disadvantages include a relatively small behavioral database in which few species are well understood; a concentration on single species and single stressors; a lack of wide acceptance in conservation, habitat management, or toxicology; a need for more standardized observation protocols and operating procedures; and variability in response data that can make the interpretation of results difficult. Moreover, present political and financial pressures to reduce the perceived burden of environmental regulations on business may limit the frequency with which highly sensitive behavioral testing is required by regulatory agencies.

Several methods used in behavioral toxicology to evaluate the potential impact of biotic and abiotic stressors were presented in this chapter. Each method has both the advantages and disadvantages listed above. We recommend that the use of behavior to understand a species' response to a stressor should not be the only approach used for decision-making because of the state of the science. The longer-term acute and chronic tests should be integrated with behavioral observations in the laboratory and in the field, and data on abiotic factors should also be obtained from the study site.

Recommendations for future work

The discipline of behavioral toxicology has progressed technically to take advantage of new computer and video technologies and socially to be recognized, albeit to a limited degree, in federal regulations as a viable approach to determining the effects of stressors in natural habitats. Additional progress is needed to make behavioral toxicology even more useful to ecology. The following suggestions for areas for future work should help provide insight for restoration and conservation biologists and site managers into the current state of the discipline.

The progress should continue toward standardizing testing and operating procedures, apparatus and terminology, and quality control and assurance. The development of inter- and intralaboratory calibration and round-robin testing between laboratories should be expanded. Methodologies, particularly statistical methods of analysis, need improvement. A much better understanding of the relationships between laboratory responses and risk to field populations should be developed. To achieve this, better capabilities to observe behavior in the field while meeting the goals of standardization are needed. In conventional toxicity testing, early life stages are often the most sensitive, and in ecology they are important to population dynamics. Given this, the study of behavior of early life stages should be expanded. A better understanding of the role of rhythms in organisms and their possible effect on the organism's response to a stressor should be developed. Efforts to understand and use multispecies behavioral tests should be expanded. The behavioral data base on important species (whether for biological, societal, or political reasons) should be expanded.

There is a need for a closer association and exchange of information between behavioral toxicology and ethology. Sometimes behavioral toxicologists find themselves 're-inventing the wheel' if they are not trained as ethologists or do not follow the literature derived from other behavioral approaches. These researchers could benefit from the combined functional and causal types of explanations applied by ethologists. Ethologists emphasize field observations and have developed methods that would be useful to the behavioral toxicologist in understanding the implications of stress analysis on natural populations. In addition, terminology long applied in ethology could also be applied to behavioral observations used in the testing of organisms exposed to stressors. Ethologists and conservationists in turn can gain from the toxicologists' experience in government regulation and policy. The interface between ethology and behavioral toxicology becomes more important as both disciplines strive to understand their fields in the broad perspective of the individual, population, community, and ecosystem processes.

Conclusions

Behavioral toxicology, restoration ecology and conservation biology provide the scientific foundations when human societies seek to preserve or restore natural ecosystems. The range of such endeavors includes the conservation of endangered species and exploited natural resources, the determination of risks of populations exposed to various forms of stress, and the restoration of contaminated or altered environments. In practice, such endeavors are complex and may involve trade-offs. For example, the restoration of a disturbed ecosystem may mean the alteration or destruction of selected components of the biotic community. What kinds of information should be used to determine which components will be altered or destroyed to satisfy the driving forces of the restoration plan, forces that may be societal or political? Cairns (1991) provided a perspective that restoration might be defined as a resetting of an 'ecological clock.' A difficult decision in restoration is where to set the clock. Should it be: (1) the time at or before the disturbance to the ecosystem; or (2) the present time, but with an ecologically superior condition compatible with the landscape in which the damage occurred? Where the stressors that necessitated the restoration include contaminants, behavioral toxicologists can provide sensitive tests and endpoints that can help link the stressors to the ecological effects. Thus, the behavioral tests can help determine where to set the clock.

While many sites of concern to conservation biologists may not be contaminated, many are subject to anthropogenic stress. The foraging or migratory ranges of rare or endangered species to be preserved, for example, may extend beyond the boundaries of a conservation area into contaminated sites in need of clean-up or restoration. In such cases, preservation and restoration are essential partners in a comprehensive conservation strategy (Noss 1991), and multidisciplinary approaches to problem solving are essential. In this example, behavioral studies could be useful in understanding the impact of the existing anthropogenic stressors or in detecting potential stress gradients.

Will the application of techniques from behavioral toxicology allow restoration and conservation biologists to understand broad ecological issues at the community and ecosystem levels? Not at the present state of the discipline, for as Brown (1995) points out, most behavioral ecologists and toxicologists are reductionists, working at the level of the individual. Yet some ecologists are proposing that addressing community or ecosystem problems is artificial (Brown 1995) because the definitions of these higher-level systems are arbitrary and dependent upon the interest of the scientist working on them. Higher-level systems are often defined simply in terms of the organisms that occur together at some arbitrary study site for some arbitrary time period. While behavioral toxicology does not focus on these higher levels of organization, whether real or arbitrary, it can provide specific answers to well-defined problems and information relevant to understanding higher-level interactions. When the behavioral methods are combined with other toxicological and ecological approaches, the combination can make a powerful tool for the restoration and conservation biologist.

Literature cited

ASTM (American Society for Testing and Materials). 1994. Standard guide for behavioral testing in aquatic toxicology. E 1604–94. Pages 1510–1517 in *1994 Annual book of ASTM standards*. Volume 11.04. American Society for Testing and Materials, Philadelphia, Pennsylvania.

Bardach, J. E., M. Fujiya, and A. Moll. 1965. Detergents: effects on chemical senses of the fish *Ictalurus natalis*. *Science* **148**: 1605–1607.

Batschelet, E. 1965. *Statistical methods for the analysis of problems in animal orientation and certain biological rhythms*. American Institute of Biological Sciences, Washington, DC.

Benfield, M. C., and D. Aldrich. 1991. A laminar-flow choice chamber for testing the response of postlarval penaeids to olfactants. *Contributions to Marine Science* **32**: 73–88.

Brown, J. H. 1995. *Macroecology*. University of Chicago Press, Chicago, Illinois.

Buckler, D. R., L. Cleveland, E. E. Little, and W. C. Brumbaugh. 1995. Survival, sublethal response, and tissue residues of Atlantic salmon exposed to acidic pH and aluminum. *Aquatic Toxicology* **32**: 203–216.

Cairns, J. Jr. 1991. The status of the theoretical and applied science of restoration ecology. *Environmental Professional* **13**: 76–84.

Cleveland, L., E. E. Little, R. H. Wiedmeyer, and D. R. Buckler. 1989. Chronic no-observed effect concentrations of aluminum for brook trout exposed in low calcium diluted acidic water. Pages 229–246 in J. E. Lewis, editor. *Environmental chemistry and toxicology of aluminum*. Lewis Publishers, Chilsea, Michigan.

Cowan, J. H., K. A. Ross, E. S. Rutherford, and E. D. Houde. 1993. Individual-based model of young-of-the-year striped bass population dynamics. 2. Factors affecting recruitment in the Potomac River, Maryland. *Transactions of the American Fisheries Society* **122**: 439–458.

Davenport, J., and A. Manley. 1979. The detection of heightened sea-water copper concentrations by the mussel, *Mytilus edulis*. *Journal of the Marine Biological Association of the United Kingdom* **58**: 843–850.

DeLonay, A. J., E. E. Little, J. Lipton, D. F. Woodward, and J. A. Hansen. 1996. Behavioral avoidance as evidence of injury to fishery resources: applications to natural resource damage assessments. In T. W. LaPoint, F. T. Price, and E. E. Little, editors. *Environmental toxicology and risk assessment*. No. 4 ASTM STP 1262. American Society for Testing and Materials, Philadelphia, Pennsylvania.

Dicks, B. 1976. The importance of behavioral patterns in toxicity testing and ecological prediction. Pages 303–319 in J. M. Baker, editor. *Marine ecology and oil pollution*. Wiley, New York.

Dodson, J. J., and C. I. Mayfield. 1979. Modification of the rheotropic response of rainbow trout (*Salmo gairdeneri*) by sublethal doses of the aquatic herbicides diquat and simozine. *Environmental Pollution* **18**: 147–157.

Drummond, R. A., C. L. Russom, D. L. Geiger, and D. L. DeFoe. 1986. Behavioral and morphological changes in fathead minnows, *Pimephales promelas*, as diagnostic endpoints for screening chemicals according to modes of action. Pages 415–435 in T. M. Poston and R. Purdy, editors. *Aquatic toxicology*. Vol. 9. ASTM STP 921. American Society for Testing and Materials, Philadelphia, Pennsylvania.

Dunning, A., and C. W. Major. 1974. The effects of cold seawater extracts of oil fractions upon the blue mussel, *Mytilus edulis*. Pages 349–366 in F. J. Vernberg and W. B. Vernberg, editors. *Pollution and physiology of marine organisms*. Academic Press, New York.

Eisler, R. 1986. *Dioxin hazards to fish, wildlife, and invertebrates: A synoptic review*. US Fish and Wildlife Service Biological Report *85*(1.8). Patuxent Wildlife Research Center, Laurel, Maryland.

Fairchild, J. F., T. W. La Point, J. L. Zajicek, M. K. Nelson, F. J. Dwyer, and P.A. Lovely. 1992. Population, community, and ecosystem-level responses of aquatic mesocosms to pulsed doses of a pyrethroid insecticide. *Environmental Toxicology and Chemistry* **11**: 115–129.

Fingerman, S. W., C. Van Meter, and M. Fingerman. 1979. Increased spontaneous locomotor activity in the fiddler crab *Uca pugilator*, after exposure to a sublethal concentration of DDT. *Bulletin of Environmental Contamination and Toxicology* **21**: 11–16.

Giattina, J. D., and R. R. Garton. 1983. A review of the avoidance-preference responses of fish to aquatic contaminants. *Residue Reviews* **87**: 43–90.

Goodrich, M. S., and J. J Lech. 1990. A behavioral screening assay for *Daphnia magna*: a method to assess the effects of zenobiotics on spatial orientation. *Environmental Toxicology and Chemistry* **9**: 21–30.

Gruber, D. S., and J. M. Diamond. 1988. *Automated biomonitoring: Living sensors as environmental monitors*. John Wiley and Sons, New York.

Howard, T. E. 1975. Swimming performance of juvenile coho salmon (*Oncorhynchus kisutch*) exposed to bleached kraft pulpmill effluent. *Journal of the Fisheries Research Board of Canada* **32**: 789–793.

Kittredge, J., F. T. Takahashi, and F. O. Sarimana. 1974. Bioassays indicative of some sublethal effects of oil pollution. Pages 891–897 in *Proceedings of the tenth annual conference of the Marine Technical Society*. Washington, DC.

Krebs, C. T., and K. A. Burns. 1977. Long-term effects of an oil spill on populations of the salt marsh crab *Uca pugnax*. *Science* **179**: 484–487.

Larrick, S. R., K. L. Dickson, D. S. Cherry, and J. Cairns, Jr. 1978. Determining fish avoidance of polluted water. *Hydrobiologia* **61**: 257–265.

Lipton, J., E. E. Little, J. C. A. Marr, and A. J. DeLonay. 1996 (in press). Use of behavioral avoidance testing in natural resource damage assessment. In D. A. Bengtson and D. S. Henshel, editors. *Biomarkers and risk assessment*, Vol. 5. ASTM STP 1306. American Society for Testing and Material, Philadelphia, Pennsylvania.

Little, E. E., F. J. Dwyer, J. F. Fairchild, A. J. DeLonay, and J. L. Zajicek. 1993*b*. Survival of bluegill and their behavioral responses during continuous and pulsed exposures to esfenvalerate, a pyrethroid insecticide. *Environmental Toxicology and Chemistry* **12**: 871–878.

Little, E. E., J. F. Fairchild, and A. J. DeLonay. 1993*a*. Behavioral methods for assessing impacts of contaminants on early life stage fishes. *American Fisheries Society Symposium* **14**: 67–76.

Little, E. E., and S. E. Finger. 1990. Swimming behavior as an indicator of sublethal toxicity in fish. *Environmental Toxicology and Chemistry* **9**: 13–19.

Little, E. E., B. A. Flerov, and N. N. Ruzhinskaya. 1985. Behavioral approaches in aquatic toxicology: a review. Pages 72–98 in P. M. Mehrle, Jr., R. H. Gray, and R. L. Kendall, editors. *Toxic substances in the aquatic environment: An international aspect*. American Fisheries Society, Water Quality Section, Bethesda, Maryland.

Logan, D. T. 1985. Environmental variation and striped bass population dynamics: a size- dependent mortality model. *Estuaries* **8**: 28–38.

Logan, D. T. 1986. Use of size-dependent mortality models to estimate reductions in fish populations resulting from toxicant exposure. *Environmental Toxicology and Chemistry* **5**: 769–775.

Marcucella, H., and C. I. Abramson. 1978. Behavioral toxicology and teleost fish. Pages 33–77 in D. I. Mastofskey, editor. *The behavior of fish and other aquatic animals*. Academic Press, New York.

McKim, J. M., S. P. Bradbury, and G. J. Niemi. 1987. Fish acute toxicity syndromes and their use in QSAR approaches to hazard assessment. *Environmental Health Perspectives* **71**: 171–186.

Mehrle, P. M., D. R. Bucker, E. E. Little, L. M. Smith, J. D. Petty, P. H. Peterman, D. L. Stalling, G. M. Degraeve, J. J. Cayle, and W. J. Adams. 1988. Toxicity and bioconcentration of 2, 3, 7, 8-tetrachlorodibenzodioxin and tetrachlorodibenzofuran in rainbow trout. *Environmental Toxicology and Chemistry* **7**: 47–62.

Miller, D.C. 1980. Some application of locomotor response in pollution effects monitoring. *Rapports et Proces-Verbaux des Reu'nions Commission International pour l'Exploration Scientific de la Mer Mediterranee Monaco* **179**: 154–161.

Noss, R. F. 1991. Wilderness recovery: thinking big in restoration ecology. *Environmental Professional* **13**: 225–234.

O'Brien, U. J. 1979. The predator–prey interaction of planktivorous fish and zooplankton. *American Scientist* **67**: 572–581.

Olla, B. L., W. H. Pearson, and A. L. Studholme. 1980. Applicability of behavioral measures in environmental stress assessment. *Rapports et Proces-Verbaux des Reu'nions Commission Internationale pour l'Exploration Scientifique de la Mer Mediterranee Monaco* **179**: 162–173.

Rand, G. M. 1977*a*. The effect of exposure to a subacute concentration of parathion on the general locomotor behavior of the goldfish. *Bulletin of Environmental Contamination and Toxicology* **18**: 259–266.

Rand, G. M. 1977*b*. The effects of subacute parathion exposure on the locomotor behavior of the bluegill sunfish and large mouth bass. Pages 253–268 in F. L. Mayer and J. L. Hamelink, editors. *Aquatic toxicology and hazard evaluation*. ASTM STP 634. American Society for Testing and Materials. Philadelphia, Pennsylvania.

Rand, G. M. 1985. Behavior. Pages 221–256 in G. M. Rand and S. R. Petrocelli, editors. *Fundamentals of Aquatic Toxicology*. Hemisphere Press, New York.

Rose, K. A., and J. H. Cowan. 1993. Individual-based model of young-of-the-year striped bass population dynamics. 1. Model description and base-line simulations. *Transactions of the American Fisheries Society* **122**: 415–438.

Slobodkin, L. B., and A. Rapoport. 1974. An optimal strategy of evolution. *Quarterly Review of Biology* **49**: 181–200.

Smith, E. H., and H. C. Bailey. 1988. Development of a system for continuous biomonitoring of a domestic water source for early warning of contaminants. Pages 182–206 in D. S. Gruber and J. M. Diamond, editors. *Automated biomonitoring: Living sensors as environmental monitors*. Ellis Horwood, Chichester, England.

Smith, E. H., and H. C. Bailey. 1989. The application of avoidance/preference testing in aquatic toxicology. Pages 34–45 in M. Cowgill and L. R. Williams, editors. *Toxicology and hazard assessment*. Vol. 12. ASTM STP 1027. American Society for Testing and Materials, Philadelphia, Pennsylvania.

Smith, E. H., and H. C. Bailey. 1990. Preference/avoidance testing of waste discharges on anadromous fish. *Environmental Toxicology and Chemistry* **9**: 77–86.

Smith, E. H., and D. T. Logan. 1993. Invertebrate behavior as an indicator of contaminated water and sediments. Pages 48–61 in J. W. Gorsuch, F. J. Dwyer, C. G. Ingersoll and T. W. LaPoint, editors. *Environmental toxicology and risk assessment*. Vol. 2. ASTM STP 1173. American Society for Testing and Materials, Philadelphia, Pennsylvania.

Spencer, W. P. 1939. Diurnal activity in fresh water fish. *Ohio Journal of Science* **39**: 119–132.

Swartz, R. C. 1987. Toxicological methods for determining the effects of contaminated sediment on marine organisms. Pages 183–198 in K. L. Dickerson, A. W. Maki, and W. W. Brungs, editors. *Fate and effects of sediment-bound chemicals in aquatic systems*. Pergamon Press, New York.

Swartz, R. C., W. A. DeBen, J. K. Jones, J. O. Lamberson, and F. A. Cole. 1985. Phoxocephalid amphipod bioassay for marine sediment toxicity. Pages 284–307 in R. D. Cordwell, R. Purdy, and R. C. Bohner, editors. *Aquatic toxicity and hazard assessment*. Vol. 7. ASTM STP 854. American Society for Testing and Material, Philadelphia, Pennsylvania.

USEPA (United States Environmental Protection Agency). 1980. *Ambient water quality criteria for phenol. EPA–400/5–80–066*. Environmental Criteria and Assessment Office, Cincinnati, Ohio.

USEPA (United States Environmental Protection Agency). 1986. *Quality criteria for water. EPA 440/5–86–001*. Office of Water, Washington, District of Columbia.

USEPA (United States Environmental Protection Agency). 1992. *National study of chemical residues in fish*. Vol. II. EPA 823-R–92–008b. Office of Science and Technology, Washington, District of Columbia.

Wahl, R. A. 1992. Body-size dependent antipredator mechanisms of the American lobster. *Oikos* **65**: 52–60.

Ward, D. V., B. L. Hawes, and D. F. Ludwig. 1976. Interactive effects of predation pressure and insecticide (Temefos) toxicity on populations of the marsh fiddler crab *Uca pugnax*. *Marine Biology* **39**: 119–126.

Ware, D. M. 1975. Relation between egg size, growth, and natural mortality of larval fish. *Journal of the Fisheries Research Board of Canada* **32**: 2503–2512.

Chapter 13

The problem of photopollution for sea turtles and other nocturnal animals

BLAIR E. WITHERINGTON

One of the most conspicuous ways we alter the natural world is to light the darkness. To an observer in space looking down at the night-time half of the Earth, a patchwork glow of artificial lighting reveals the extent of our presence on the planet. Beyond providing a space-age index of human habitation, this errant light is known to have its own unique and profound ecological effects. These effects have the potential to create fundamental shifts in the selective pressures on living organisms and should be of keen interest to conservation biologists.

Verheijen (1985) used the term 'photopollution' to describe the introduction of detrimental artificial light into the environment. To be sure, light is not among the more familiar pollutants. Unlike most other pollutants, light is energy rather than substance and the measurement of its biological effects must account for varying sensitivities (among animals and instrumentation) to light's many properties (intensity, wavelength, directivity, polarity, form, periodicity). The most important peculiarity of light as a pollutant is in its effects upon animals, the most harmful of which are upon behavioral systems, such as those controlling visual orientation and the timing of periodic behavior. In either case, artificial light is best described, not as a toxic material, but as misinformation. With its great potential to disrupt behaviors that rely on correct information, artificial lighting can have profound effects on survival.

In the following discussion I present examples showing the consequences of behaviors disrupted by photopollution, offer mechanisms by which this disruption may occur, and list a number of strategies with which to address this conservation problem. A focus of this discussion is upon disrupted visual orientation, and in particular, the well-studied example of the effects of artificial lighting on the orientation of hatchling sea turtles. The research of lighting problems for sea turtles and the pursuit of solutions to these problems can provide a model to shape similar conservation efforts for other species.

Examples

Disruption of visual orientation

Night-flying insects are commonly seen circling and colliding with point sources of light and are perhaps the best known organisms to have their orientation disrupted by artificial lighting (Schacht & Witt 1986). Workers studying insect phenology capitalize on this phenomenon to capture insects *en masse* in light-traps (Takahashi & Haruta 1986; Wada *et al.* 1987; Pedgley & Yathom 1993). Commercially available light-traps sold for pest-control ('bug zappers') are engineered to attract large numbers of insects and kill them with electricity (Syms & Goodman 1987). Insects die in great numbers at other light sources; however, their mortality is generally only reported as a concern when the density of dead insects interferes with the functioning of light sources or is otherwise a nuisance (Ali *et al.* 1994). Although the loss of insects to artificial lighting can be extensive, the ecological implications of this mortality are unknown.

The aggregation of fish and squid at artificial light sources has been discussed primarily as a phenomenon that assists harvesting (Ben-Yami 1976). Although the principal harm of artificial lighting for fishes may be to simply facilitate over-exploitation, the use of light in fisheries also may have some potential to conserve resources. Because fishes have species-specific responses to light intensity and color, light may be used to reduce the catch of non-target species in net fisheries that are otherwise indiscriminate (Clarke *et al.* 1986).

Sea turtles are affected profoundly by artificial lighting, making photo-pollution at sea turtle nesting beaches an important conservation issue for the six sea turtle species (of seven extant species) listed as threatened or endangered. Lighting deters sea turtles from emerging from the sea to nest on otherwise preferred beaches (Witherington 1992*a*). Where nesting on lighted beaches does occur, hatchlings emerging nocturnally from nests are unable to locate the sea and wander inland toward light sources. This sea-finding disruption (often termed disorientation) results in high hatchling mortality. Conspicuous examples include cases where hatchlings have been lured onto roadways and crushed (McFarlane 1963; Philibosian 1976; Peters & Verhoeven 1994) or burned to death in the flames of an abandoned fire (Mortimer 1979). More commonly, however, 'lost' hatchlings are depredated by beach crabs or birds or become exhausted and dehydrated deep in dune vegetation. This mortality is most often inconspicuous, as in an example from Melbourne Beach, Florida, USA, where hundreds of dead Loggerhead (*Caretta caretta*) hatchlings were discovered beneath a single mercury vapor luminaire (L. M. Ehrhart, personal communication). At this site, the number of hatchlings and their varied conditions indicated that the light had attracted hatchlings for many nights. The dead hatchlings were only discovered after a close inspection of the grounds near the light source. In Florida each year, approximately one million hatchling sea turtles are misdirected by lighting which results in hundreds of thousands of hatchling deaths.

Mortality of nocturnally migrating birds at artificial light sources is well documented, and accounts show that many species are affected. The most acute bird-kills occur during periods when the moon provides little light or when cloud cover is heavy (Verheijen 1981). Literature accounts of specific bird-kill events show that the light sources responsible are varied. Examples include an oil-industry flare in northwestern Alberta, Canada, that killed approximately 3000 passerines representing 24 species during a period of a few days (Bjorge 1987). On a single night, flood lights at a park on Padre Island, Texas, USA, killed approximately 10 000 birds representing 39 species (James 1956). On a lighted boat off Alaska, approximately 6000 Crested auklets (*Aethia cristatella*) collided with the vessel nearly sinking it (Dick & Donaldson 1978). Lighthouses (Verheijen 1981), lighted towers (Kemper *et al.* 1966), airport ceilometers (Howell *et al.* 1954), and area lighting (James 1956; Reed *et al.* 1985) also mark areas where heavy bird mortality is known to occur. It is estimated that lighted television towers kill roughly a million migrating birds each year (Aldrich *et al.* 1966).

In endangered Hawaiian populations of petrels and shearwaters, fledgling birds leaving burrows in mountain nesting colonies fly toward the lighting of nearby resorts and are killed as they strike poles and buildings. This attraction to lighting is known to occur in 21 species of procellariiform birds (Reed *et al.* 1985).

Disruption of rhythmic behaviors

The natural periodicity of light intensity, both the differentiation of day and night and the change of day length with season, is critical to the timely expression of many behaviors. One effect of artificial light sources is to alter the environmental periodicity on which rhythmic behaviors depend. Thus, migrations may occur during an improper season, reproductive behavior and territoriality in individuals may occur out of phase with the rest of the population, and behaviors selected for during the day are expressed at night.

There is ample laboratory evidence for the change of rhythmic behavior under artificial lighting conditions. Artificial light-periodicity has been shown to affect behaviors as diverse as the timing of death feigning in domestic chicks (Stahlbaum *et al.* 1986), emergence of wasps from pupae (Dahiya *et al.* 1993), reproductive swarming of polychaetes (Fong 1993), and the accuracy of the sun compass used by orienting birds (Foa & Saviozzi 1990).

Although the effects on captive animals are well documented, few field studies have been initiated to examine the light-periodicity effects on wild populations. There are some anecdotal observations. At street lighting, normally diurnal birds are known to call at night, and in lighted harbors marine polychaetes are sometimes seen to swarm outside the normal new-moon period. There are likely to be additional important but less conspicuous effects, yet it is presently unknown how widespread these phenomena may be.

Does photopollution matter?

Abundance and life history traits of animals in a population will determine whether it can withstand a specific level of additional mortality. For instance, many insects may appear to be locally or seasonally abundant and to have a high reproductive potential that would offset additional mortality. In such a simple population model, photopollution-related mortality would have to be high relative to population size for there to be a significant effect on population growth or decline. Yet, for many insect species the phenomenon of local or seasonal abundance is critical for reproduction. High mortality during these critical periods of abundance would have large effects on population stability, selection of behaviors, and therefore, the expression of aggregation phenomena. As aggregation phenomena diminish, predators that depend on insect aggregations would suffer in turn.

Additional ecological effects may ensue as artificial lighting affects insects that occupy specific niches, such as pollinators of night-flowering plants. Again, important effects are not only limited to reductions in insect abundance but also include evolutionarily abrupt changes in the insect behavior patterns on which other organisms may depend. The ecological consequences of insect mortality at artificial light sources have yet to be adequately measured.

Compared with insects, there would seem to be a greater potential for photopollution-related mortality to cause population declines in turtles and birds. A population model for the Hawaiian dark-rumped petrel (*Pterodroma phaeopygia*) has shown that fledgling mortality caused by lighting near nesting colonies is an important factor in predicting whether the population is eventually extirpated (Simons 1984). For many sea turtle and bird populations, historic declines may be readily demonstrated, but underlying life history traits are poorly understood. Even without complete demographic information, however, it stands to reason that any level of mortality related to photopollution can be significant to a population that is threatened with extinction.

Artificial light sources and orientation systems

The best studied evidence of the effects of photopollution comes from animals that have trouble with visual orientation tasks near artificial light sources. In understanding this effect, it is important to understand the fundamental differences between light from artificial sources and light from celestial sources.

Clearly, it is celestial light rather than artificial light that has had the greatest opportunity to shape the evolution of visual orientation systems; but how does artificial light differ? Much of the answer lies in the nature of the light fields produced by the two kinds of sources. A light field is produced by a light source (or sources) but is measured from the perspective of an observer. It is a directional picture of all the light that can be detected by an observer. Light fields for celestial sources are only moderately directed, which means that although there may be one brightest direction, this direction is not tremendously brighter than other competing directions. These 'natural' light fields are moderated because both the observer and the illuminated features that the observer can see are similarly distant from the light source. Celestial light has a distant origin and reaches an observer not only directly but also indirectly as it is scattered in the atmosphere and reflected from the features on the Earth's surface (other competing directions). As a result, a celestial light field has clearly visible brightness from many directions.

Artificial light fields are produced by sources that are less intense than celestial sources, although artificial sources appear bright to an observer due to their proximity. Other illuminated features that would contribute to an artificial light field (sky, clouds, landscapes) are distant relative to the proximity of the source, and the light reflected from them, comparatively dim. Consequently, an observer near an artificial light source experiences a highly directed light field that is overwhelmingly dominated by the light source.

Verheijen (1958) cited the high directivity of artificial light fields as the reason why artificial lighting elicits an abnormal 'light-trapping' response in many animals. He reasoned that as an orienting animal approaches an artificial source, the light field the animal perceives becomes less multidirectional (more directed) until a point is reached where the light source may be the only visible feature of the animal's world. At this point the animal would only be able to orient relative to this single feature in its environment.

An artificial source need not functionally blind an animal in order to disrupt its orientation: it need only disrupt the balance of visual information that an animal perceives. Proper spatial orientation is critical, especially to animals that migrate. Insects, amphipods, lobsters, fish, turtles, and birds all accomplish orientation by using multiple sensory modalities. As specific sensory cues fail or become contradictory, additional or 'backup' modalities assist in maintaining proper orientation (Schöne 1984). In such multimodal systems, sensory input is often weighted so that the relative intensity of a cue determines its influence on orientation. An animal near an artificial light source experiences high light intensity, and may be compelled to orient toward the light (use positive phototaxis) although this orientation may be counter to information from other directional cues.

The behaviors of animals affected by photopollution should not be thought of as so abnormal as to have no relation to behaviors that have been shaped by selection. Light cues are important to visually orienting animals, so it is reasonable to assume that before an animal is close enough to a light source to be overwhelmed by it, the animal may be using that light to guide (or in actuality, misguide) a visual orientation task that is critical for survival. Hatchling sea turtles are an example of this. When placed beneath a lighted luminaire on a beach at night, Loggerhead, Green turtle (*Chelonia mydas*), Hawksbill (*Eretmochelys imbricata*), and Olive ridley (*Lepidochelys olivacea*) hatchlings circle and appear unable to orient in a constant direction. When only 10 to 20 m away from the light source, however, hatchlings crawl on a constant bearing toward the light source. This positive phototaxis is just as directed and deliberate as the orientation of hatchlings toward the sea on a naturally lighted beach. Hatchlings also show movement toward artificial light sources that are hidden from view, where light fields are not as directed as those produced by visible sources. I have often observed Loggerhead hatchlings crawling directly toward a wall of a beachfront building that is lit with a hidden floodlight. Although phototaxis alone would seem to explain this effect, there are additional mechanisms shaped by selection that influence the behavior of hatchlings near lighting. As I will discuss later, hatchlings do not always move in the brightest direction and can rely on additional visual information to lead them seaward.

In addition to positive phototaxis, many animals may be drawn to artificial light sources as a result of using them as reference cues in light-compass orientation systems. Bees, moths, and many bird species can maintain critical course bearings by using the sun, moon, an individual bright star, or star groups as reference points (Schöne 1984; Baker 1987). An animal that selects an artificial light source as a reference point for its light compass might not be able to maintain a constant orientation. To keep a straight course, an animal using a light compass keeps the bearing to a light-reference at a constant angle with the course bearing it has selected. When an earth-bound, artificial source is used as a light-reference, the deviation between the initial course bearing and the light-reference bearing will change with the position of the animal and defeat the function of the light compass. For example, a bird or insect may select an artificial light source to indicate a reference bearing at azimuth 45° (right of center). Initially, the animal's path may be relatively straight but as its flight takes it closer to the artificial source, it will begin to arc to the right toward the source until a point is reached when the animal actually circles the light source. From the animal's perspective, the light source will remain at a 45° angle to its path throughout the flight. Close to the light source, the animal may collide with the source or become light-trapped by high light-field directivity. Aircraft warning lights on tall towers would appear to be above the horizon to a flying bird and may deceptively mimic celestial point sources. This may explain high bird mortalities at these structures.

Searching for solutions

An understanding of species-specific visual orientation systems or light-mediated rhythmic behavior is fundamental to the correct application of specific remedies to photopollution problems. One common-sense remedy, however, would seem to be universal: turn off the lights; but which lights? It may be that only some at a site contribute to the photopollution hazard. Extinguishing lights indiscriminately may not gather much support from property owners; therefore, a more prudent conservation strategy would be to selectively manage the light rather than prohibit it.

For a light management program to be guided successfully, certain questions should be answered. For instance, how important is light to the behavior in question? How much of a given light source is too much? Is there a compromise that will allow light required for human activity yet minimize light that can disrupt orientation? Is there a differential between the properties of light required for human use and the properties of light that can disrupt animal behaviors? Can disruption of orientation be minimized by introducing other cues that may guide multimodal systems? Where and when is light management needed (where and when do critical behaviors take place)?

Below I describe studies on hatchling sea turtle orientation that have begun to provide answers to these questions and that have resulted in some novel solutions to photopollution problems at sea turtle nesting beaches. Many solutions, however, are not novel and come instead from light management practices meant to solve problems that directly affect humans (e.g., energy efficiency, aesthetics, astronomical observation). As might be the case for other animals, sea turtle conservation has benefited from this synthesis of basic biological understanding and practical field techniques.

Light and hatchling sea turtle orientation

Like other animals affected by photopollution, sea turtles rely on accurate visual orientation to accomplish critical nocturnal tasks. For sea turtle hatchlings this critical task is the expeditious movement from nest to sea (sea-finding). Even moderate delay of sea-finding can be deadly for hatchlings.

Photopollution also has been shown to deter nesting and thus limit the extent of nesting habitat for sea turtles (Witherington 1992*a*). However, lighting that is bright enough to disrupt sea-finding in hatchlings often is not bright enough to deter females from nesting. Hatchling mortality from lighting is a significant conservation problem because nests are commonly placed on beaches that are too bright to allow hatchlings to escape.

Sea turtle hatchlings emerge from their sand nests almost exclusively at night (Witherington *et al.* 1990) and, under natural lighting conditions, move immediately and directly from nest to sea (Fig. 13.1). The inability of hatchlings to orient seaward when wearing bilateral blindfolds or when their view of the ocean is blocked by light shields or dune topography indicates that the principal ocean-recognition cues are visual (Mrosovsky & Shettleworth 1968, 1975; Mrosovsky 1972). Although hatchling sea turtles can respond to some non-visual cues, specifically beach slope (Rhijn 1979; Salmon *et al.* 1992), these cues appear to have a minor influence on directional movement even when visual cues are not present.

Hatchlings show a strong tendency to orient toward the brightest direction in a way that is predicted well by a complex phototropotaxis model (Mrosovsky & Kingsmill 1985). In this model, hatchlings turn to balance the visual input of multiple comparators at the retina of each eye

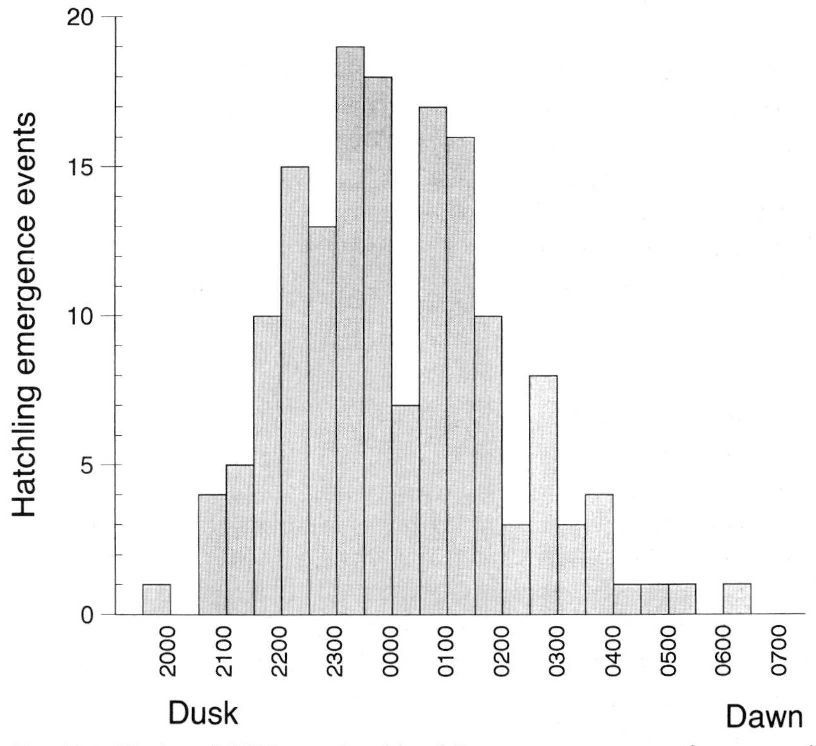

Fig. 13.1. Timing of 157 Loggerhead hatchling emergence events from natural nests on a Florida beach between 29 July and 1 September 1988. An emergence event was defined as the movement of ten or more hatchlings from nest to sea. Data from Witherington *et al.* (1990).

so that the hatchling is oriented toward the brightest horizon. Such a mechanism is thought to facilitate sea-finding by orienting hatchlings toward the horizon with the greatest portion of visible sky.

To assess lighting that may disrupt sea-finding, it is important to understand the criteria with which hatchlings 'measure' brightness. Some understanding of the spectral properties of this measure has been gained through determinations of relative electrical potential across retinas exposed to spectral light (electroretinography, ERG). The ERG spectrum for the dark-adapted Green turtle shows that the animal's eye is

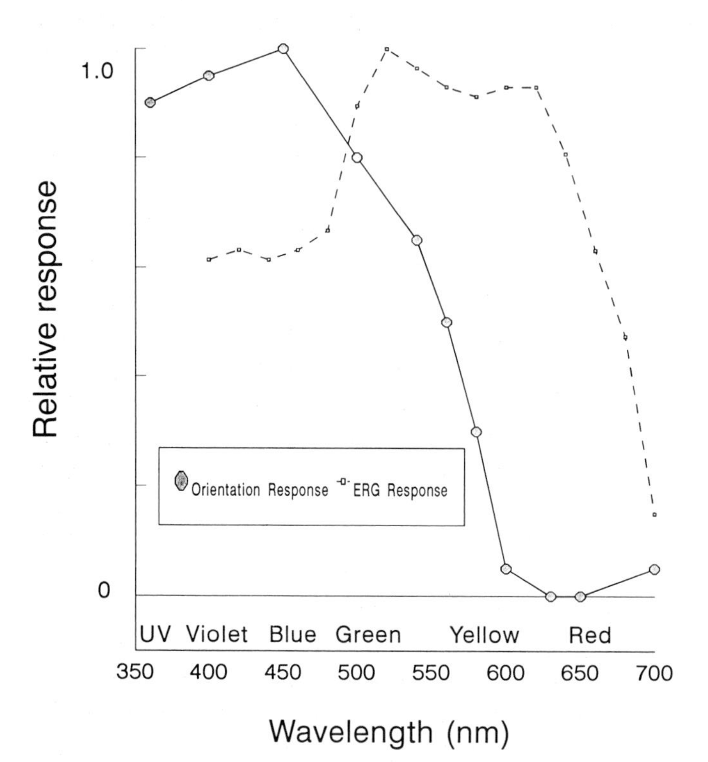

Fig. 13.2. Physiological and behavioral responses to spectral light in dark-adapted Green turtle hatchlings. The behavioral response curve represents the proportion of green turtle hatchlings in a modified Y-maze choosing a window lighted with colored light at the wavelengths indicated. The opposing window was lighted with an unvarying source (Witherington 1992*b*). Although wavelength at the colored-light window varied, radiance remained at 1.44 x 10^{17} photons s^{-1} m^{-2} nm^{-1} (approximate radiance of a full-moon night sky). The electroretinography (ERG) response curve is adapted from Granda & O'Shea (1972) and represents log sensitivity among wavelengths. Absolute intensity of the ERG light source was reported to be 2.7 x 10^{17} photons s^{-1} m^{-2} at 580 nm.

most sensitive in the blue-green to orange region of the visible spectrum, from 500 to 640 nm (Fig. 13.2; Granda & O'Shea 1972).

In addition to measuring what spectral light animals can see, it is important to measure how they respond behaviorally to spectral light. To measure the influence of spectral light on orientation behavior, Witherington & Bjorndal (1991*a*) conducted a series of two-choice light-preference experiments. Using a modified Y-maze, we measured hatchling attraction to high-intensity (greater than the illuminance of a full-moon night sky) diffuse light of various wavelengths presented relative to a second source of constant intensity and color (peak emission at 520 nm). Wavelengths that Green turtle hatchlings were most likely to orient toward did not coincide with their peak ERG sensitivity (Fig. 13.2). Although Green turtles can see yellow light comparatively well, they are only weakly attracted to yellow sources.

Among the four species of sea turtle hatchlings tested, there are two distinct patterns of spectral light preference (Fig. 13.3; Witherington 1992*b*). When dark-adapted Green turtle, Hawksbill, and Olive ridley hatchlings are presented high-intensity, diffuse spectral light and the option to orient toward or away from the source, hatchlings show attraction to short-wavelength light (ultraviolet-yellow) and are indifferent to long-wavelength light (red; Fig. 13.3). In contrast, Loggerhead hatchlings are attracted to ultraviolet-green light, are indifferent to green-yellow and red light, and move away from yellow light. In additional trials with Loggerhead hatchlings, neither blue-green nor red light prompted the aversion response seen for yellow light even when intensities were up to 100 times the intensity of yellow light. This suggests that the aversion to yellow may involve color perception rather than a response to brightness mediated by spectral sensitivity.

The Loggerhead's unique aversion to yellow light and the Green turtle's apparent discounting of yellow light occur only at what are probably photopic light levels (intensity sufficient for cone vision). To measure loggerheads' orientation responses to low-intensity spectral light, I conducted an orientation titration (Witherington 1992*b*). The apparatus was an enclosed circular arena with 72 pitfalls at the perimeter and a single light window in the perimeter wall. In this titration, light intensity of a given wavelength at the window was increased incrementally with each of several hatchling groups released at the arena center, until a hatchling group showed statistically directed orientation. At these low, threshold light-levels, all orientation was toward the light source for all wavelengths, including yellow. As in experiments with higher-

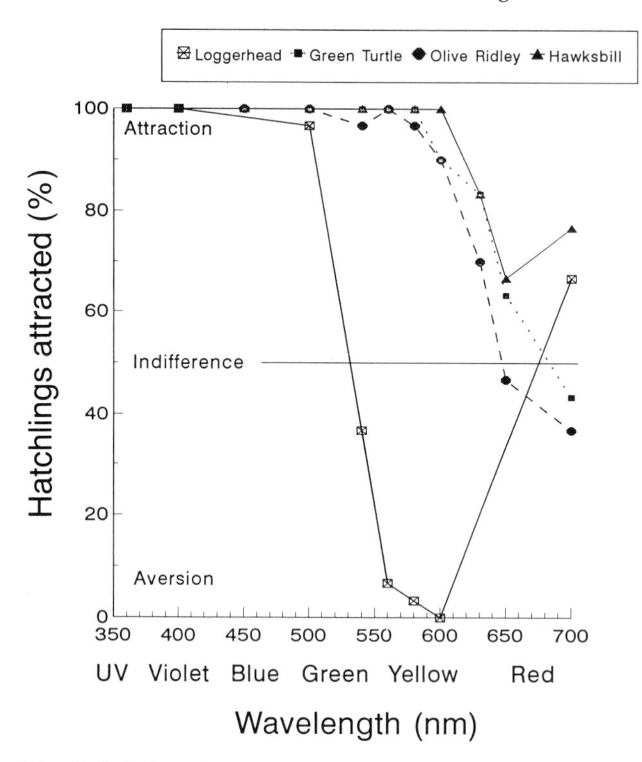

Fig. 13.3. Orientation responses to colored light from hatchlings from four species of sea turtles. Responses were measured as the proportion of hatchlings in a modified Y-maze choosing a window lighted with colored light at the wavelengths indicated. The opposing window was not lighted (Witherington, 1992*b*). Radiance of the lighted window remained at 1.44 x 10^{17} photons s^{-1} m^{-2} nm^{-1}. Aversion indicates that hatchlings predominately chose the dark window over the lighted window.

intensity light, ultraviolet-green light was most attractive (i.e., required the lowest intensity to elicit directed orientation; Fig. 13.4). Threshold intensities of yellow and red light were approximately one and five orders of magnitude higher than the threshold intensities for the shorter wavelengths studied. Loss of color perception with scotopic vision (night vision depending upon rod photoreceptors) may explain the lack of yellow aversion at these low light levels. It is at these low light levels that sea turtle hatchlings most often must locate the sea.

In addition to the weighing of spectral information, hatchlings also factor the relative direction of light into their measure of brightness. Sensitivity to directional light in hatchlings can be characterized by a specific 'cone of acceptance,' which describes how much of the world a

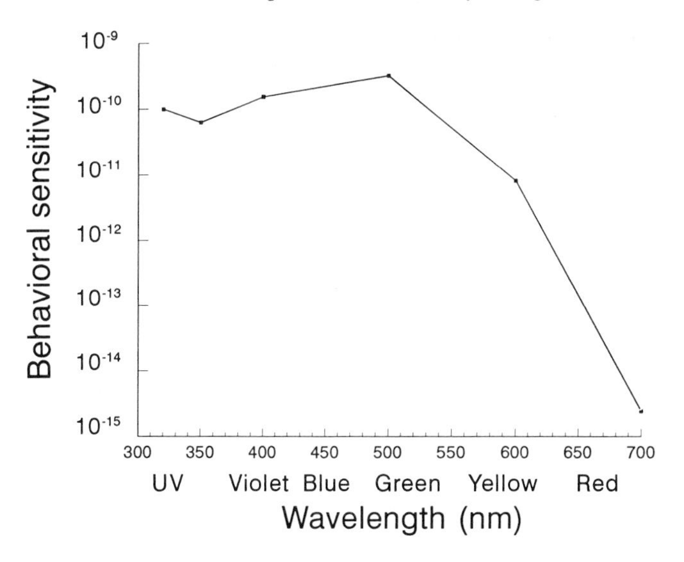

Fig. 13.4. Behavioral sensitivity of loggerhead hatchlings to low-intensity spectral light. Behavioral sensitivity is represented as the inverse of the light-source radiance required to evoke directed orientation in groups of individually released hatchlings (n=30 per group). Orientation at light thresholds was toward the light source. The ordinate is a log scale of the units (photons s^{-1} m^{-2} nm^{-1} $sr)^{-1}$. Data are from Witherington (1992*b*).

hatchling measures at any one instant (Fig. 13.5). An acceptance cone encompasses the angular span over which a brightness measurement is integrated and may be different from the turtle's field of view. The height and breadth of a hatchling's (or light detector's) acceptance cone critically influence its assessment of brightest direction and can determine whether point sources or a broad glow will have the greatest influence on measured brightness.

The horizontal component of the acceptance cone for Green turtle, Olive ridley, and Loggerhead hatchlings has been deduced through studies of hatchling orientation in controlled light fields (Verheijen & Wildschut 1973; Witherington 1992*b*). In these studies, hatchlings oriented within a vertically striped cylinder having many 'brightest directions' that varied according to width of the acceptance cone. On average, hatchlings of each species chose directions that corresponded with the brightest direction as measured with a wide acceptance cone (approximately 180° horizontally) (Fig. 13.5). To determine the vertical component of the hatchling's acceptance cone, researchers observed orientation in hatchlings that were presented light sources at varying vertical angles.

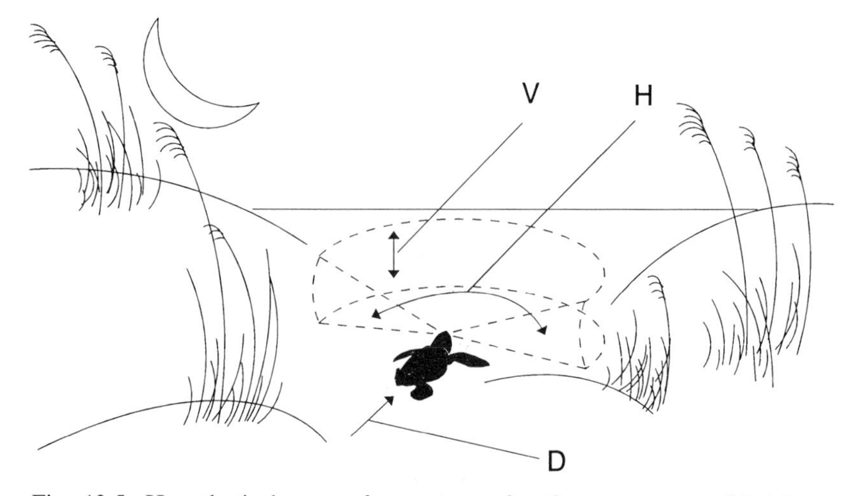

Fig. 13.5. Hypothetical cone of acceptance for the assessment of brightest direction by a sea turtle hatchling. The vertical component of the cone (*V*) is approximately 10° to 30° from the horizontal plane. The horizontal component of the cone (*H*) is approximately 180°. Light sources within this cone of acceptance are thought to be integrated into a measure of brightness for the direction D.

The vertical component of the acceptance cone was found to be a small span (approximately 30° or less) and centered near the horizon (Verheijen & Wildschut 1973; Salmon *et al.* 1992; Witherington 1992*b*). This vertically flat acceptance cone indicates that light closest to the horizon plays the greatest role in determining orientation direction. The horizontal breadth of the acceptance cone means that irradiance (light reaching the hatchling) is a better measure of relevant brightness than radiance (light emanating from a source). Integrating light measurement over a broad range may be important in mitigating the effects of the sun and moon on orientation direction (Fig. 13.6).

Although the brightest direction has an important influence on hatchling sea-finding orientation, hatchlings also respond to shape cues. Both Limpus (1971) and Salmon *et al.* (1992) have presented convincing evidence that Loggerhead and Green turtle hatchlings tend to orient away from darkened silhouettes. On most beaches this tendency would direct hatchlings away from the profile of the dune and toward the ocean. However, this response could be an epiphenomenon that results from brightest-direction orientation. Dune silhouettes by their nature darken the portion of the horizon within a hatchling's cone of acceptance in addition to presenting a characteristic shape.

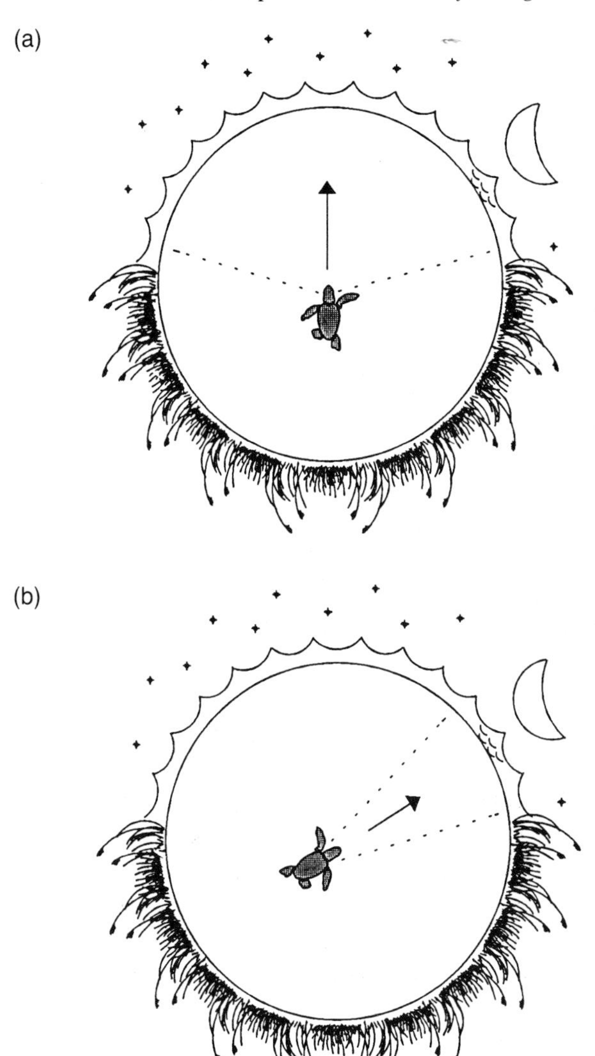

Fig. 13.6. Orientation consequences of brightness assessments using two hypothetical angles of acceptance. Hatchlings (a) and (b) both orient toward the center of the brightest horizon as measured within their respective angles of acceptance (dotted lines). Hatchling (a) uses a broad angle of acceptance and has the most accurate seaward orientation. Hatchling (b), having a narrow angle of acceptance, is directed toward the rising moon.

Elucidating the effects of shape on orientation, independent of the effects of brightness, is not a straightforward task. One's confidence in distinguishing shape-cue orientation from brightness-cue orientation should only be as great as one's confidence in measuring brightness as hatchlings do. Although our understanding of a hatchling's response to brightness remains at a basic level, there are clear indications that shape does influence orientation direction.

The effects of shape and brightness on hatchling orientation have been studied in cue-conflict experiments in which brightness was measured with instruments having the broad, flat acceptance cone described above. Both Green turtle and Loggerhead hatchlings tend to orient away from sets of vertical black and white stripes and toward an opposing, uniformly illuminated direction (open horizon) in the laboratory (Rhijn & Gorkom 1983; Witherington 1992*b*). In Loggerhead hatchlings, orientation away from contrasting stripes persists even when the striped direction is approximately twice the brightness of the open horizon (Witherington 1992*b*). When the striped horizon is brightened to three times the brightness of the open horizon, however, orientation of groups of hatchlings is randomly directed, and only after the striped horizon is made approximately five times brighter do hatchlings significantly orient in that direction.

In Loggerhead hatchlings, orientation, either away from contrasting shapes irrespective of brightest direction or toward the brightest direction irrespective of contrasting shapes, depends upon the directivity of the light field from the perspective of the hatchling. As this directivity increases, the brightest direction becomes more pronounced, less ambiguous perhaps, and apparently a greater orientation stimulus to the hatchling. An aversion to portions of the horizon that have visible structure (i.e., shapes as opposed to openness, variable spatial patterns of light as opposed to a uniform pattern) would benefit sea-finding. Commonly, nesting beaches have a vegetated dune that provides a structured profile in contrast with a flat and featureless ocean.

In nature and under artificial conditions, cues do conflict. Brightness measurements (with the spectral and directional considerations outlined above) made on nesting beaches indicate that the brightest direction is often not seaward (Rhijn 1979; Witherington 1992*b*). Measurements made under varying natural conditions show that the brightest direction depends largely on the direction of the moon during moonset and moonrise (Witherington 1992*b*). Under natural lighting conditions, regardless of the brightest direction, released hatchlings orient to the sea.

Measured light fields are most directed when artificial lighting is present, and on these lighted beaches, hatchlings are most likely to orient in the brightest direction, toward lighting (generally inland).

By presenting an unambiguous brightest direction to hatchlings, artificial lighting may be perceived as a supernormal sea. At such high levels of light directivity, brightness cues may be weighted so that shape cues are essentially ignored. Although hatchlings possess a multimodal visual orientation system that directs seaward movement under a variety of natural conditions, the function of this system is defeated in the presence of artificial lighting.

Lessons from sea turtles in understanding photopollution

From studies of hatchling orientation come a number of observations that may be common to other animals affected by photopollution.

(1) Although ERG spectral curves can be used to approximate the sensitivity of a behavior to spectral light, the most direct way to determine this is to measure the spectral sensitivity of the behavior itself. In doing so, one should measure the behavior where and when the behavior is normally expressed (such as the measurement of seafinding in hatchlings taken directly from the nest at night).

(2) Changing the spectral properties of an artificial light source can be an important way to reduce the effect it has on animal behavior. In some sea turtles, there is a reduced sensitivity to yellow and red light.

(3) Spectral sensitivity of a behavior and the extent to which a behavior is disrupted may be closely tied to light intensity. Behavioral experiments should be conducted at more than one light intensity level.

(4) Just as an animal may have a specific spectral sensitivity to light, it also may have a specific directional sensitivity. This directional sensitivity may be described by a cone of acceptance with specific horizontal and vertical dimensions. This acceptance cone will determine how the position, number, and distribution of artificial sources will affect an animal's orientation. Dim sources that are numerous and widely separated may have the same effect on orientation as a single bright source because hatchling sea turtles tend to move in the brightest direction as measured with a broad, flat acceptance cone.

(5) An animal may depend upon variously weighted multiple cues for correct orientation. Although highly directed artificial light fields can overwhelm these cues, it is important to understand how additional cues can help correct orientation that is affected by lighting. In hatchling sea turtles, the addition of shape cues can sometimes override their tendency to move in the brightest direction.

Conservation strategies for photopollution

An initial step in a conservation strategy is to identify areas where the potential for photopollution problems exist. Some of these areas have been identified: (1) where lighting is visible from sea turtle nesting beaches during the nesting and hatching season; (2) where lighting is located near the seaward route taken by fledging procellariiform birds; (3) where high-intensity lighting, especially up-lighting, is located along fall and spring migration routes of passerines; (4) at towers with aircraft warning lighting; (5) at lighthouses; and (6) at industrial oil flares. There are many other areas where nocturnal animals are affected by photopollution (e.g., insects at street lights, polychaetes in lighted harbors), although the consequences of these events are poorly understood. Research is sorely needed.

Mortality from photopollution can be both cryptic and variable in time. Consequently, the lack of discovery of dead animals at a lighted site does not exculpate a site of photopollution problems and it may be necessary to initiate light management at sites where photopollution problems are merely suspected.

After photopollution problems are identified, a light management plan can be enacted. Light management is more readily accepted as a conservation strategy than is simply prohibiting light within an area of concern. With proper light management, light sources are altered so that human needs for lighting are met and disruption of animal behavior is minimized. The following are some important considerations for light management.

Timing

Many animals are affected by photopollution only during critical periods. For sea turtles, the nesting and hatching period is critical. In the tropics this period varies, but at more temperate latitudes it usually occurs from late spring through early fall. For shearwaters and petrels autumn fledging is a vulnerable time. For passerines most mortalities occur during spring and fall migrations. Restrictions on lighting during these periods are most important. Both birds and sea turtles are most vulnerable on cloudy and moonless nights (Verheijen 1981; Salmon & Witherington 1995).

Reducing perceived directivity and intensity

Other than turning light sources off, the principal way to reduce both intensity of a photopollution source (which may have many individual light sources) and its contribution to light field directivity is to reduce the power of constituent light sources. Lowering the wattage of luminaires will usually lower their contribution to a photopollution hazard but it will also reduce light that may be required for human activity. A way to reduce the photopollution hazard from a source while maintaining adequate levels of useful light is to restrict the distribution of light from the sources. Focusing a light source toward the ground can reduce the amount of errant light and decrease the directivity of the resulting light field as it is perceived by animals outside the immediate vicinity. Other useful techniques to control the escape of light include lowering pole-mounted luminaires, recessing lights into eaves, and shielding them. The use of light shields has reduced mortality at sea turtle nesting beaches and near shearwater and petrel nesting colonies. Although shields can be easily fashioned from any opaque material, luminaires specifically designed to eliminate stray light (e.g., cut-off luminaires) are a more efficient solution. Diffusing light from point sources by adding globes and other diffusers may decrease their attractiveness to insects but does not reduce their attractiveness to sea turtles.

Using light screens

In some situations, a 'non-point-source' glow affecting a specific area can be addressed by the use of light screens. For example, blocking distant city light from urban sea turtle nesting beaches has been shown to be helpful (Salmon *et al.* 1995). Although constructing opaque screens on the dune would suffice, preserving natural dune vegetation is a more permanent solution. Problems with photopollution are exacerbated at nesting beaches with a low dune profile and little dune vegetation.

Altering spectral properties

Often, the spectral sensitivity profiles of species that are affected by artificial lighting are different from the spectral requirements of humans. Birds are not sensitive to ultraviolet light and bird mortality at airport ceilometers has been essentially eliminated where ultraviolet sources have replaced white-light sources in ceilometer beams. Unfortunately, birds have a spectral range that is broader than ours (Norren 1979), leaving it difficult or impossible to select a colored light source that humans see well but that birds see poorly. Compared with humans, sea turtle spectral sensitivity is lower in the yellow-red end of the spectrum. Yellow, low-pressure-sodium-vapor luminaires, therefore, are good alternatives to broad-spectrum lighting on sea turtle nesting beaches (Witherington 1992*a*; Witherington & Bjorndal 1991*b*). Incandescent lamps that have been tinted to reduce blue emission (marketed as 'bug lights' because of their reduced attraction to insects) are also good alternatives to white sources. These colored sources, however, can affect sea turtles to some degree and should be shielded from beaches. Both low-pressure-sodium-vapor lighting and yellow bug lights kill substantially fewer nocturnal insects than other commercial sources. As a bonus, low-pressure-sodium-vapor sources are the most energy-efficient of all commercial sources, and they provide little interference with astronomical observation (because of the ease with which their monochromatic light can be filtered).

Reducing light duration

The motion-detector switch has been an important technological innovation for the management of security lighting. These switches activate a light source only when it is approached and can reduce the time that the light source affects animals. Work with sea turtles has shown that flashing lights are much less attractive than constant light sources, which suggests that hatchlings may integrate their measure of brightness over time (Mrosovsky 1978). In Australia, flashing lights have been deployed to mark paths to sea turtle nesting beaches. On many communications towers, constant or slow-pulse blinking lights for warning aircraft have been replaced by flashing strobes which has resulted in reductions in bird mortalities (Verheijen 1985). Similar reductions in bird kills have been observed at lighthouses where a flashing light has replaced a rotating beam.

Providing additional orientation cues

Ensuring that dunes at sea turtle nesting beaches remain vegetated will help provide additional shape cues to hatchlings and may help hatchlings correct orientation that is affected by artificial lighting. Verheijen (1985) suggests that using downward-directed flood lighting to brighten obstacles near towers may help reduce bird collisions by decreasing the directivity of the light field perceived by birds. A similar strategy of lighting the ocean at sea turtle nesting beaches would not be a solution to photopollution problems there. Light sources on the water could attract swimming hatchlings and cause hatchlings on adjacent stretches of beach to move parallel to the shore (Witherington 1991).

Public enlightenment

Photopollution problems on Florida sea turtle nesting beaches began to be recognized when workers in the field published details of sea-finding disruption and hatchling mortality they had observed. Awareness spread as biologists working at various nesting beach sites shared their concerns with the news media and members of the public. A biological interpretation of hatchling mortality events allowed the media to correctly interpret hatchling misdirection as the result of artificial lighting, rather than the result of more fanciful causes. Local residents and governmental officials in Florida became impressed when orientation and mortality problems were demonstrated in their areas. At an important Loggerhead and Green turtle nesting beach in Brevard County, Florida, public recognition of hatchling mortality became widespread following numerous news articles and editorials in the local media in the early 1980s. By 1985 Brevard County had promulgated the first local ordinance restricting lighting of the beach during the nesting season. Other coastal communities in Florida followed suit, and at present most of the Florida counties and municipalities near nesting beaches have ordinances designed to protect sea turtles from the effects of artificial lighting.

In some areas of heavily developed coastal Florida, substantial photopollution problems persist. In Dade and Broward counties (located in southeast Florida) most sea turtle nests are relocated to hatcheries that confine hatchlings as they emerge so that they can be gathered and released at dark-beach sites. Most sea turtle biologists believe that this strategy creates problems for hatchling survivorship. An important problem stemming from this technique is the illusion that the artificial lighting problem has been resolved.

In Florida and South Carolina, USA, involvement of local power companies has proven to be an important way to reach the public. Florida power companies have funded educational pamphlets and have begun the practice of including educational materials in the power bills sent to coastal consumers. In the USA, Costa Rica, Greece, and Australia, seaside resorts distribute information on how beach visitors can minimize the effects of lighting on sea turtles.

Bird kills at lighted structures are poorly known by the public considering the widespread nature of the phenomenon. Some specific cases, however, have enjoyed considerable attention such as the mortality of Hawaiian petrels and shearwaters (Reed *et al.* 1985). Deaths of these birds were substantially reduced after each of the entities responsible for the harmful lighting became convinced that their light management efforts would result in reduced bird mortality.

Elsewhere, serendipity more than design has been responsible for bird-friendly changes in light sources. The flashing strobes on modern communication and navigation towers are used primarily for their effectiveness in attracting human attention (thus deterring aircraft strikes and ship groundings). If experimental evidence confirms what anecdotal observations suggest, replacement of constant or blinking sources with strobes may be an important effort for migrating bird conservation. Many tall structures remain lighted with sources that can cause bird kills. Many of these are public facilities, so it may be most useful to concentrate efforts to resolve these conservation problems at a governmental level rather than at individual facilities.

Light management is likely to be most successful when initiated at the project-planning level. Light management is cheaper, more effective, and generally more palatable to developers when solutions are engineered into plans rather than retrofitted. Nonetheless, methods to reduce the effects from existing light sources abound and include options that generally are of minimal cost and inconvenience.

Continued research aimed at understanding behaviors affected by photopollution is essential for the development and testing of methods to mitigate the effects of artificial lighting. Solutions to lighting problems will need to be supported by research if they are to be embraced by development planners, governmental officials, and the public.

The role of conservation behaviorists

Although some conspicuous examples of behavioral modification and mortality from photopollution are known, many important effects are not likely to be discovered without considerable effort. Studies are needed that would elucidate the ecological effects of photopollution. Effects are sure to exist, whether large or small, almost anywhere that artificial light sources are positioned outdoors. Dark areas where artificial sources are scheduled to be introduced would provide unique opportunities for experimental discovery. This field provides an opportunity for behaviorists to make important contributions to conservation biology.

As natural areas are developed, the potential for biological photopollution is likely to increase. Of the many effects of human habitation on organisms, the effects from photopollution may be among the most readily managed. With the many techniques available for using light while minimizing its effects on nocturnal animals, the control of biological photopollution should be among the most prominent conservation success stories. The work of behavioral conservation biologists can help make this so.

Literature cited

Aldrich, J. W., R. R. Graber, D. A. Munro, G. J. Wallace, G. C. West, and V. H. Cahalane. 1966. Report of committee on bird preservation. *Auk* **83**: 465–467.

Ali, A., G. Ceretti, L. Barbato, G. Marchese, F. D'Andrea, and B. Stanley. 1994. Attraction of *Chironomus salinarius* (Diptera: Chironomidae) to artificial light on an island in the saltwater lagoon of Venice, Italy. *Journal of the American Mosquito Control Association* **10**: 35–41.

Baker, R. R. 1987. Integrated use of moon and magnetic compasses by the heart-and-dart moth, *Agrotis exclamationis*. *Animal Behaviour* **35**: 94–101.

Ben-Yami, M. 1976. *Fishing with light*. Fishing News Books, London.

Bjorge, R. R. 1987. Bird kill at an oil industry flare stack in northwest Alberta. *Canadian Field Naturalist* **101**: 346–350.

Clarke, M. R., P. L. Pascoe, and L. Maddock. 1986. Influence of 70 watt electric lights on the capture of fish by otter trawl off Plymouth. *Journal of the Marine Biological Association of the United Kingdom* **66**: 711–720.

Dahiya, A. S., W. B. Tshernyshev, and V. M. Afronia. 1993. Diurnal rhythm of emergence from pupae in parasitic wasp *Trichogramma evanescens* Westw. (Insecta: Hymenoptera). *Journal of Interdisciplinary Cycle Research* **24**: 162–170.

Dick, M. H., and W. Donaldson. 1978. Fishing vessel endangered by crested auklet landings. *Condor* **80**: 235–236.

Foa, A. and G. Saviozzi. 1990. Effects of exogenous melatonin on sun-compass orientation of homing pigeons. *Journal of Biological Rhythms* **5**: 17–24.

Fong, P. P. 1993. Lunar control of epitokal swarming in the polychaete *Platynereis bicanaliculata* (Baird) from central California. *Bulletin of Marine Science* **52**: 911–924.

Granda, A. M., and P. J. O'Shea. 1972. Spectral sensitivity of the green turtle (*Chelonia mydas mydas*) determined by electrical responses to heterochromatic light. *Brain Behavior and Evolution* **5**: 143–154.

Howell, J. C., A. R. Laskey, and J. T. Tanner. 1954. Bird mortality at airport ceilometers. *Wilson Bulletin* **66**: 207–215.

James, P. 1956. Destruction of warblers on Padre Island, Texas, in May, 1951. *Wilson Bulletin* **68**: 224–227.

Kemper, C. A., D. G. Raveling, and D. W. Warner. 1966. A comparison of the species composition of two TV tower killed samples from the same night of migration. *Wilson Bulletin* **78**: 26–30.

Limpus, C. J. 1971. Sea turtle ocean finding behaviour. *Search* **2**: 385–387.

McFarlane, R. W. 1963. Disorientation of loggerhead hatchlings by artificial road lighting. *Copeia* **1963**: 153.

Mortimer, J. A. 1979. Ascension Island: British jeopardize 45 years of conservation. *Marine Turtle Newsletter* **10**: 7–8.

Mrosovsky, N. 1972. The water-finding ability of sea turtles. *Brain Behavior and Evolution* **5**: 202–225.

Mrosovsky, N. 1978. Effects of flashing lights on sea-finding behavior of green turtles. *Behavioral Biology* **22**: 85–91.

Mrosovsky, N., and S. F. Kingsmill. 1985. How turtles find the sea. *Zeitschrift für Tierpsychologie* **67**: 237–256.

Mrosovsky, N., and S. J. Shettleworth. 1968. Wavelength preferences and brightness cues in the water finding behaviour of sea turtles. *Behaviour* **32**: 211–257.

Mrosovsky, N., and S. J. Shettleworth. 1975. On the orientation circle of the leatherback turtle, *Dermochelys coriacea*. *Animal Behaviour* **23**: 568–591.

Norren, D. van. 1979. Two short wavelength sensitive systems in pigeon, chicken, and daw. *Vision Research* **15**: 1164–1166.

Pedgley, D. E., and S. Yathom. 1993. Windborne moth migration over the Middle East. *Ecological Entomology* **18**: 67–72.

Peters, A., and K. J. F. Verhoeven. 1994. Impact of artificial lighting on the seaward orientation of hatchling loggerhead turtles. *Journal of Herpetology* **28**: 112–114.

Philibosian, R. 1976. Disorientation of hawksbill turtle hatchlings, *Eretmochelys imbricata*, by stadium lights. *Copeia* **1976**: 824.

Reed, J. R., J. L. Sincock, and J. P. Hailman. 1985. Light attraction in endangered procellariiform birds: reduction by shielding upward radiation. *Auk* **102**: 377–383.

Rhijn, F. A. van. 1979. Optic orientation in hatchlings of the sea turtle, *Chelonia mydas*. I. Brightness: not the only optic cue in sea-finding orientation. *Marine Behavior and Physiology* **6**: 105–121.

Rhijn, F. A. van and J. C. van Gorkom. 1983. Optic orientation in hatchlings of the sea turtle, *Chelonia mydas*. III. Sea-finding behaviour: the role of photic and visual orientation in animals walking on the spot under laboratory conditions. *Marine Behavior and Physiology* **9**: 211–228.

Salmon, M. and B. E. Witherington. 1995. Artificial lighting and seafinding by loggerhead hatchlings: evidence for lunar modulation. *Copeia* **1995**: 931–938.

Salmon, M., M. G. Tolbert, D. P. Painter, M. Goff, and R. Reiners. 1995. Behavior of loggerhead sea turtles on an urban beach. II. Hatchling orientation. *Journal of Herpetology* **29**: 568–576.

Salmon, M., J. Wyneken, E. Fritz, and M. Lucas. 1992. Seafinding by hatchling sea turtles: Role of brightness, silhouette and beach slope as orientation cues. *Behaviour* **122**: 56–77.

Schacht, W., and T. Witt. 1986. Warum nachtaktive Insekten kuenstliche Lichtquellen anfliegen (Insecta). *Entomofauna* **7**: 121–128.

Schöne, H. 1984. *Spatial orientation*. Princeton University Press, Princeton, New Jersey.

Simons, T. R. 1984. A population model of the endangered Hawaiian dark-rumped petrel. *Journal of Wildlife Management* **48**: 1065–1076.

Stahlbaum, C. C., C. Rovee-Collier, J. W. Fagen, and G. Collier. 1986. Twilight activity and antipredator behavior of young fowl housed in artificial or natural light. *Physiology and Behavior* **36**: 751–758.

Syms, P. R., and L. J. Goodman. 1987. The effect of flickering U-V light output on the attractiveness of an insect electrocutor trap to the house-fly, *Musca domestica*. *Entomologia Experimentalis et Applicata* **43**: 81–86.

Takahashi, F., and A. Haruta. 1986. A portable automatic light trap and its application. *Kontyu* **54**: 373–380.

Verheijen, F. J. 1958. The mechanisms of the trapping effect of artificial light sources upon animals. *Archives Neerlandaises de Zoologie* **13**: 1–107.

Verheijen, F. J. 1981. Bird kills at tall lighted structures in the USA in the period 1935–1973 and kills at a Dutch lighthouse in the period 1924–1928 show similar lunar periodicity. *Ardea* **69**: 199–203.

Verheijen, F. J. 1985. Photopollution: artificial light optic spatial control systems fail to cope with. Incidents, causations, remedies. *Experimantal Biology* **44**: 1–18.

Verheijen, F. J., and J. T. Wildschut. 1973. The photic orientation of sea turtles during water finding behaviour. *Netherlands Journal of Sea Research* **7**: 53–67.

Wada, T., H. Seino, Y. Ogawa, and T. Nakasuga. 1987. Evidence of autumn overseas migration in the rice planthoppers, *Nilaparvata lugens* and *Sogatella furcifera*: analysis of light trap catches and associated weather patterns. *Ecological Entomology* **12**: 321–330.

Witherington, B. E. 1991. Orientation of hatchling loggerhead turtles at sea off artificially lighted and dark beaches. *Journal of Experimental Marine Biology and Ecology* **149**: 1–11.

Witherington, B. E. 1992*a*. Behavioral responses of nesting sea turtles to artificial lighting. *Herpetologica* **48**: 31–39.

Witherington, B. E. 1992*b*. Sea-finding behavior and the use of photic orientation cues by hatchling sea turtles. Ph.D. dissertation, University of Florida, Gainesville. UMI Dissertation Information Service, Ann Arbor.

Witherington, B. E., and K. A. Bjorndal. 1991*a*. Influences of wavelength and intensity on hatchling sea turtle phototaxis: implications for sea-finding behavior. *Copeia* **1991**: 1060–1069.

Witherington, B. E., and K. A. Bjorndal. 1991*b*. Influences of artificial lighting on the seaward orientation of hatchling loggerhead turtles (*Caretta caretta*). *Biological Conservation* **55**: 139–149.

Witherington, B. E., K. A. Bjorndal, and C. M. McCabe. 1990. Temporal pattern of nocturnal emergence of loggerhead turtle hatchlings from natural nests. *Copeia* **1990**: 1165–1168.

Chapter 14

Light, behavior, and conservation of forest-dwelling organisms

JOHN A. ENDLER

Animals use vision in many behavioral contexts that are essential for their survival and reproduction. These include finding and attracting a mate, mating, defending a territory, detecting food and oviposition sites, and luring in prey (Edmunds 1974; Hailman 1977). Their color patterns are also used in these contexts as well as for hiding from predators, startling or warning predators, hiding from ambushed prey, and thermoregulation (Wickler 1968; Edmunds 1974; Hailman 1977, 1979; Endler 1978; Burtt 1986). Plants use visual signals (flowers) for attracting pollinators, hiding from or confusing herbivores (Gilbert 1975; Rausher 1978; Williams & Gilbert 1981), hiding unripe fruit from dispersers and frugivores, and attracting fruit and seed dispersers (Willson & Whelan 1990). Plants also use the color of ambient light to determine rates of growth, stem elongation and branching direction, flowering and fruiting time, and position of fruits and flowers in the forest (Hart 1988; Salisbury & Ross 1992). Any factor that alters the function, reception, and perception of color can have dire fitness consequences at the population, species, or subspecies levels.

When an animal looks at a colored patch on an animal, flower, or fruit, its perception of the patch depends upon the interaction between the color of the ambient light, the intrinsic color of the patch, effects of fog or dust in the light path, the capture and processing of the light by the eyes and brain, and the properties of the other patches on the object and in the adjacent visual background (Hailman 1977, 1979; Lythgoe 1979; Endler 1978, 1986, 1990, 1991, 1993*a*,*b*; Burtt 1986). As a result, what is perceived may depend strongly upon the color of the ambient light. The color of ambient light in vegetation depends upon vegetation geometry (Endler 1993*a*), and can change as a result of habitat disturbance. If the light changes, perception of the object and its visual background can change, affecting the ability of the viewer to detect the object against the background. If the perception of the object and background change differently with changing light, then what was conspicuous in the old light environment can now be inconspicuous and vice versa (examples in Lythgoe 1979 and Endler 1991). Changes in the light environment may therefore make it more difficult to find food, predation may be more likely, and fewer matings and appropriate ovipositions may be achieved.

Changes in the light environment can have direct effects on many different species, but indirect effects may involve many more species regardless of whether their fitness depends upon vision or visual signals. Species live in interacting networks rather than in isolation. A change in one species can affect many other species in the food web even though only one of them is affected directly (Cohen *et al.* 1990; Kawanabe *et al.* 1993) by a change in ambient light. Consider, for example, a *Heliconius* butterfly living in the understory of a tropical forest. It obtains food as an adult by feeding on nectar and pollen, and its ability to detect flowers will affect the amount of time it can search for *Passiflora* vines on which to lay eggs, as well as the number of eggs laid. The butterfly's ability to detect flowers at the appropriate stages for feeding will depend upon the light environment. It uses vision for detecting the appropriate species of host plant (for example, Gilbert 1975), and also to determine whether or not eggs have already been laid on the plant (avoids cannibalism and crowding by the larvae), and whether or not the egg-shaped objects are actually eggs or egg-mimics used by the plant to reduce oviposition (Williams & Gilbert 1981). (Chemical cues are used for final decisions at short distances once the butterfly is within contact distance). Given that host plant recognition and egg mimicry involves various shades of green and yellow, the light environment is even more critical for oviposition behavior. A change of light environment can therefore have a direct affect on lifetime reproductive success through both survivorship and fecundity changes in the butterfly. In addition, there will be indirect effects on *Passiflora* and other plants whose flowers are pollinated by *Heliconius*. Reduced pollination would result in reduced or zero seed set and, subsequently, in less food for frugivores, seed predators, and seed dispersers. These effects would then affect other plant species and the insects and vertebrates that feed upon them. Specialists would be even more strongly affected. There is a similar chain starting with *Passiflora* and its herbivores and pollinators. Of course, changes in the light environment are likely to affect more than one species simultaneously, leading to the entire ecological web being affected. The main point is that a change in the conditions under which vision and visual signals are used can have repercussions throughout the food web and involve species that do not use vision as well as those that do. We can think of a change in light in a forest as likely to give rise to a chain reaction throughout the food web. Such cascading effects can lead to instability and extinction of many members of the ecological web (Cohen *et al.* 1990; Kawanabe *et al.* 1993).

Forests have a complex mosaic of light environments that differ in color and brightness (Endler 1993*a*). The conspicuousness or crypsis of visual signals will depend upon which light environment is used to illuminate the animal, flower, or fruit (Endler 1990, 1991, 1993*a*). These environments also provide valuable and specific cues for plant growth, flowering, and fruiting (references in Hart 1988; Salisbury & Ross 1992; Endler 1993*a*). Thus, for animals and plants the spectral composition of light can have significant fitness consequences (Burtt 1986; Endler 1986, 1990, 1991, 1992, 1993*a*,*b*). Alteration of the physical structure of a forest will affect the distribution of light environments, and hence the ability of species to use them. The purpose of this paper is to describe the light environments of forests, to examine the effects of forest disturbance on these light environments, and to suggest how a consideration of the quality of forest light environments may allow predictions about how certain species will respond under various kinds of disturbances.

Light spectra basics

A light spectrum is the physical manifestation of what we and animals perceive as color and brightness. In conjunction with a knowledge of the optical and neural properties of the eye and brain, we can use the light spectrum coming from an object to predict many aspects of the object's perceived color and brightness, even though some animals have very different vision than we do (Lythgoe 1979; Lythgoe & Partridge 1989; Endler 1990, 1991). A spectrum is simply a plot of light intensity versus wavelength, measured in units appropriate for vision (see Endler 1990 for detailed discussion). A rainbow is equivalent to our perception of the spectrum of sunlight, where brightness is intensity and color is proportional to wavelength.

There are four different kinds of spectra – ambient, reflectance, radiance, and absorbance – and all are needed to understand the function of color. An ambient light spectrum, such as in sunlight or shade, is the spectrum of ambient light striking the animal, flower, fruit, or visual background. A reflectance spectrum summarizes the 'intrinsic' color of an object and gives the fraction of ambient light that is reflected from the object towards the viewer at each wavelength. An example of a reflectance spectrum would be an intrinsically red object that reflects only long-wavelength light. A radiance spectrum is the spectrum of the beam of light from the animal, flower, fruit, or background to the viewer. A radiance spectrum from a given colored patch is a product of the ambient light spectrum and the reflectance spectrum of the patch. A white object illuminated by blue light, for example, will yield a blue radiance spectrum. A blue object illuminated by white light, however, will also yield a blue radiance spectrum, which is why we need to know something about both ambient light and intrinsic color. An absorbance spectrum summarizes the ability of the eye or one of its components (photoreceptors) to capture light at each wavelength from a radiance spectrum. Each species has its own set of absorbance spectra: the visual systems of animals are as diverse as the animals themselves and range from those with poorer color vision to those with color vision much superior (birds) to that of humans (Lythgoe 1979; Levine & MacNichol 1979; Jacobs 1981; Laughlin 1981; Goldsmith 1990). For further details, including the effects of water color for aquatic organisms and discussions of the effects of varying vision on perception, see Lythgoe (1979) and Endler (1986, 1990, 1991). This paper will be primarily concerned with the joint effects of ambient light and reflectance spectra because the vision of most animals (especially endangered ones) has not been studied, and because many of the effects of ambient light are independent of vision.

Spectra contain information that predicts how the light will be perceived. What we and animals perceive as 'brightness' or 'value' is the average height of the curve on the Y-axis (intensity), or more correctly, the area under the curve over a defined interval of wavelengths (X-axis). What is perceived as 'color' has two components: hue and chroma. Hue is 'color' in the vernacular sense of red, green, blue, etc., and is proportional to wavelength. In terms of human-perceived colors, the correspondences to wavelengths are approximately Ultra-Violet (UV), 320–400 nm; Violet, 400–440 nm; Blue, 440–490 nm; Green, 490–530 nm; Yellow, 530–570 nm; Orange, 570–620 nm; and Red, 620–710 nm. Hue, however, corresponds to wavelength only in the artificial situation of light of a 'pure' color, or a spectrum consisting of a very narrow and high peak at a particular wavelength and zero elsewhere. Natural colors in forests, woodlands, and grasslands never have spectra consisting of a single narrow peak but generally consist of broad curves. In this case, hue is roughly determined by the location on the curve where the intensity changes most rapidly with wavelength. An orange fruit, for example, will show little light below 600 nm, then a rapid increase in intensity around

600 nm and high intensity above 600 nm. A red fruit will have the same spectral shape, but the transition will be around 660 nm. Chroma, also known as 'purity' or 'saturation,' is a measure of how strong the color is. Pink, for example, is a low-chroma red and 'baby-blue' is a low-chroma blue. Chroma is proportional to the difference in total light intensity above and below the part of the spectra where intensity changes most rapidly. A pink flower, for example, will show a smaller difference in intensity above and below 660 nm than will a red flower. For detailed discussion and examples and the relationship to visual processes, see Endler (1990).

In summary, the perception of a given colored patch depends upon the product of the ambient light spectrum, the reflectance spectrum of the patch, and the absorption spectrum of the eye. Consequently, if the ambient light spectrum changes, then so too does the radiant spectrum reaching the eye, hence the perception of that patch. This is why a consideration of light environments, and the effects of changed light environments, is so important.

Light environments

Forests contain five light environments: Forest Shade, Woodland Shade, Small Gaps, Large Gaps, and Early/Late. These habitats are found in a variety of different kinds of forests, both temperate and tropical, broad-leaf and coniferous (Endler 1993*a*). Woodlands are forests in which most of the crowns do not touch and have many canopy gaps, whereas forests proper have continuous canopy and fewer gaps. All light environments are found in both woodlands and forests. The first four light environments occur during the day and the fifth occurs early and late in the day at sun angles less than 10 degrees from the horizon. The spectra of the light habitats vary and are shown in Fig. 14.1. Forest Shade is rich in middle wavelengths and appears yellow-green to humans. Woodland Shade is rich in shorter wavelengths and appears bluish. Small Gaps are rich in longer wavelengths and appear yellow-orange. Large Gaps are relatively even over all but the shortest wavelengths and appear whitish. Early/Late, which is deficient in middle wavelengths, occurs near dawn and dusk and appears purplish. When it is cloudy, the spectra converge on that of large gaps (Endler 1993*a*) and appear whitish.

The spectra of the five light environments depend upon the geometry of vegetation, the sun angle, and the fraction of the sky occupied by clouds (Endler 1993*a*). The ambient light striking the surface of an

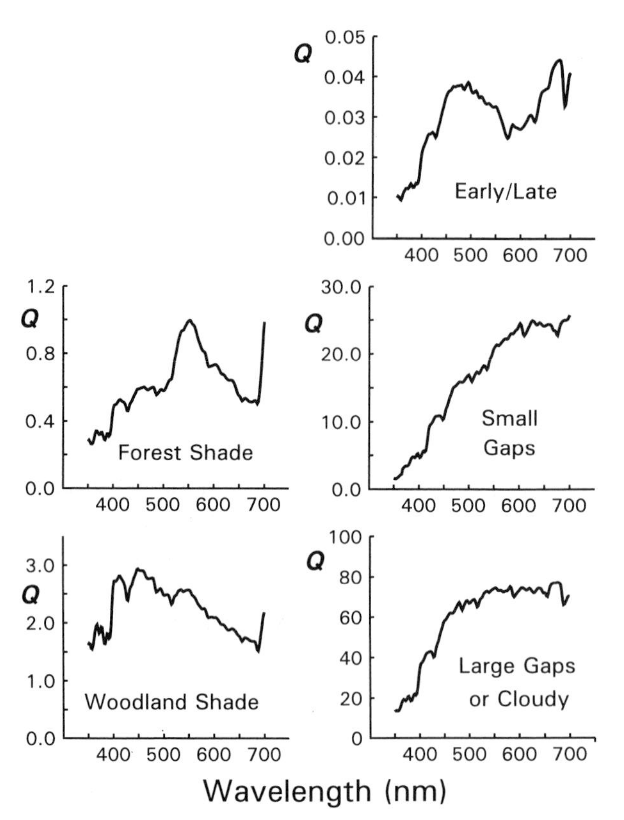

Fig. 14.1. The five light environments found in undisturbed forests. Each plot displays the irradiance Q (light reaching a surface from a 180° solid angle) as a function of wavelength (nanometers). To humans, below 400 nm is invisible (ultraviolet or UV), and 400–700 nm ranges from violet, through blue, green, yellow, orange to red. Q is measured in $\mu mol\ m^{-2}\ s^{-1}\ nm^{-1}$ rather than watts m^{-2}, lux, or other units because only the former is relevant to how animal and plant photoreceptors work. Forest Shade is rich in middle wavelengths (yellow-green to humans), Woodland Shade in shorter wavelengths (bluish), Small Gaps in longer wavelengths (yellowish-orange), Large Gaps roughly equal except the shortest wavelengths (whitish), and Early/Late deficient in middle wavelengths (purplish). The spectrum of large gaps is almost identical to that of open areas in full sunlight (Endler 1993*a*).

animal, flower, or fruit consists of light arriving from various light sources with different spectra. The principal sources are the sun (yellowish), the cloudless part of the sky (blue, if there is no air pollution or dust; white, brownish, to brownish yellow if dusty or polluted), clouds (white), live leaves (greenish), and bark and dead vegetation (various shades of gray, brown, and reddish-brown). The light environment in a particular place depends upon the relative contribution of these light sources to the ambient light spectrum, or technically, the product of their intensity and the solid angle that they subtend on the illuminated surface (for details, see Endler 1993*a*). Canopy holes affect the mixture of light from the sun and sky and the light reflected from vegetation (Fig. 14.2). Forest Shade and Small Gaps are associated with closed forest with canopy holes that subtend angles similar in size to that of the sun. Woodland Shade and Large Gaps are associated with large canopy holes. Forests with open crowns, such as pine and eucalyptus forests, tend to have less Forest Shade and Small Gaps, and more Woodland Shade and Large Gaps than

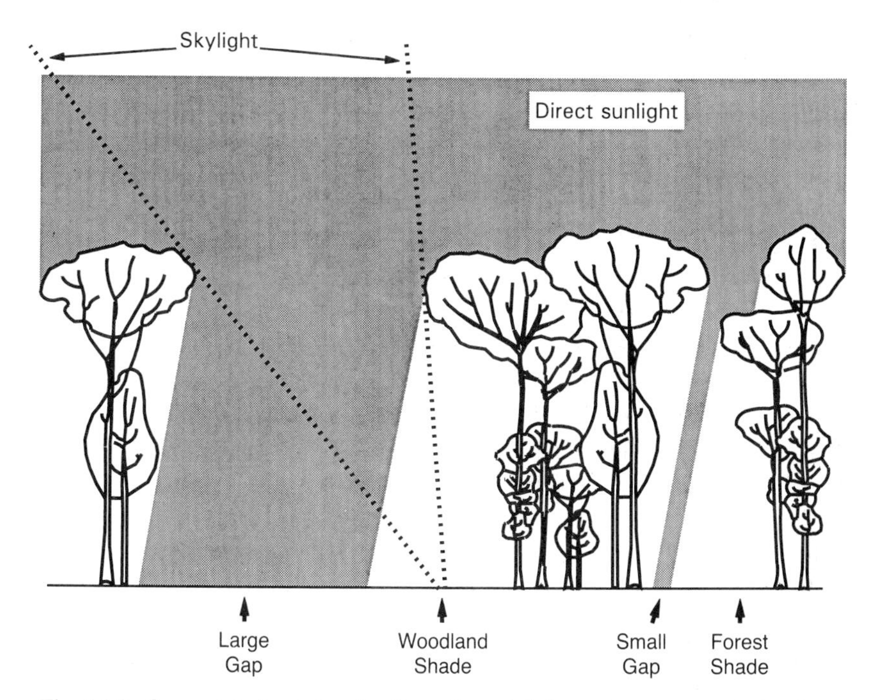

Fig. 14.2. Geometric basis for the four diurnal light environments. Animals or plants in Large Gaps are illuminated by direct sunlight and skylight from large holes in the canopy. Individuals in Woodland Shade are illuminated by skylight from large and/or many holes in the canopy but not direct sunlight. Individuals in Small Gaps are illuminated by direct sunlight but the canopy hole is too small to admit much skylight. Individuals in Forest Shade are mostly or entirely illuminated by light reflected by and transmitted through leaves and other vegetation. The color of the light at any one location is controlled by the size of the canopy holes, whether the sun is in a canopy hole, and by the fraction of holes that are filled with clouds. This geometry affects the mixture of light coming from the sun (yellowish), blue sky, white clouds, green leaves, and dead vegetation (see Endler 1993*a* for details).

do denser forests. Family or species composition, however, is not a good predictor of the frequency of light habitats; a dense redwood or spruce forest will consist mostly of forest shade and small gaps, and a tropical forest with unstable soil and frequent tree falls will have proportionally more Woodland Shade and Small Gaps than a redwood forest.

The spectrum of Forest shade (Fig. 14.1) is rich in middle wavelengths because most of the light is from sunlight transmitted and reflected from leaves, little or none comes from the sky, and none comes from the sun (Fig. 14.2). Forest Shade spectra are similar to that of leaves. The spectrum of small gaps is yellowish to yellowish-orange (Fig. 14.1) because virtually all of the light comes directly from the sun, almost none comes from the open sky, and the light from vegetation is so much less intense than direct sun that its contribution is minor (Fig. 14.2). Small Gap spectra are almost identical to the radiance spectrum of the sun. The spectrum of Woodland Shade is bluish (Fig. 14.1) because most of the light comes from the open sky (when unpoliuted), none from the sun, and comparatively little from vegetation (Fig. 14.2). Woodland Shade spectra are similar to, but not as chromatic (saturated), as blue sky. If the canopy gaps that cause Woodland Shade have clouds instead of blue sky, then the spectrum of Woodland Shade will be whitish (like Large Gaps) rather than bluish. In general, the degree of blueness is directly proportional to the fraction of the canopy hole or holes occupied by blue sky rather than clouds. With dust or air pollution, the spectrum of Woodland Shade will be brownish or whatever color the pollution induces in the open sky. The spectrum of Large Gaps is whitish (Fig. 14.1) because the intense spectrum of the sun at a small fraction of the sky area is counteracted by the complementary color of the blue sky at lower intensity but much larger total area (Fig. 14.2). Clouds also contribute to the whiteness of the light in Large Gaps. When the sun is blocked by clouds, and in the case of Woodland Shade, when a large fraction of the sky is covered by clouds, then the ambient light spectra of Forest Shade, Small Gaps, and Woodland Shade converge on that of Large Gaps or become whitish (Endler 1993*a*). The Early/Late light environment (Fig. 14.1) depends upon different phenomena. At low sun angles the light from the sun must pass a long way through the earth's atmosphere before illuminating an animal or plant. On its way it is absorbed by ozone. Ozone absorbs light most efficiently at 602 nm, resulting in the deficiency of middle wavelengths (Fig. 14.1), or purplish light. The depth of the central trough is proportional to the ozone concentration in the atmosphere (Endler 1993*a*). If there are a lot of clouds in the atmosphere, they will also be illuminated by the long path-length light. On the way, the shorter wavelengths are scattered (causing blue skies at high sun angles), leaving the longer wavelengths to go tangentially through the atmosphere. By the time the light strikes the clouds, most of it consists of long wavelengths, causing reddish sunsets and red-illuminated clouds. The radiance from the red-illuminated clouds has the effect of increasing the intensity of light to the right of the 600 nm trough, making the ambient light reddish-purple rather than purple, until the sun falls low enough to no longer illuminate the clouds (for details, see Endler 1993*a*).

In summary, the spectral properties, or color, of ambient light in forests varies in space and time. The spectrum at a particular place depends upon the sizes of canopy gaps, the percentage of the gaps covered by clouds, whether or not the sun is in the gap, whether or not the sun is covered by a cloud, and the sun's angle from the horizon. These light environments are found in forests with very different species composition because the environments are dependent on vegetation geometry. The differences between the light environments are much greater than differences due to variation in leaf color. As a result, the same light habitats are found, for example, in broadleaf, pine, and eucalyptus forests (Endler 1993*a*). Although the same light habitats are found in all forests, the proportion of the habitats depends upon the forest and canopy geometry. For example, Woodland Shade is proportionally more common in eucalyptus and pine forests, which have more open canopies, and Forest Shade is more common in forests with closed or dense canopies. A color pattern can be displayed anywhere in the forest, but the ambient light illuminating the color pattern will be a predictable function of the time and place in which it is displayed. In addition to daily changes, there will also be seasonal changes, because seasons differ both in cloud cover and in sun angle timing (Hailman 1977; Endler 1992, 1993*a*). Species have evolved to use particular seasons, and their color patterns and behavior are adjusted accordingly.

Crypsis and conspicuousness in light environments

The color patterns of animals, flowers, fruit, or their visual backgrounds may be regarded as mosaics of patches differing in reflectance spectra (color, brightness), size, and shape (Endler 1978). Color patterns can be described by the statistics of the distributions of spectral and other properties of the component patches – pattern statistics. A color pattern is cryptic (inconspicuous) if it represents a random sample of the visual background at the time and place it is viewed (Endler 1978). The degree of crypsis can be measured by the similarity between the pattern statistics of the visual background and animal (or plant) color patterns (Endler 1978, 1984, 1991). Crypsis may take on more specific forms in which the animal mimics inedible objects (Wickler 1968), but the same principles apply. There are fewer ways to be cryptic than to be conspicuous because a deviation between animal and background in any one or more of the pattern statistics will make the pattern more conspicuous, whereas all statistics must be similar in order to be cryptic. Nevertheless, most natural visual backgrounds (especially in forests) are sufficiently complex that there are a very large number of ways of being equally cryptic on the same visual background (Endler 1978, 1984, 1988). Ambient light (Fig. 14.1) affects the appearance of color patterns regardless of the color vision of the viewer because the light traveling between the color pattern and the viewer (radiance) is the product of the ambient light spectrum and the reflectance spectrum of each patch (Endler 1990). It is also affected by fog or dust between the object and viewer, but I will neglect these effects here (see Lythgoe 1979 for details). There are two effects of changes in the ambient light spectrum: brightness and color.

A change in the ambient light spectrum can affect the brightness (radiance) of a color patch, and hence, the brightness contrast of the entire color pattern. If a patch reflects only long wavelength light (orange-red, say above 600 nm), and it is illuminated by Woodland Shade or Forest Shade, then that patch will not be very bright because the ambient light is not rich in long wavelength light. On the other hand, if the same patch is illuminated by Gaps or Early/Late, then it will be brighter, especially in Small Gaps where it will reflect much more of the ambient light (Table 14.1). In contrast, a yellow-green patch (reflecting only at intermediate wavelengths, say 520–620 nm) will be brightest in Forest Shade and least bright in Early/Late light environments (Table 14.1; more examples in Endler 1986, 1990). The effect will be strongest if one light environment has a similar spectral shape to the patch's reflectance spectrum and the other is very different. In general, a shift between light environments will cause a shift in the relative brightnesses of the components of color patterns. The shift will be stronger if some patches have similarly and other patches have differently shaped spectra from the light environments. Differences among light environments affect brightness contrast within color patterns and between the color pattern and the visual background, hence the degree of conspicuousness (Hailman 1977; Endler 1978, 1986, 1991, 1992, 1993*a*; Burtt 1986).

Table 14.1 *Percentage total reflectance (350–700 nm) of ambient light by two color patches illuminated by six different light environments.*

	Forest Shade	Small Gaps	Woodland Shade	Large Gaps	Early/ Late	White[a] (for comparison)
Orange-red	29.3	39.8	23.8	35.7	34.26	29.0
(600–700)[b]	0.0%[c]	+35.8%	−18.7%	+21.8%	+16.9%	
Yellow-green	39.4	37.2	29.9	35.5	28.7	29.0
(520–620)[b]	0.0%[c]	−5.6%	−24.1%	−9.9%	−27.2%	

[a] White light has the same intensity at all wavelengths.
[b] Each color patch reflects 100% of the ambient light between the indicated wavelengths and no light outside those limits.
[c] Percentage change in total reflectance when illuminated spectrum shifts from Forest Shade.

A change in the ambient light spectrum can affect the color (spectral shape) of a color patch and hence the color contrast of the entire pattern. Consider a flower with three shades of blue. The reflectance spectra of the three patches are shown in the upper-left panel of Fig. 14.3. They have similar brightness because the areas under their curves are similar; if they were illuminated by white light, they would also reflect a similar amount of light (similar total radiance). The three patches have the same hue (blue-violet to human eyes, also reflecting in the UV) because they all reflect more light below 450 nm and less light above 450nm. The three patches have different chroma. Patch A reflects a lot more light below 450 nm than do B and C, and B reflects more below 450 than C, so the chroma of A is greater than B and both are greater than C (for more examples, see Endler 1990). Now consider the effects of illuminating patches A, B, and C with the five light environments found in forests (Fig. 14.3); this is obtained by multiplying the irradiance spectra of each light environment (Fig. 14.1) by the reflectance spectrum of each patch (Fig. 14.3). Patch A, with high chroma, is relatively insensitive to varying light environments; its radiance spectrum always shows more light below 450nm, so it will always be perceived as blue-violet-UV. Patch C with low chroma is very sensitive to the ambient light. In Woodland Shade it will still be perceived as bluish, but this will not be true in the other environments. In fact, in Small Gaps, Patch C actually appears yellowish because it is radiating more long-wavelength light than it radiates in the short wavelengths. In Large Gaps, or when the sun goes behind a cloud, Patch C will actually appear close to colorless (gray). Patch B, with intermediate chroma, has an intermediate sensitivity to the ambient light.

In summary, patches with low chroma will change chroma (patches B and C) and may also change hue (Patch C) if displayed in different light environments, whereas those with high chroma are less sensitive to changes in ambient light.

Because patches in natural color patterns vary in hue and chroma, changes in the light environment can have very strong effects on color contrast within the color pattern, as well as between the color pattern and the visual background. For example, if some patches change hue and the others do not as light changes, a cryptic pattern will suddenly be conspicuous, and the rate of predation may increase. Visual backgrounds in forests will also change in hue, chroma, and brightness because, aside

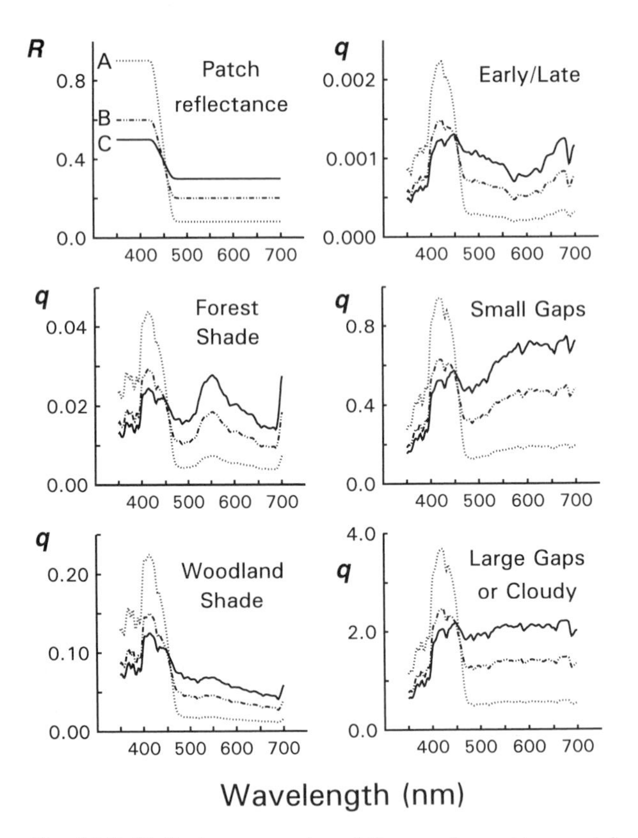

Fig. 14.3. Reflectance spectra of three color pattern patches (upper left panel) and their resulting radiance spectra when illuminated by each of the five light environments in forests. Note how the shape (color) of the radiance spectrum of patch A changes little with ambient light, whereas that of C changes almost as much as the ambient light spectra. A gray patch (flat reflectance spectrum) would always radiate the same color (spectral shape) as that of the ambient light.

from live healthy leaves and some flowers, all other patches have low chroma and varying brightness (Endler 1990, 1991, 1992, 1993*a*). Species, therefore, that must be cryptic in a variety of light environments must have patches that change with ambient light in the same way as the background patches. Species that require conspicuous colors in sexual, territorial, or warning (aposematic) displays must either choose low chroma colors for specific light environments, or high chroma colors that retain their contrast and hue independently of light environment. Too few species have been studied to know whether or not these hypotheses are true.

In summary, the degree of inconspicuousness or conspicuousness depends strongly upon the light environment in which the color pattern is illuminated. Changes in light environment alter the relative brightness as well as the chroma of each colored patch, and the effects are strongest with low-chroma colors. In addition, the hue of the patches, especially low chroma patches, can actually shift in different light environments. As a consequence of the dependence of the visibility of a color pattern upon the light environment, microhabitat choice and choice of light environments can either enhance or decrease visibility.

Examples of the effects and use of the light environments

The interaction between visibility and light environments has been investigated in detail in a fish (*Poecilia reticulata*) and three lekking bird species (*Rupicola rupicola*, *Corapipo gutturalis*, and *Lepidothrix serena*). Leks are sites where groups of males congregate and display jointly to females, females choose and mate with a male, and females receive only gametes from males (Bradbury 1981). Leks are microsite specific and long-lasting and are ideal systems to start investigations of the effects of ambient light environments.

Guppies (*Poecilia reticulata*, Poeciliidae) are small fish that live in the clear mountain streams of northeastern South America and adjacent islands. The streams are shallow (generally less than 1 m deep), have clear water, gravel bottoms, and run through tropical rain forests. Guppies are genetically polymorphic for male color patterns, and the color patterns of a particular local population represent a balance between predation (crypsis) and sexual selection. Females prefer the brighter and more colorful males (summaries in Endler 1983, 1995). (Guppies do not form leks.) Although all five light environments are available to all guppy populations in undisturbed forest streams, guppies do most of their conspicuous displays to females in the Early/Late light environment, whereas most predation takes place in the other four light habitats (Endler 1987, 1991). The color patterns consist of one or more UV, blue, bronze-green, yellow, orange, red, purple, silver, white, brown, and black patches. Although all of these color classes are found in most guppy populations, most guppies have color patterns that are predominantly blue (and UV), orange, purple, brown, and black with some silver. The genes controlling yellow are rare, and bronze-green tends to be limited to a small surface area, if present at all. The result is that guppies have the highest brightness and color contrast in the Early/Late light environment and the lowest brightness and color contrast in the other light environments, especially Forest Shade (see Endler 1991 for details). This effect is very interesting because it increases visibility to potential mates during courtship and decreases visibility to potential predators at other times and places (Endler 1991). The rarity of yellow patches, which are known to be genetically determined, is also striking because yellow would be most conspicuous during times and places of maximum predation risk (especially in Forest Shade), and least conspicuous in the Early/Late courtship light environment. The differences in color and brightness contrast in the different light environments are made even greater by the differences in color vision between guppies and their predators (Endler 1991). Microhabitat choice of light environments allows guppies to reduce the conflict of selection pressures on color patterns so that the same color pattern appears conspicuous when used in courtship, but relatively less conspicuous at other times and places when predation risk is high.

The Guianian cock-of-the-rock (*Rupicola rupicola*, Cotingidae) lives in the tropical forests of the Guianas, northeastern South America. Leks are located in patches of forest that are qualitatively different from the surrounding forest. The species composition of this forest is a highly select subset of the species found in the surrounding forest, consisting mostly of trees whose fruit *R. rupicola* eats and then defecates the seeds on the lek site (Théry & Larpin 1993). This forest has a thin low canopy with relatively more transparent leaves and produces Forest Shade with more chroma than usual (Endler & Théry 1996). Small Gaps are abundant but Large Gaps and Woodland Shade are rare and distant from the lek site. Males display on horizontal perches about 2 m above a 1 m^2 area of ground cleared of leaves, and fly to the cleared patch when a female arrives. Each male 'owns' a particular perch and cleared area, but males visit the perches of males in the same aggregate of perches (Gilliard 1962;

Trail 1985, 1987). Males are mostly orange- red with yellowish-orange wing 'strings' (silky fringe feathers with reduced barbs emerging from the innermost secondary feathers) and brown-and-white wing bars. Males display only when the beams of light from Small Gaps pass over their horizontal perches, and display with part of the body in the Small Gap and the rest of the body in Forest Shade. Males fight for the position that allows simultaneous illumination by Small Gaps and Forest Shade. When the present perch loses its sunbeam, the males move to new perches that have sunbeams in the appropriate position (Endler & Théry 1996). All displays stop when the sun goes behind a cloud (Gilliard 1962; Endler & Théry 1996), which converts both Forest Shade and Small Gap spectra to that of Large Gaps (Fig. 14.1; Endler 1993*a*).

The combination of behavior, habitat choice, and light environment is similar to that of guppies. When *R. rupicola* is entirely within forest shade, it is relatively cryptic, in fact surprisingly so if one has only seen this bird in a field guide or museum. The reflectance spectrum of most of the bird is deep orange; there is little reflectance below 650 nm and high reflectance above that wavelength (Endler & Théry 1996). When this color is illuminated by Forest Shade, there is very little radiance from the bird because forest shade has relatively few photons above 650 nm (Fig. 14.1). In addition, the total reflectance (brightness or luminance) of the bird is not significantly different from that of most elements of the visual background in Forest Shade. The bird is therefore cryptic with respect to brightness, even if it does mismatch in hue and chroma (Endler & Théry 1996). On the other hand, the light of Small Gaps is very bright at the same wavelengths that *R. rupicola* reflects strongly, so the brightness in Small Gaps is very high indeed, and in fact, much brighter than any object in the background. The effect of the extremely high intensity and chroma of the part of the bird in the Small Gap, adjacent to the very low intensity in the adjacent Forest Shade, is to make the displaying male extraordinarily conspicuous, as well as considerably more conspicuous than a male on the same perch who did not win the position allowing partial illumination by the Small Gap. The male cock-of-the-rock is thus maximizing its visual contrast during its display, but minimizing it off the display perch. Minimizing conspicuousness when not displaying is important because these leks last for a hundred years or more and predators learn their locations (Trail 1987). The higher than average chroma of the Forest Shade of the lek forest compared with the forests in which these birds forage also means that they are relatively more cryptic in the lek forest when not displaying than in the foraging areas. This difference in conspicuousness between lekking and foraging light environments is important because the lek site is a more predictable potential food source for predators than the foraging site. Note that these birds cannot use the border between Forest Shade and Large Gaps because such a border is geometrically impossible (large canopy holes produce Large Gaps with Woodland Shade surrounding them; Forest Shade cannot start until many tens of meters away from the large canopy hole, far enough away for the hole to subtend a very small angle). The boundary between Large Gaps and Woodland Shade is probably not used for two reasons: (1) it is more difficult to find a perch with a sunbeam edge and a sunbeam small enough to illuminate only part of the bird; and (2) the brightness contrast and color contrast of the birds is much greater in Forest Shade than Woodland Shade. In summary, the cock-of-the-rock, like the guppy, maximizes its conspicuousness to potential mates while minimizing its conspicuousness to predators by selecting light habitats that generate these differences in conspicuousness (Endler & Théry 1996).

The White-throated manakin (*Corapipo gutturalis*, Pipridae) is a small bird native to the tropical forests of northern South America. Mature males display on moss-covered fallen logs that they keep clear of fallen leaves and twigs, and there is also an aerial display above the log. Like *R. rupicola*, each male 'owns' a given log but visits adjacent logs in the cluster (Théry 1990*a,b*). The surrounding forest consists of Forest Shade and Small Gaps, but the size of the sun patches is larger than those illuminating the *R. rupicola* lek, and the spectra of these gaps is intermediate between that of Small Gaps and Large Gaps (more white than yellowish). The logs are sufficiently old that the canopy holes created by their fall are almost completely filled in, so there are neither Large Gaps nor Woodland Shade light environments nearby (Endler & Théry 1996). As with *R. rupicola*, male displays occur and female arrivals are more common when a sun patch illuminates part of the log. Successful males also have much darker visual backgrounds away from the log than unsuccessful males (Théry 1990*a,b*). Males have a bluish-black back, flanks and wings, black cap, white throat, white chest, and white wingbars. Male displays consist of a high velocity swoop from a high perch in Forest Shade to the log. The male lands at the edge of the sun patch with the dark back in the shade and the white chest and throat in the sun with the head pointed up. The landing is more of a bounce, as the male then jumps above the log and through the sun patch to the opposite sun patch edge and lands again with his dark parts in Forest Shade and his light parts in the sun patch. If a female is present he will back up towards her, displaying his conspicuous white wing bars (Théry 1990*a*,*b*; Endler & Théry 1996). Displays rarely occur without a patch of the proper size (around 1 m of the log illuminated), and will stop if the patch disappears, either because the sun is obstructed by a cloud, obstructed by the experimenter, or if the earth's rotation moves the patch away from the log (M. Théry, unpublished data).

Once again, the combination of the light environments, the animal color pattern, and the behavior interact to maximize the visual contrast when contrast is needed, such as in displays to conspecifics. Unlike the previous two species, color is not so important, although there may be an unknown effect of the green reflection of the moss off the white chest and throat. The contrast between the sunlit chest and throat with the shaded dark head, back and wings, is very high. At other times when a male is wholly in Forest Shade, his contrast will not be as great. The light intensity of the larger-than-average Small Gaps is 10–30 times brighter than Forest Shade, and so light-adaptation to the Gap may further reduce perceived contrast when the bird is wholly in Forest Shade. By analogy, an underexposed or overexposed photograph of a black and white object will have lower contrast than a properly exposed photograph of the same object. The blue iridescence of the back increases the difference in effects of the light environments. Blue iridescence is a result of structural colors, which require directional light to work best. As a result, the blue is brighter and has more chroma in direct sunlight (the Gap) than in the diffuse light of Forest Shade. In addition, the structural blue of *C. gutturalis* is weak, and so only shows when the bird is in the Gap. The result is that the blue shows only during the passage through the Gap during the display. If a cloud blocks the sun, then the boundary between Forest Shade and the Gap disappears, the intensity is the same in both places, and the diffuse light of the clouds obliterates the visibility of the structural blue. In summary, the color patterns and behavior of *C. gutturalis* increase the bird's conspicuousness to potential mates while reducing its conspicuousness to predators by selecting light habitats that generate these differences in conspicuousness (Endler & Théry 1996).

The White-fronted manakin (*Lepidothrix* [=*Pipra*] *serena*, Pipridae) also lives in the tropical forests of northern South America. It prefers the denser parts of forests with nearly complete canopy and few canopy holes. Its lek display is different from most lekking birds in that the lek location is not constant, and it breeds more intensely in the wet season. The lek moves downhill as the sun rises above the hill at the opposite side of the valley. The location stays in the shadow of the mountain just in front of the advancing zone of Forest Shade and Small Gaps (Théry 1990*a,b*; Endler & Théry 1996). Virtually all displays occur when the sun goes behind a cloud, and very few displays occur when the sun is unobstructed. When the sun comes out, the birds immediately start foraging for insects, which may be more active at that time. Thus, *L. serena* displays in the whitish light of cloudy conditions (Fig. 14.1), but does not display in Forest Shade (Endler & Théry 1996). The lek sites are in areas of densest vegetation, and so Small Gaps are comparatively rare, and Large Gaps and Woodland Shade are not available. The total light intensities (Q_T) of the display conditions are significantly lower than the average for Forest Shade under cloudy conditions. *L. serena* males have a white cap, black and bluish-black back, turquoise-blue rump, and yellow-orange chest. The white cap has the highest reflectance of any natural color I have ever measured, the bluish-black back is a structural color as in *C. gutturalis*, but the turquoise-blue rump appears to be a mixture of pigment and structural colors.

L. serena behavior and color patterns are affected by the light environments as in the other three species. The remarkably reflective white cap maximizes conspicuousness in the unusually dark display conditions, but also will be conspicuous in other light environments. The black back will make most of the bird inconspicuous against the background in any light it uses, because both Forest Shade and Cloudy conditions are diffuse light, minimizing the chroma of the pale structural blue. (The function of this color is obscure as the birds avoid direct sunlight.) The turquoise blue rump patch changes the most when the sun goes behind a cloud and then re-emerges. The reflectance spectrum of this patch shows high reflectance from UV to about 500 nm, and it declines rapidly above 500 nm. Forest Shade has few photons in the part of the spectrum that is reflected most strongly by turquoise, and has more photons where the turquoise patch does not reflect. Consequently, the rump patch will reflect relatively little light in Forest Shade, and its contrast with the rest of the back and tail will be reduced. Cloudy conditions are rich in all wavelengths, and the intensity of a surface illuminated by Forest Shade changes little when the sun becomes blocked and the spectrum is converted to the white cloudy light environment (Endler 1993*a*). When the sun is blocked by clouds, a greater proportion of the ambient light is reflected by the turquoise patch than occurred in Forest Shade, while there is little change in the spectrally flat back and tail. Consequently, when the sun is blocked by clouds, the brightness and chroma of the turquoise patch increases, and contrast with the back and tail is increased. So by displaying only during cloudy conditions, the color contrast of the bird goes up during displays and down when it is sunny and the bird is foraging. Note that this effect of the light environment on visual contrast will work only under the nearly dense canopy that results in Forest Shade when it is sunny; it will not work in Woodland Shade. The spectrum of the yellow-orange chest is complementary to the rump patch and helps to increase color contrast, but is not affected as much as the turquoise patch by changes in light environments. In summary, the color and brightness contrast of *L. serena* increases during the time and at the places in which males display to females, but is lower at other times and places (Endler & Théry 1996).

In all four species that have been analyzed for the effects of light environments on visual contrast, the color and/or brightness contrast is greater at the time and places of displays to potential mates, and is smaller at other times and places. The appropriate habitats and light microhabitats must be present, not only to yield the appropriate radiance spectra from the patches, but also to give the appropriate cues to elicit the appropriate display behavior.

This conclusion applies to plants as well. Plants use the spectral composition of ambient light to grow towards or away from light gaps, time their flowering and fruiting, and adjust their growth rates (Hart 1988; Salisbury & Ross 1992), so the presence of the appropriate light habitats is essential for survival and reproduction. Although plants do not 'behave' in the animal sense, over longer time scales they do have choices of where to display their fruits and flowers, where to place leaves and branches, and even when seeds should germinate. Certain locations will experience different light environments or proportions of light environments depending on the time of day and weather. In addition, the wet season will have different proportions of the light environments than the dry season. This means that forest plants could potentially take advantage of changes in visibility with light environment in a way similar to animals, albeit less precisely.

Effects of forest disturbance on light environments

The presence and abundance of the light environments will be affected by any kind of damage to the forest that modifies the geometry of the vegetation (Fig. 14.4). Figure 14.4a is a simplified view of the geometry of a forest. Note the presence of Small Gaps (s) and Forest Shade (f) on the forest floor, and Large Gaps (l) and Woodland Shade (w) in the vicinity of

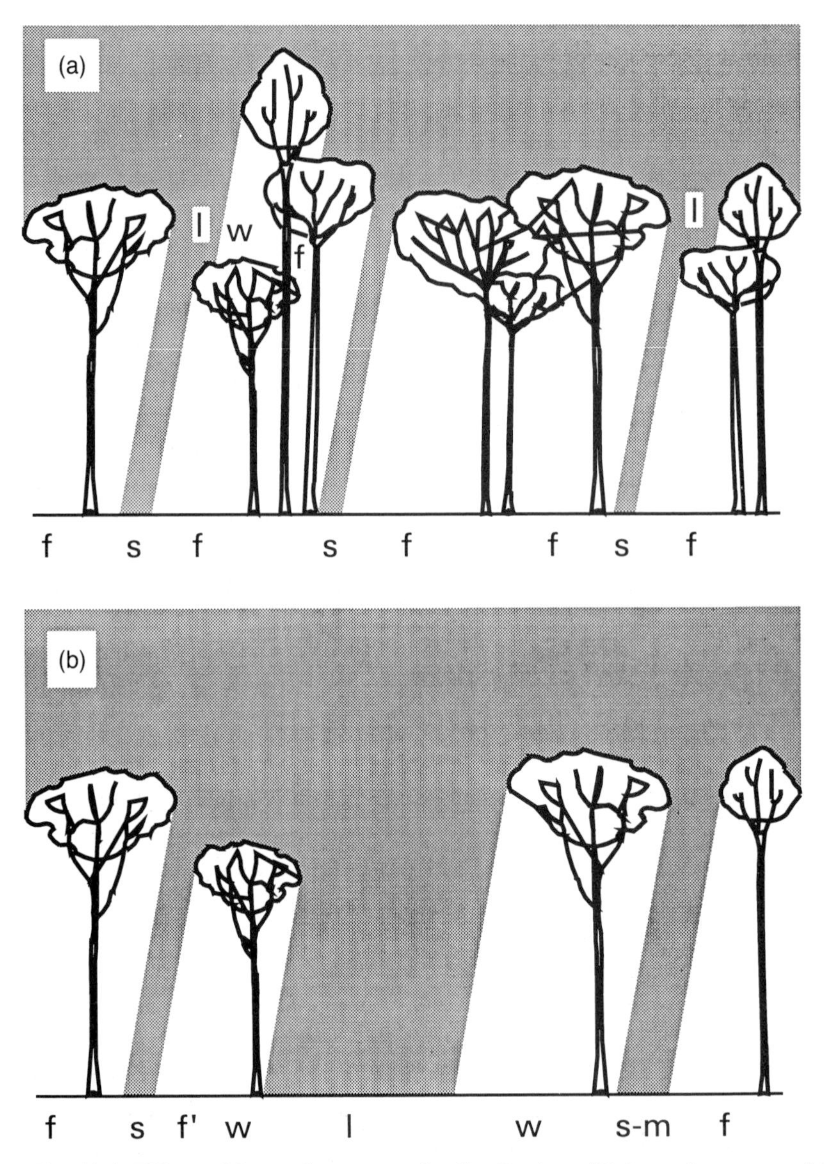

Fig. 14.4. Effects of forest damage on the distribution of light environments. (a) Before damage. (b) After damage. Light environment codes: f, Forest Shade; s, Small Gaps; l, Large Gaps; w, Woodland Shade; f', modified Forest Shade, transitional to Woodland Shade; s-m, modified Small Gaps, transitional to Large Gaps. Note the presence of l, w, f in the canopy whenever there are canopy emergents or other complex structure (s is also possible, but not shown). The positions of these habitats change as the sun angle changes. For example, the f in the canopy of (a) can change to s with a sun angle of about 60° to the left of the illustrated angle.

canopy emergent trees. Large Gaps and Woodland Shade are also found on the forest floor at and near tree fall openings (Fig. 14.2), and all habitats can be found at any height, depending upon the geometry of the canopy (details in Endler 1993*a*).

Now consider the effect of disturbance (Fig. 14.4b). Even the removal of a single tree can have a significant effect. On the far right a single subcanopy tree was removed, causing a larger-than-normal small gap (s-m), and a smaller area of forest shade. The ambient light spectrum in the larger Small Gap would be more whitish and less yellowish, and the spectrum of the adjacent Forest Shade would have an increased contribution from the open sky and would be more blue-green than green, or, if cloudy, much less green than normal. The effect of converting forest to parkland (rest of Fig. 14.4b) has a drastic effect on the distribution of light environments by greatly reducing the frequency of Small Gaps and Forest Shade from the forest floor, and completely eliminating them from the canopy. The clump of trees, which has a canopy gap small enough to produce a regular Small Gap, and adjacent Forest Shade may result in only a modified version of Forest Shade (Fig. 14.4b, f′, blue-green rather than green) because of the proximity of the adjacent large canopy hole, and hence Woodland Shade. Removing even more trees would result in the complete elimination of Forest Shade and Small Gaps. It is clear that the greater the disturbance, the lower the frequency of Forest Shade and Small Gap light environments. The elimination of these light environments occurs at a higher rate in the canopy than in the understory, and still less on the forest floor. The removal of trees, therefore, should result in a greater effect on canopy than on subcanopy species, and still less effect on forest floor species.

Generally, we can conclude that any degree of disturbance to a forest can have profound effects on the distribution and abundance of light environments. Species that require the light environments most affected by disturbance (Forest Shade and Small Gaps) will be more strongly affected than species requiring Woodland Shade and Large Gaps. Species that use Large Gaps, however, or only display under cloudy conditions, will not be as affected by forest disturbance as those that display when the sun is unobstructed and there is much blue sky. Species (such as guppies) that display in the Early/Late light environment will be less affected by forest disturbance than species that display in the middle of the day because the Early/Late environment is found in all parts of the forest at low sun angles.

Effects of air pollution on forest light environments

Although the Early/Late environment is relatively insensitive to disturbance of forest geometry, it is affected by air pollution. Early/Late species are sensitive to ozone depletion in the atmosphere because ozone is responsible for the purplish light of the Early/Late environment (Endler 1993*a*); the more ozone, the deeper (more chroma) the purple. Ozone depletion will convert the purplish light to gray, drastically changing the visual contrast of visual signals used at low sun angles. A conversion of the purplish to gray light, for example, will drastically reduce the visual contrast of courting guppies (Endler 1991) and, ultimately, their reproductive success and the population's ability to recruit enough offspring to keep up with predation (Endler 1995). This could happen to any species like guppies that primarily use the Early/Late light environment for reproduction or territoriality. For example, it will affect birds that use visual as well as vocal displays during the dawn chorus. It may also affect those species that forage visually in this environment.

Air pollution is also likely to affect species using Woodland Shade because this light environment is primarily determined by the open sky. In an atmosphere free of dust or air pollution, Tyndall scattering makes the sky blue, hence giving Woodland Shade its bluish spectrum. If there is much dust, such as from natural sources in arid or other environments undergoing construction or road building, the air will be filled with larger dust particles. Larger dust particles cause scattering of all wavelengths (Mie scattering), making the sky grayish rather than blue (discussion in Endler 1993*a*). This scattering of all wavelengths will convert Woodland Shade to the whitish light of Cloudy conditions, and will reduce the visual contrast of animal or plant signals specialized for maximum contrast under bluish light. Cryptic patterns that are specialized for Woodland Shade will become more conspicuous as reddish-browns and reds increase in brightness and chroma. These changes can have significant effects on both signaling and predator–prey relationships.

On a longer time scale, any climatic change that alters the cloud cover and its seasonal characteristics will shift the proportions, and hence availability, of the light environments. If other cues for seasonality do not change, or change at a different rate, then the unchanged color patterns will be exposed at the wrong time and place, yielding lower reproduction, poorer territorial defense, and upset the delicate balance of natural selection on color patterns between signaling and predator avoidance.

Conclusions

Forests contain five light environments that are available to organisms for use in signaling to potential mates; territorial infringers; pollinators; fruit and seed dispersers; warning or hiding from potential predators, herbivores, and parasites; finding food; and for plants, the pattern and direction of growth and timing of flowering, fruiting, and germination. All of these factors affect survival and reproductive success, and some of the interactions are shown in Fig. 14.5. Each species uses a different combination of particular light environments and, for the four species examined in detail, is specifically adapted in behavior and color patterns for these environments. The success in each of these functions depends upon the appropriate light environment being present. If the appropriate light environment is very rare or absent, then the efficiency of these functions is reduced or destroyed, endangering the population, subspecies, or species.

The light environments depend upon weather, sun angle, and forest geometry, particularly the size and frequency of canopy holes. Human-induced disturbance alters the geometry of forests, and as geometry determines the presence and relative abundance of the light environments, disturbance must alter the presence or absence and relative abundance of the light environments. It is therefore likely that disturbance will significantly affect all the components of fitness that depend upon light for both animals (Fig. 14.5) and plants. As mentioned in the introduction, the effects will be direct on the species using light, but also will be indirect throughout all parts of the food web that interact with the light-dependent species. Consequently, altered light environments can have potentially catastrophic effects on individual species as well as ecosystems. Note that these effects are in addition to the direct effects of disturbance, such as those found in physically destroyed habitats, but can also occur even when there appears to be some 'appropriate' habitat left, when measured by conventional methods (shelter, food availability, etc.).

Although the importance of the light environments is large, very few species have been studied. In fact, the function and importance of environmental light has been an active area of research for less than 20 years (Hailman 1977; Lythgoe 1979), and the forest light environments have only recently been disentangled and characterized (Endler 1993*a*). The entire subject is ripe for major discoveries, and we can only guess how important and widespread are the effects described in all four species

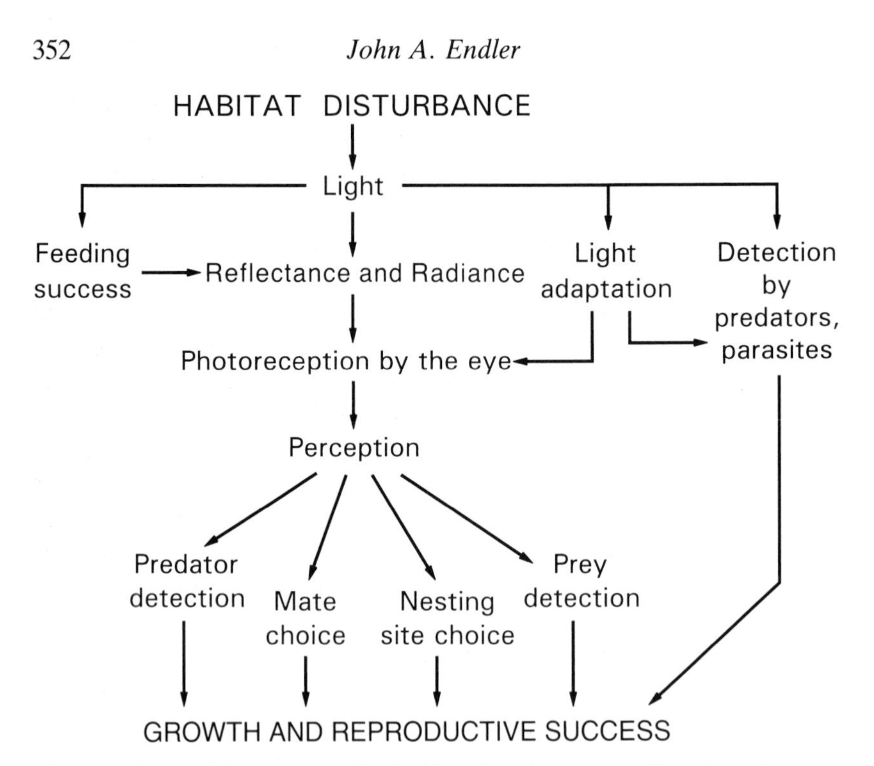

Fig. 14.5. Some of the ways in which habitat disturbance can affect the well-being of an animal species through its interaction with light environments. Plants will experience additional effects because they use the color of light as cues for growth, development, and timing of flowering, fruiting, and germination. Feeding success influences the input of pigments (such as carotenoids) or pigment precursors which will directly affect the reflectance spectra of some color pattern patches. Light adaptation will affect the ability of an eye to perceive objects as well as to perceive brightness and color contrast. Light adaptation is the equivalent of choosing the correct f-stop of a camera to give the proper exposure; too high or too low a setting will reduce contrast, but the same setting will not work well in another location with more or less total light available. Light adaptation also affects the ability to detect different colors because photoreceptors sensitive to different sets of wavelengths will light adapt differently if the ambient light is not white.

that have been studied in detail (Endler 1991; Endler & Théry 1996). Other aspects of light and color patterns, such as thermoregulation and glare reduction (Burtt 1986), may also be significantly affected by disturbance. Given that this entire subject is so little explored, I expect that there will be a certain amount of hesitation to consider these phenomena. Our reluctance is exacerbated by the fact that we are so visually-oriented and have unconsciously trained ourselves to ignore these effects. A little observation with a white card in any forest during a sunny and comparatively cloud-free day will demonstrate the light environments, and it will be difficult to look at a forest in the same way once the light environments are noted. In any case, we cannot tell whether or not a new phenomenon is important without seriously studying it.

The highly light-specific behavior of the four species discussed here, and the potential significance of these effects on their fitness and persistence, points out the need to investigate the use and importance of light environments on a large variety of species and ecosystems. The kinds of questions we need to ask are: (1) in the undisturbed habitat of a species, what light environments are present and which ones are used? (2) How frequently are they used? (3) For what purposes are the environments used? (4) Which light environments are essential for (a) foraging and finding food, (b) avoiding predation, parasitism, herbivory, nectar robbers, fruit predators/parasites, (c) defense of territory and/or young, (d) attracting a mate or pollinator, and (e) raising young, maturing fruit, or germination? (5) What is the effect of various kinds of forestry, lumbering, or other physical disturbance on the light environments at various levels in the forest? (6) What is the effect of various kinds of disturbance on the uses (item 3) and essential uses (item 4) of each light environment? (7) Which light environments are absolutely critical for continued existence of the population, and how are they affected by disturbance? (8) Can any modification in forestry practices be made to maintain critical light environments as well as the more conventional requirements of the species? (9) What are the relationships between the light environments and the conventional needs of species? (10) What kinds of disturbances are more likely to affect many species? (11) What kinds of disturbances are most likely to affect 'keystone' species in ecological webs? We essentially know nothing of the answers to any of these questions, yet they are potentially very important to understanding a species' requirements, and hence preventing its extinction.

The general conclusion is that an attempt to maintain an endangered species in its natural habitat will require information on the light environments and natural geometry of its habitat, and not just upon the more conventional ecological parameters such as habitat, food, and microclimate. We need detailed knowledge of the species in its natural environment in order to understand what goes wrong when its habitat is disturbed.

Literature cited

Bradbury, J. W. 1981. The evolution of leks. Pages 138–169 in R. D. Alexander and D. W. Tinkle, editors. *Natural selection and social behavior: Recent research and new theory*. Chiron Press, New York.

Burtt, E. H., Jr. 1986. An analysis of physical, physiological, and optical aspects of avian coloration with emphasis on wood-warblers. *Ornithological Monographs* **38**: 1–125.

Cohen, J. E., F. Briand, and C. M. Newman. 1990. *Community food webs: Data and theory*. Biomathematics, Vol. 20. Springer, New York.

Edmunds, M. 1974. *Defense in animals: A survey of anti-predator defenses*. Longman, London.

Endler, J. A. 1978. A predator's view of animal color patterns. *Evolutionary Biology* **11**: 319–364.

Endler, J. A. 1983. Natural and sexual selection on color patterns in Poeciliid fishes. *Environmental Biology of Fishes* **9**: 173–190.

Endler, J. A. 1984. Progressive background matching in moths, and a quantitative measure of crypsis. *Biological Journal of the Linnaean Society of London* **22**: 187–231.

Endler, J. A. 1986. Defense against predation. Pages 109–134 in M. E. Feder & G. V. Lauder, editors. *Predator–prey relationships, perspectives and approaches from the study of lower vertebrates*. University of Chicago Press, Chicago.

Endler, J. A. 1987. Predation, light intensity, and courtship behaviour in *Poecilia reticulata*. *Animal Behaviour* **35**: 1376–1385.

Endler, J. A. 1988. Frequency-dependent predation, crypsis, and aposematic coloration. *Philosophical Transactions of the Royal Society of London B* **319**: 505–523.

Endler, J. A. 1990. On the measurement and classification of color in studies of animal colour patterns. *Biological Journal of the Linnaean Society of London* **41**: 315–352.

Endler, J. A. 1991. Variation in the appearance of guppy color patterns to guppies and their predators under different visual conditions. *Vision Research* **31**: 587–608.

Endler, J. A. 1992. Signals, signal conditions, and the direction of evolution. *American Naturalist* **139**: s125-s153.

Endler, J. A. 1993*a*. The color of light in forests and its implications. *Ecological Monographs* **63**: 1–27.

Endler, J. A. 1993*b*. Some general comments on the evolution and design of animal communication systems. *Philosophical Transactions of the Royal Society of London B* **340**: 215–225.

Endler, J. A. 1995. Multiple trait coevolution and environmental gradients in guppies. *Trends in Ecology and Evolution* **10**: 22–29.

Endler, J. A., and M. Théry. 1996. Ambient light specificity of leks, lekking behavior, and color patterns in three tropical forest-dwelling birds. *American Naturalist* **148**: 421–452.

Gilbert, L. E. 1975. Ecological consequences of a coevolved mutualism between butterflies and plants. Pages 210–240 in L. E. Gilbert and P. E. Raven, editors. *Coevolution of animals and* plants. University of Texas Press, Austin.

Gilliard, E. T. 1962. On the breeding behavior of the cock-of-the-rock (Aves, *Rupicola rupicola*). *Bulletin of the American Museum of Natural History* **124**: 31–68

Goldsmith, T. H. 1990. Optimization, constraint, and history in the evolution of eyes. *Quarterly Review of Biology* **65**: 281–322.

Hailman, J. P. 1977. *Optical signals: animal communication and light*. Indiana University Press, Bloomington.

Hailman, J. P. 1979. Environmental light and conspicuous colors. Pages 289–354 in E. H. Burtt, Jr., editor. *The behavioral significance of color*. Garland STPM Press, New York.

Hart, J. W. 1988. *Light and plant growth*. Unwin Hyman Publishers, London.

Jacobs, G. H. 1981. *Comparative color vision*. Academic Press, New York.

Jacobs, G. H. 1993. The distribution and nature of colour vision among the mammals. *Biological Reviews* **68**: 413–471.

Kawanabe, H., J. E. Cohen, and K. Iwasaki, editors. 1993. *Mutualism and community organisation: Behavioural, theoretical, and food-web approaches*. Oxford University Press, Oxford.

Laughlin, S. B. 1981. Neural principles in the peripheral visual system of invertebrates. Pages 133–280 in H. Autrum, editor. *Invertebrate visual centers and behaviour; handbook of sensory physiology*. Vol. VII/6b. Springer, New York.

Levine, J. S., Jr, and E. F. MacNicol. 1979. Visual pigments in teleost fishes: effects of habitat, microhabitat, and behavior on visual system evolution. *Sensory Processes* **3**: 95–131.

Lythgoe, J. N. 1979. *The ecology of vision*. Oxford University Press, Oxford.

Lythgoe, J. N., and J. C. Partridge. 1989. Visual pigments and the acquisition of visual information. *Journal of Experimental Biology* **146**: 1–20.

Rausher, M. D. 1978. Search image for leaf shape in a butterfly. *Science* **200**: 1071–1073.

Salisbury, F. B., and C. W. Ross. 1992. *Plant physiology*. Fourth edition. Wadsworth Publishing Co., Belmont.

Théry, M. 1990*a* . Influence de la lumière sur la choix de l'habitat et le comportement sexuel des Pipridae (Aves: Passeriformes). *Guyane Franéaise Revue d'Ecologie (La Terre et la Vie)* **45**: 215–236.

Théry, M. 1990*b*. Display repertoire and social organization of the White-fronted and White-throated Manakins. *Wilson Bulletin* **102**: 123–130.

Théry, M. and D. Larpin. 1993. Seed dispersal and vegetation dynamics at a Cock-of-the-Rock lek in the tropical forest of French Guiana. *Journal of Tropical Ecology* **9**: 109–116.

Trail, P. 1985. Territory and dominance in the lek-breeding Guianan Cock-of-the-Rock. *National Geographic Research* **1**: 112–123.

Trail, P. 1987. Predation and antipredator behavior at Guianan Cock-of-the-Rock leks. *Auk* **104**: 496–507.

Wickler, W. 1968. *Mimicry in plants and animals*. World University Library, New York.

Williams, K. S., and L. E. Gilbert. 1981. Insects as selective agents on plant vegetative morphology: egg mimicry reduces egg laying by butterflies. *Science* **212**: 467–469.

Willson, M. F., and C. J. Whelan. 1990. The evolution of fruit color in fleshy-fruited plants. *American Naturalist* **136**: 790–809.

Chapter 15

On becoming a conservation biologist: Autobiography and advice

KATHERINE RALLS

'If not us, who? If not now, when?'
Jewish Proverb

Knowledge of a species' behavior can be helpful, even key, to successful conservation efforts. The discovery, for example, that some Sea otters (*Enhydra lutris*) travel much farther offshore over deeper water than previously suspected facilitated the passage of legislation prohibiting set-net fishing in waters shallower than 30 fathoms throughout most of the species' range in California (Ralls *et al.* 1995). Many Sea otters had been accidentally drowned in these nets, and the restriction of this fishery led to an increase in the Sea otter population, which is now growing at about 5% per year (Estes *et al.* 1995).

Conversely, ignorance of behavior can lead to conservation failures. For example, a major effort to establish a new population of Sea otters failed because of a lack of knowledge about homing behavior (National Biological Service, unpublished data). From 1987 to 1991, the US Fish and Wildlife Service translocated 139 otters from the central California coast to San Nicolas Island off southern California. Most of the otters left the island and at least 36 of them returned to their capture location. In 1995, there were only about 17 otters at the island, a colony size that has persisted since about 1992.

In spite of the importance of behavioral studies for conservation planning, however, animal behaviorists have been relatively slow to join the ranks of conservation biologists. This chapter tells how I made the transition from purely academic research to conservation-related work. I have written the chapter in the hope that my experiences will prove helpful to others who would like to make the same transition. Many animal behaviorists may be interested in contributing professionally to conservation (after all, biodiversity is what inspires and motivates many of us) but be uncertain as to how to begin.

The majority of the scientists who founded the Society for Conservation Biology in 1985 were ecologists; I was the only one with a background in behavior. This disparity in participation between ecologists and animal behaviorists persists today. While recent presidents of the Ecological Society of America (ESA) have stressed the importance of ecological research to the development of conservation policy (e.g., Lubchenco *et al.* 1991; Lubchenco 1995) and the ESA has facilitated the publication of applied work by founding two new journals, the Animal Behaviour Society (ABS) has not been active in conservation policy, and its journal *Animal Behaviour* rarely publishes papers related to conservation.

There are, however, welcome signs that more animal behaviorists are becoming interested in conservation. ABS appointed an ad hoc Conservation Committee in 1995. In addition, both the 1995 meeting of ABS and the 1995 International Ethological Congress had symposia on the applications of behavior to conservation.

From neurophysiologist to conservation biologist

At the beginning of my career, I had no intention of becoming a conservation biologist. Conservation biology was not even a separate field when I received my Ph.D. in 1965. In retrospect, however, I can identify five key events that helped me make the transition from purely academic work to conservation biology: becoming expert on a particular taxon (mammals), getting a job in a zoo, serving on national-level committees that dealt with conservation issues beyond the comparatively limited concerns of my own institution, learning federal conservation policy and associated sources of funding, and making a personal commitment to conservation.

I have always been interested in mammals and have had the good fortune to be able to study many aspects of their biology. I decided to become a mammalogist in high school but did not discover animal behavior until I was a graduate student at Harvard. E. O. Wilson's marvelous lectures inspired me to pursue a career in behavior. In 1960 he had not yet discovered that mammals could be interesting, and I could not be convinced that ants were the most fascinating of living creatures! I became a student of Donald Griffin, who, although working on the neurophysiology of hearing in bats, was at least studying mammals.

My thesis was on high frequency hearing in mice as indicated by auditory evoked potentials (Ralls 1967). After a post-doc with Peter Marler at Berkeley in 1967, where I shifted my emphasis from neurophysiology to animal behavior, especially chemical communication in mammals, I was a guest investigator at Rockefeller University in New York, where I studied scent marking in small antelopes at the Bronx Zoo (Ralls 1971).

My first 'real job' was teaching at Sarah Lawrence College, where I became interested in sexual dimorphism. In 1972 several women in the faculty organized a course on 'Women: myth and reality' and invited me to give the lectures on sex differences in mammals (including humans). As I read about sexual dimorphism in mammals, the theories seemed too simple to account for the facts. I applied for and was awarded a Smithsonian fellowship to study sexual dimorphism in antelopes. I did write some papers on sexual dimorphism (Ralls 1976, 1977) but never did write one on sexual dimorphism in antelopes.

I did not return to Sarah Lawrence because, in 1976, John Eisenberg hired me to work in the research department at the Smithsonian's National Zoo. He was looking for an expert on the behavior of captive ungulates to replace the late Hal Buechner, who had studied the Uganda kob, *Kobus kob*, (e.g., Buechner 1974) and other species. I qualified for the job because of my previous work at the Bronx Zoo. Eisenberg and I discussed a prospective research program centering on the behavior of small antelopes. I soon diverged from this plan, however, and took my first inadvertent step towards becoming a conservation biologist.

Part of my job was to advise the zoo curators responsible for ungulates. The pedigree of the zoo's Eld's deer (*Cervus eldii*) herd was relatively simple and I noticed that all seven fawns resulting from father–daughter matings had died. To avoid more father–daughter matings and consequent increases in fawn mortality, I suggested that the curator exchange the breeding males between our two enclosures with these deer. The curator refused, saying that deer might be injured during efforts to move them and that there was no evidence that inbreeding was a problem. I decided to collect more data on the effects of inbreeding by examining the zoo's records. Without realizing it, I had started down the path towards becoming a conservation biologist.

Much of my next ten years was devoted to learning population genetics, collaborating with Jon Ballou, then my assistant but now a population geneticist at the National Zoo, on a series of papers documenting the higher mortality rates of inbred young in captive mammals (Ralls *et al.* 1979; Ralls & Ballou 1982*a*,*b*), and organizing a workshop to lay the foundations for the genetic management of captive populations (Ralls & Ballou 1986). Prior to this pioneering workshop, zoos managed captive animals as individuals, not as populations. In the process of organizing the workshop, Ballou and I met many first-rate geneticists who greatly contributed to our understanding of conservation genetics. We also introduced many of these geneticists, such as Richard Frankham of Macquarie University, and Russ Lande of the University of Oregon, to practical management questions, and they later made important scientific contributions to conservation genetics (e.g., Frankham 1995; Lande 1995).

Our genetic studies of captive populations, together with the guidelines developed at the workshop for the genetic management of captive populations, played a major role in inducing zoos to manage captive breeding programs at the national level. Zoos are now cooperating on a large scale to manage captive populations of threatened and endangered species to maximize the preservation of genetic diversity (Ralls & Ballou 1992).

Michael Soulé (now at the University of California at Santa Cruz) invited me to speak at the Second Conference on Conservation Biology in 1985 (Ralls *et al.* 1986) because of my work on inbreeding. The participants in this conference voted to establish the Society for Conservation Biology and Soulé immediately drafted several of us to help with the work involved in establishing a new professional society. I soon found myself, as a member of the fledgling society's first board of directors, involved with such matters as writing bylaws and finding an editor and publisher for the Society's new journal, *Conservation Biology*, which began publication in 1987. Through my involvement with this Society, I met many leading conservation biologists who again influenced my thinking and expanded my knowledge of the field.

Another step towards becoming a conservation biologist was serving on the Marine Mammal Commission's (MMC) Committee of Scientific Advisors from 1979 to 1982. Established by the Marine Mammal Protection Act (MMPA), the MMC provides scientific oversight for US Government activities involving marine mammals. John Twiss, Executive Director of the MMC, asked me to serve because he believed that mammalian social behavior should be one of the disciplines represented on this advisory group. Two colleagues who were experts on marine mammals, Robert Brownell, Jr, now at the National Marine Fisheries Service's Southwest Fisheries Science Center, and Donald Siniff of the University of Minnesota had recommended me to him.

I was certainly not an expert on either marine mammals or their conservation in 1979 but serving on the committee did a lot to remedy these deficiencies. I learned much about the MMPA, the Endangered Species Act, the permit process, and federal research on marine mammals. My work on this committee led to my involvement in many other scientific advisory committees on conservation issues in later years, such as the Southern Sea Otter Recovery Team and workshops and program reviews of various marine mammals. Participation in these committees continued my education on how the US Federal Government works and my knowledge of marine mammal conservation issues, such as the high degree of dolphin mortality resulting from the practice of setting nets around the schools of dolphins often associated with large tuna.

One of the species within the purview of the MMC was the Sea otter (*Enhydra lutris*). Each year, we advised the US Fish and Wildlife Service (USFWS) and the California Department of Fish and Game (CDFG) about Sea otter issues, including needed research on the threatened California Sea otter population. As our advice was often ignored, Donald Siniff, who had studied Sea otters in Alaska, jokingly told me that the only way to get this research done would be to do it ourselves.

In 1983, the joke became reality when Siniff and I began to study sea otters under a contract from the Minerals Management Service, which needed a Sea otter population model to help predict the effects of oil spills on the California sea otter population. I originally saw it primarily as a long-awaited opportunity to conduct field research on a large mammal. This opportunity was timely because my three daughters had left home for college and graduate school, I sensed that I had contributed what I could to captive breeding issues, and I was still physically capable of demanding field work. In addition to finally getting to do field work, however, I learned more about the intricacies of the state and federal permit processes, small boats, radiotelemetry, and Sea otter biology and politics.

Furthermore, as I explained in the introduction to this chapter, our work eventually proved very helpful to the California Sea otter population by providing the data needed to restrict the use of gill and trammel nets. This population now consists of over 2350 individuals, up from approximately 1250 when we began our work.

When funding for our Sea otter work ended in 1989, due to a moratorium on oil leasing along the California coast, I looked for another field project in central California. This time, I wanted to do a study with the explicit purpose of producing information that would be useful in conservation planning. By the end of the Sea otter study, I had finally realized I had become a conservation biologist. When confronted with a new opportunity, I was asking myself not only 'Is this scientifically interesting?' but also 'Is this likely to be useful in some conservation context?' I had consciously decided to join the relatively small group of scientists attempting to apply conservation biology to real-life problems in the USA. I resolved to concentrate my efforts in California. As a native Californian, the biodiversity of this state mattered more to me than biodiversity anywhere else. I wanted to help preserve this biodiversity for future generations, including my four granddaughters, the first of whom was born in 1988 and all of whom live in California.

At this time, The Nature Conservancy (TNC), CDFG, and the Bureau of Land Management were just combining forces to start a new reserve, the Carrizo Plain Natural Area, a short distance inland. I visited TNC's state ecologist in San Francisco to ask about research needs in the new reserve. Learning that he had funding for a study of the impact of Coyotes (*Canis latrans*), on the endangered San Joaquin kit foxes (*Vulpes macrotis mutica*), in the reserve, I wrote a successful proposal to do the work. Although I had never worked on either Kit foxes or Coyotes, I had what I needed to get the project going: experience in getting state and federal permits, helpful colleagues who were willing to teach me how to trap and handle foxes and Coyotes, and a basic knowledge of radiotelemetry. After the first year, I found a graduate student, P. J. White, who soon became my main collaborator on this project.

Our work on Kit foxes (Mercure *et al.* 1993; White & Ralls 1993; White *et al.* 1994, 1995; Ralls & White 1995) has been widely used in recovery planning for this species. Although we found that Coyotes were the principal source of mortality to Kit foxes in the reserve (Ralls & White 1995), we advised against coyote control. We discovered that a third canid, the non-native Red fox (*Vulpes vulpes*), was present in the reserve and that coyotes were probably suppressing the red fox population. Red foxes have apparently totally replaced Kit foxes in some parts of California, and as Coyotes tend to exclude Red foxes, but not Kit foxes (White *et al.* 1994) from their territories, Coyotes may actually be beneficial to kit foxes.

In the course of our fox work, I also learned about other endangered species in the same area, such as the Giant kangaroo rat (*Dipodomys ingens*). In fact, Dan Williams, a kangaroo rat expert, Rob Fleischer, the molecular geneticist in my department, and I are currently collaborating to work out the genetic relationships among various populations of this kangaroo rat.

My training in basic research did not prepare me for one aspect of conservation biology that I now consider among the most interesting challenges – how to make decisions in the face of uncertainty: uncertainty about the science and uncertainty about the future. This interest grew out of my experience on various scientific advisory committees concerned with conservation issues. These committees are often faced with the need to make recommendations despite uncertainty (e.g., Ralls *et al.*, in press). My main collaborator in this work has been Anthony Starfield, an applied mathematician and modeler at the University of Minnesota. We recently published our first papers, which concern the 'mobbing' problem in the endangered Hawaiian monk seal, *Monachus schauinslandi* (Starfield *et al.* 1995; Ralls & Starfield 1995). When adult sex ratios become male-biased, groups of adult male monk seals 'mob' and kill adult females and immature seals. Our work contributed to the National Marine Fisheries Service's decision to remove males from the Laysan Island monk seal population in 1995. This population now has an equal adult sex ratio, which we hope will lead to a reduction in mobbing behavior, a decrease in mortality, and resumed population growth.

Although my career path was unique, elements of it are shared with other conservation biologists. Expertise on a particular taxon and committee service seem to be a common route to conservation involvement. Andrew Smith, for example, in his 'A sabbatical odyssey at the SSC' (Smith 1995), relates how his knowledge of pikas (*Ochotona* spp.) led to his involvement first with the Lagomorph Specialist Group (one of the IUCN's many international scientific committees concerned with particular taxa) and then to a sabbatical leave at IUCN's headquarters in Switzerland. A personal commitment to conservation is another common element among conservation biologists. Richard Primack, for example, dedicated his text, *A Primer of Conservation Biology* (Primack 1995), to his children 'in the hope that they will experience the undiminished richness of nature' and Gary Meffe and C. Ronald Carroll are donating a third of the royalties from their text, *Principles of Conservation Biology* (Meffe & Carroll 1995), to major conservation organizations.

Why haven't more academics become involved with conservation?

Conserving biodiversity is obviously of importance to anyone employed as an animal behaviorist. So why have more academics not gotten involved in conservation? In my case, it is clear that the zoo environment has facilitated my involvement in conservation. In a zoo, opportunities for conservation-related work are obvious and one's colleagues tend to agree that such work is a worthwhile activity. Furthermore, there are so few scientists employed in zoos that the 'if not us, who?' factor is important. When I became involved in inbreeding research, for example, it was clear that I had found an important problem and that if I did not follow through with it (even though I had a background in behavior rather than genetics), no one else was likely to. It seems no accident that Devra Kleiman, another behavioral researcher at the National Zoo, and I both became involved in conservation-related research, even though her studies have remained more focused on behavior than mine (e.g., Kleiman 1983, 1994).

In contrast, I suspect that some aspects of academia make it more difficult for scientists to get involved in conservation. In an academic department without colleagues interested in conservation, it may be less clear as to how one might begin some conservation-related research. In addition, one's colleagues may be less encouraging – some departments may have an overly critical attitude that tends to deter people from moving in completely new directions. Moreover, some academics feel that funding from National Science Foundation (NSF) is superior to funding from elsewhere. Surely the worth of a research program is best judged from the resulting publications, rather than the funding source?

Other academics may feel that they cannot afford the risk that the scientific payoff, i.e. the number or type of publications resulting from applied research, may be smaller than that resulting from a similar amount of effort devoted to more conventional academic studies. Progress in conservation-related work may be slow, especially if the study requires capture and handling of an endangered species. Sample sizes may be small, experiments may be precluded, and permit problems are likely to consume time and energy. When we started the otter project, for example, the prospects for the successful completion of the fieldwork in California were so precarious that we never dared to involve a graduate student in that part of the research; however, three of Siniff's students did receive their Ph.D.s on other aspects of this project (Brody 1988; Monnett 1988; Rotterman 1992). Although we eventually published several papers on the California work (Ralls *et al.* 1989, 1992, 1995; Ralls & Siniff 1990; Siniff & Ralls 1991), events beyond our control could have made it impossible to complete the work successfully.

Other academics seem overly concerned with whether or not their colleagues see conservation biology as real science. The idea that 'basic' science is superior to 'applied' science may still be a problem for some. Before his sabbatical with IUCN, for example, Smith asked himself

"Was this pure folly – to devote my hard-earned sabbatical leave to slave from dawn to dusk on issues that had real-world application, rather than to the basic science preferred by my academic colleagues?" (Smith 1995). Personally, I have found that there is often a fruitful interaction between basic and applied research. Thinking about an applied question can be very stimulating and suggests new areas of basic research: for example, my concern with juvenile mortality in captive ungulates eventually led to a paper that estimated the 'cost of inbreeding' in various species of mammals (Ralls *et al.* 1988). Although the cost of inbreeding figures in a variety of theoretical models (e.g., Dawkins 1976; Bengtsson 1978; Smith 1979), this cost had not been estimated for mammals other than humans prior to our work (May 1979).

My view is that "Conservation biology will become what we collectively make it. If enough of us do it and do it well, conservation biology, at least those parts of it created by individuals trained as scientists, will certainly become a science if it is not one already" (Ralls 1995).

Who is best suited to make the transition to applied work?

I think the transition is relatively easy for well-established scientists, like myself. Many of them are motivated by a concern with the quality of life their descendants will experience. And if they choose to devote some of their energies to conservation, they can safely ignore the negative opinions of some colleagues because they are secure in their jobs. In addition, established scientists can afford the risk that a project may not yield impressive numbers of publications.

Of course, one does not have to begin with a major conservation-related research project; those who are very concerned about the likelihood of publications could begin on a smaller scale, perhaps by joining a committee, lecturing in a conservation biology course, or attending meetings of scientific societies, such as SCB, ESA, or the Wildlife Society, that are concerned with conservation issues. The meetings of local chapters of these professional societies may be quite informative, as they are often involved in local conservation problems, and agency personnel may be among the members. I think that any conservation-related activities that tend to expand one's horizons would probably be helpful, because the first step may be the hardest. After that, one thing tends to lead to another and you end up involved with conservation issues.

Personality is another important consideration. Conservation research is often issue-driven rather than discipline-driven (Meffe & Viederman 1995). Solving issue-driven problems is likely to require the combined efforts of experts in a variety of disciplines. Thus, many conservation biologists are good collaborators who are willing to take the time to learn about academic areas outside of their primary discipline. Patience is helpful too – one is likely to attend some meetings that seem to be moving very slowly. Any personality type can make a useful contribution to conservation – the important thing is to pick something that suits your personality as well as your scientific interests. As Meffe & Carroll (1995) say in the final sentence of their conservation biology text: "Where do you think you could make the best contributions, given your own talents, interests, and experiences?"

Unpleasant consequences of conservation-related work

Animal behaviorists who venture into conservation-related research must be prepared for the political/management consequences of their work. They may encounter resistance to change from managers and intense scrutiny from non-scientists. For example, I encountered a great deal of resistance to the implications of my studies on inbreeding depression in captive populations for several years. When I distributed my first in-house report on the effects of inbreeding on juvenile mortality in captive ungulates in 1977, several zoo curators objected to the zoo's research department (i.e., me) meddling in animal management. Many people maintained that my findings were false; it could not possibly be true that inbred young had higher mortality rates than outbred young. The director of the zoo even called me into his office and forbade me from publishing my results. He ordered me to search the zoo's records and examine every possible alternative hypothesis, such as the possibility that more inbred young were born to inexperienced mothers. When I explained how much work such a project would be (the zoo's records were not computerized), he gave me some money for an assistant, and I was able to hire Jon Ballou. So Ballou spent a year or so extracting data from the zoo's records, and we wrote a paper that excluded all the alternative hypotheses. This paper (Ballou & Ralls 1982) was politically useful but perhaps not scientifically necessary because the evidence I had presented in my original study was quite strong.

Similarly, when working on our sea otter project, Siniff and I encountered intense scrutiny from non-scientists such as the Friends of the Sea Otter. Our research was controversial because we proposed to use intraperitoneally implanted transmitters (Ralls *et al.* 1989) in California sea otters, which are federally listed as threatened. Although this technology had been used successfully in river otters, it had not yet been used in Sea otters. We succeeded by developing the technique on Sea otters in Alaska, which are not threatened or endangered, before we requested permits to use it with California otters. Even after we had demonstrated that the procedure was not harmful to Alaskan otters, we were allowed to begin with only five California otters. We were required to monitor them intensively for an entire year before we could request permission to implant transmitters in more otters.

Members of the Friends of the Sea Otter and other conservation organizations vigorously opposed our project and attempted to have both our federal permit and our state permit denied at several points. They almost succeeded at one meeting of the California State Fish and Game Commission – we were granted permission to continue our project by only one vote. Siniff and I spent a good deal of time answering questions from the federal permit office, some of which were rather uninformed because the person responsible for sea otter permits had been trained as a botanist. At one point in 1985, the permit office ordered us to stop implanting transmitters because some captive otters in someone else's project had died following the use of a completely different type of implanted transmitter. We were allowed to resume work, however, in a couple of weeks and eventually completed the project with a sample of 47 otters. Our study was so beneficial to the California Sea otter population that even the Friends of the Sea Otter ultimately became supporters of the work (Fulton 1989).

So my experience has been that even though there may be some unpleasant experiences along the way, good scientific studies ultimately can have an impact on management.

From science to policy

That said, researchers beginning studies relevant to conservation policy must be prepared to find that science is not automatically converted to policy. One problem is that lawyers and scientists have different conceptions of the 'Truth' and often communicate poorly. Furthermore, conservation biologists often must give advice in the face of uncertainty, which is very difficult for us because it goes against our scientific training. Data may be sparse and possibly inaccurate. Decisions are usually based, at least in part, on estimates of relative risk: the risk of extinction (Ralls &

Taylor, 1996), the risk of acting on imperfect information, and the risk of waiting to collect more information before acting. Often, we are attempting to predict the future, an endeavor at which humans have had very limited success.

Lawyers want yes or no answers right now while scientists offer probabilities or call for more research. We must find ways to make prudent decisions in the face of scientific uncertainty. Effective policies under conditions of uncertainty are possible but they must take uncertainty into account (Ludwig *et al*. 1993). Furthermore, we will be able to make better decisions in the future if we design management efforts so that they will yield information that will improve our understanding of the problem. This sort of dynamic, or adaptive management, helps reduce uncertainty over time (Walters 1986; Walters & Holling 1990).

My current collaboration with Tony Starfield aims at improving methods for making decisions under uncertain conditions; we are using both computer models and ideas from the field of operations research. Our initial work on Hawaiian monk seals (Starfield *et al*. 1995; Ralls & Starfield 1995) appealed to the government managers responsible for this species, but they are scientists who have acquired management responsibilities. I don't know how lawyers would react to such an approach.

Some have suggested improvements in the way the legal system handles scientific issues, such as the use of judges trained to have scientific expertise and the use of court-appointed expert witnesses who are not associated with either the prosecution or the defense. There is also room for improvement in the behavior of scientists. As long as lawyers, managers, and politicians see that scientists differ, they can easily ignore us.

Scientists need to avoid scientific arrogance and treat agency managers with respect. As Jerry Franklin, a veteran of the spotted owl controversy, has said: "Hubris is the most serious limitation of the scientist involved in policy analysis" (Franklin 1995). A scientist who adopts a superior attitude can have negative impacts that extend beyond the problem at hand, because one unpleasant experience with such a scientist may lead managers and decision-makers to conclude that all scientists are arrogant, impractical eggheads. Furthermore, in my experience, scientific arrogance is rarely justified. Relevant science may be limited or lacking and you are likely to discover that agency managers know many important things that you do not.

Scientists also need to consider the consequences of erring in various ways (NRC 1995). They need to pay particular attention to the results of making a Type II error, i.e., accepting a false null hypothesis, such as that population size has remained constant over time. This may be difficult at first because scientists are trained to minimize the Type I error, i.e., the probability of reporting an effect that does not exist. As Peter Medawar has said, "The most heinous offence a scientist can commit is to declare to be true that which is not so" (Medawar 1984). In conservation biology, however, making a Type II error can have disastrous consequences because it can result in protecting species less than necessary (Maguire 1991; Noss 1992). Statistical power is often low in studies of rare species. The power to detect a decline in abundance over time, for example, decreases as populations become smaller, and for the endangered porpoise (*Phocoena sinus*) power to detect trends is very low within the range of estimated population sizes (Taylor & Gerrodette 1993). Thus, detection of a decline should not be a requirement before conservation measures are enacted for this species. Scientists newly involved in conservation issues may call for more research before making a management recommendation, without fully realizing that delaying action is itself a management decision, and one that may have a serious impact on a population.

Scientists also need to provide consensus advice whenever possible, because managers confronting conflicting scientific opinions often tend to ignore science altogether. I first experienced this in my work on inbreeding depression in captive populations. While Ballou and I were presenting evidence that inbreeding was a problem in these populations (Ralls *et al.* 1979; Ralls & Ballou 1982*a*,*b*), others, particularly Bill Shields (Shields 1982), were maintaining that inbreeding was often beneficial. At that point, many zoo managers felt they could safely ignore genetics because the advice they got would depend on which geneticist they asked. To overcome this problem Ballou and I organized the workshop on the genetic management of captive populations. Our objective was to achieve consensus recommendations regarding the genetic management of captive populations despite disagreement among population geneticists. We convened scientists representing the range of opinions in the field, asked them what they agreed about and what they disagreed about and, given the scientific uncertainties, how we could best manage captive populations until the scientific disagreements were resolved. This approach proved quite successful and the general guidelines recommended at that workshop (Ralls & Ballou 1986) are still in use.

My impression is that small groups of scientists, such as Recovery Teams and other scientific advisory committees, are doing fairly well at providing consensus advice. If we want to have a serious impact on national and international conservation policy, we will have to do this on a larger scale.

Rewards of conservation biology

My involvement in conservation issues has been fruitful for both myself and the animals I have studied. Continued exposure to new species, problems, and academic disciplines has been conducive to my scientific productivity. Considering practical problems has contributed to my education and breadth as a scientist and has enabled me to work closely with, and learn from, stimulating colleagues in other fields. I have found it very gratifying to affect actions implemented in the real world.

My advice to anyone contemplating behavioral research relative to a conservation issue is: do it. In my experience, both the need and the rewards are great.

Literature cited

Ballou, J., and K. Ralls. 1982. Inbreeding and juvenile mortality in small populations of ungulates: a detailed analysis. *Biological Conservation* **24**: 239–271.

Bengtsson, B. O. 1978. Avoid inbreeding: at what cost? *Journal of Theoretical Biology* **73**: 439–444.

Brody, J. A. 1988. Population dynamics of California Sea otters and a model for the risk of oil spills. Ph.D. Thesis, Department of Ecology, Evolution and Behavior, University of Minnesota, St. Paul.

Buechner, H. K. 1974. Implications of social behavior in the management of the Uganda kob. Pages 853–871 in V. Geist and F. Walther, editors. *The behavior of ungulates and its relation to management*. IUCN Publications, n. s., No. 24, Morges, Switzerland.

Dawkins, R. 1976. *The selfish gene*. Oxford University Press, Oxford, UK.

Estes, J. A., R. J. Jameson, J. L. Bodkin, and D. R. Carlson. 1995. Pages 110–112 in E. T. LaRoe, G. S. Ferris, C. E. Puckett, P. D. Doran, and M. J. Mac, editors. *Our living resources: A report to the nation on the distribution, abundance, and health of US plants, animals, and ecosystems*. US Department of the Interior, National Biological Service, Washington, DC.

Frankham, R. 1995. Inbreeding and extinction: a threshold effect. *Conservation Biology* **9**: 792–799.

Franklin, J. F. 1995. Scientists in wonderland. *BioScience, Science and Biodiversity Policy Supplement* **S**: 74–78.

Fulton, C. 1989. Monterey Bay gill nets take tragic toll of sea otters and harbor porpoise. *Otter Raft* (Summer), p. 13.

Kleiman, D. G. 1983. The behavior and conservation of the golden lion tamarin, *Leontopithecus rosalia*. Pages 35–53 in M. Thiago de Mello, editor. *Annals First Congress Brasil Primatologia*, Belo Horizonte.

Kleiman, D. G. 1994. Mammalian sociobiology and zoo breeding programs. *Zoo Biology* **13**: 423–432.
Lande, R. 1995. Mutation and conservation. *Conservation Biology* **9**: 782–791.
Lubchenco, J. 1995. The role of science in formulating a biodiversity strategy. *BioScience, Science and Biodiversity Policy* Supplement **S**: 7–9.
Lubchenco, J., A. M. Olson, L. E. Brubaker, S. R. Carpenter, M. M. Holland, S. P. Hubbell, S. A. Levin, J. A. MacMahon, P. A. Matson, J. M. Melillo, H. A. Mooney, C. H. Peterson, H. R. Pulliam, L. A. Real, P. J. Regal, and P. J. Risser. 1991. The Sustainable Biosphere Initiative: an ecological research agenda. *Ecology* **72**: 371–412.
Ludwig, D., R. Hilborn, and C. Walters. 1993. Uncertainty, resource exploitation, and conservation: lessons from history. *Science* **260**: 17, 36.
Maguire, L. 1991. Risk analysis for conservation biologists. *Conservation Biology* **5**: 123–125.
May, R. M. 1979. When to be incestuous. *Nature* **279**: 192–194.
Medawar, P. B. 1984. *The limits of science*. Harper and Row, New York.
Meffe, G. K., and C. R. Carroll. 1995. *Principles of conservation biology*. Sinauer, Sunderland, Massachusetts.
Meffe, G. K. and S. Viederman. 1995. Combining science and policy in conservation biology. *Wildlife Society Bulletin* **23**: 327–332.
Mercure, A., K. Ralls, K. P. Koepfli, and R. K. Wayne. 1993. Genetic subdivisions among small canids: mitochondrial DNA differentiation of swift, kit, and arctic foxes. *Evolution* **47**: 1313–1328.
Monnett, C. 1988. Patterns of movement, post-natal development, and mortality of sea otters in Alaska. Ph. D. Thesis, Department of Ecology, Evolution and Behavior, University of Minnesota, St. Paul.
NRC (National Research Council). 1995. Science and the Endangered Species Act. National Academy Press, Washington, DC.
Noss, R. F. 1992. Biodiversity: many scales and many concerns. Pages 17–22 in H. F. Kerner, editor. *Proceedings of the Symposium on Biodiversity of Northwestern California*, October 28–30, Santa Rosa, California.
Primack, R. B. 1995. *A Primer of conservation biology*. Sinauer, Sunderland, Massachusetts.
Ralls, K. 1967. Auditory sensitivity in mice: *Mus musculus* and *Peromyscus*. *Animal Behaviour* **15**: 123–128.
Ralls, K. 1971. Mammalian scent marking. *Science* **171**: 443–449.
Ralls, K. 1976. Mammals in which females are larger than males. *Quarterly Review of Biology* **51**: 245–276.
Ralls, K. 1977. Sexual dimorphism in mammals: avian models and unanswered questions. *American Naturalist* **111**: 917–938.
Ralls, K. 1995. But is it Science? *Conservation Biology* **9**: 8.
Ralls, K., and J. Ballou. 1982*a*. Effects of inbreeding on juvenile mortality in some small mammal species. *Laboratory Animals* **16**: 159–166.
Ralls, K., and J. Ballou. 1982*b*. Effects of inbreeding on infant mortality in captive primates. *International Journal of Primatology* **3**: 491–505.
Ralls, K., and J. Ballou, editors. 1986. Proceedings of the Workshop on the Genetic Management of Captive Populations. *Zoo Biology* **5**: 81–238.
Ralls, K., and D. B. Siniff. 1990. Time budgets and activity patterns in California sea otters. *Journal of Wildlife Management* **54**: 251–259.
Ralls, K., and J. Ballou. 1992. Managing genetic diversity in captive breeding and reintroduction programs. *Transactions 57th North American Wildlife and Natural Resources Conference (1992)*, pp. 263–282.

Ralls, K., and A. M. Starfield. 1995. Choosing a management strategy: two structured decision-making methods for evaluating the predictions of stochastic simulation models. *Conservation Biology* **9**: 175–181.

Ralls, K., and B. L. Taylor. 1996. How viable is Population Viability Analysis? In S. T. A. Pickett, R. S. Ostfeld, M. Shachak, and G. E. Likens, editors. *Enhancing the ecological basis of conservation: Heterogeneity, ecosystem function, and biodiversity. Proceedings of the Sixth Cary Conference*. Chapman and Hall, New York (in press).

Ralls, K., and P. J. White. 1995. Predation on endangered San Joaquin kit foxes by larger canids. *Journal of Mammalogy* **276**: 723–729.

Ralls, K., J. Ballou, and A. R. Templeton. 1988. Estimates of lethal equivalents and the cost of inbreeding in mammals. *Conservation Biology* **2:** 185–193.

Ralls, K., K. Brugger and J. Ballou. 1979. Inbreeding and juvenile mortality in small populations of ungulates. *Science* **206**: 1101–1103.

Ralls, K., D. P. DeMaster, and J. A. Estes. Developing a delisting criterion for the Southern Sea Otter under the US Endangered Species Act. *Conservation Biology*, in press.

Ralls, K., P. H. Harvey, and A. M. Lyles. 1986. Inbreeding in natural populations of birds and mammals. Pages 35–56 in M. Soulé, editor. *Conservation biology: The science of scarcity and diversity*, Sinauer, Sunderland, Massachusetts.

Ralls, K., B. Hatfield, and D. B. Siniff. 1995. Foraging patterns of California sea otters based on radiotelemetry. *Canadian Journal of Zoology* **73**: 523–531.

Ralls, K., D. B. Siniff, A. Doroff, and A. Mercure. 1992. Movements of sea otters relocated along the California coast. *Marine Mammal Science* **8**: 178–184.

Ralls, K., T. W. Williams, D. B. Siniff, and V. B. Kuechle. 1989. An intraperitoneal radio transmitter for sea otters. *Marine Mammal Science* **5**: 376–381.

Rotterman, L. M. 1992. Patterns of genetic variability in sea otters after severe population subdivision and reduction. Ph. D. Thesis, Department of Ecology, Evolution and Behavior, University of Minnesota, St. Paul.

Shields, W. 1982. *Philopatry, inbreeding, and the evolution of sex*. State University of New York Press, Albany.

Siniff, D. B., and K. Ralls. 1991. Reproduction, survival, and tag loss in California sea otters. *Marine Mammal Science* **7**: 211–229.

Smith, A. T. 1995. A sabbatical odyssey at the SSC. *Species* **24**: 11–13.

Smith, R. H. 1979. On selection for inbreeding in polygynous animals. *Heredity* **43**: 205–211.

Starfield, A. M., J. Roth, and K. Ralls. 1995. 'Mobbing' in Hawaiian monk seals: the value of simulation modeling in the absence of apparently crucial data. *Conservation Biology 9:*166–174.

Taylor, B. L., and T. Gerrodette. 1993. The uses of statistical power in conservation biology: the vaquita and northern spotted owl. *Conservation Biology* **7**: 489–500.

Walters, C. J. 1986. *Adaptive management of renewable resources*. McGraw-Hill, New York.

Walters, C. J. and C. S. Holling. 1990. Large-scale management experiments and learning by doing. *Ecology* **71**: 2060–2068.

White, P. J., and K. Ralls. 1993. Reproduction and spacing patterns of kit foxes relative to changing prey availability. *Journal of Wildlife Management* **57**: 861–867.

White, P. J., K. Ralls, and R. A. Garrott. 1994. Coyote-kit fox spatial interactions based on radiotelemetry. *Canadian Journal of Zoology* **72**: 1831–1836.

White, P. J., K. Ralls, and C. A. Vanderbilt White. 1995. Overlap in habitat and food use between coyotes and San Joaquin kit foxes. *Southwestern Naturalist* **40**: 342–341.

Author acknowledgments

Janine R. Clemmons and Richard Buchholz, *Chapter 1*. Many of the ideas in this chapter stem from discussions with the participants in this book and of the symposium and paper session on behavior and conservation held at the 1995 ABS meetings in Lincoln, Nebraska. Helpful comments and discussion of sections or earlier drafts of this paper were provided by Tim Allen and his lunch group, Dianna Padilla, Warren Porter, and Ethan Clotfelter.

Stephen R. Beissinger, *Chapter 2*. Development of the tools concept used here benefited from discussions with Tim Clark, Bob Lacy, Ariel Lugo, Peter Stacey, and Joe Wunderle. Suggestions by Janine Clemmons, Richard Buchholz, and two anonymous reviewers greatly improved this paper.

Peter Arcese, Luke F. Keller, and John R. Cary, *Chapter 3*. Many behaviorists have had inputs into this work, directly via their comments on drafts, and indirectly in discussions about how and why one pursues careers in science. In particular, Ron Ydenberg encouraged us to develop behavioral examples of conservation practice to stimulate research. Justin Brashares, Rich Buchholz, Janine Clemmons, Don Waller and anonymous referees made constructive suggestions to improve this chapter. This work was funded by a grant from the US National Science Foundation (IBN–94568122) and the College of Agriculture and Life Sciences, University of Wisconsin-Madison.

Hugh Dingle, Scott P. Carroll, and Jenella E. Loye, *Chapter 4*. We thank B. Fisher for helpful discussion and references, and several reviewers for their constructive criticisms.

John C. Wingfield, Kathleen Hunt, Creagh Breuner, Kent Dunlap, Gene S. Fowler, Leonard Freed, and Jaan Lepson, *Chapter 5*. Unpublished research presented in this paper was supported by the US Fish and Wildlife Service, Pacific Islands Office, Grant Number 14–48–0001–92690. We are most grateful to Richard C. Wass for the permit to work at Hakalau and very special thanks for the thoughtful guidance and suggestions from Jack Jeffrey. We are most grateful to Dr Carol Terry of the Department of Land and Natural Resources for ensuring that our permits were correct and sufficient to allow us to complete the project. Jim Jacobi, Thane Pratt and Gerald Lindsey gave us valuable advice concerning field work on endemic Hawaiian birds. Lynn Erckmann gave valuable assistance when assaying the plasma samples. Bengt Silverin, David Crews and an anonymous reviewer provided useful comments on the manuscript. Finally, many thanks to Karen Rosa for the grant funds that made the work on Hawaiian forest birds possible.

Ian G. McLean, *Chapter 6*. I thank the many people who have been involved in my research on predator recognition in animals; all will have contributed in small or large ways to the points made here. In particular, I thank C. Hoelzer, D. Hume, R. Maloney and G. Rhodes. D. Anderson and H. Hayne provided some essential literature. R. Buchholz, J. Clemmons, L. Drickamer, L. Haase, D. Hume, J. Marzluff, G. White and an unidentified reviewer made many valuable and insightful comments on the manuscript. I thank R. Buchholz and J. Clemmons for their patience during manuscript preparation. Funding for my research is provided by Donaghys Industries Ltd, Fletcher Building Ltd, the New Zealand Lotteries Board and the University of Canterbury.

Scott H. Stoleson and Stephen R. Beissinger, *Chapter 7*. We thank Janine Clemmons and Rich Buchholz for having the foresight and initiative to organize the symposium that led to this volume, and for inviting us to take part. Many of the concepts in this paper resulted from collaborative work between SRB and Enrique Bucher. Our parrotlet research was funded by an American Ornithologists' Union Research Grant, the Frank M. Chapman Fund of the American Museum of Natural History, the National Geographic Society, NSF grants DEB–9503194 and IBN–9407349, Sigma Xi, the Smithsonian Institution's International Environmental Sciences program in Venezuela, and Yale University. Comments by P. Arcese, R. Buchholz, J. R. Clemmons, S. Derrickson, L. S. Forbes, J. Robinson and S. Tygielski improved this chapter.

Richard Buchholz and Janine R. Clemmons, *Chapter 8*. We thank John Carr, Frank Pezold, Carla Restrepo, Bill Stevens, and Tsunemi Yamashita for reviewing an early draft of this paper, and for giving us the systematists' perspectives on conservation and behavioral diversity. On many points they totally disagreed with one another, or with us; however their diverse and constructive critiques helped us understand the differences of opinion between the subfields of systematics, and clarified our own arguments.

Luis F. Baptista and Sandra L. L. Gaunt, *Chapter 9*. Comments on an earlier manuscript by J. Faaborg, M. Greenfield and an anonymous reviewer aided us in focusing this chapter. D. Nelson and the Conservation Biology Discussion Group at The Ohio State University served as a valuable sounding board. Important references where provided by Kelly Allman, Ken Herman, Jeff Jackson, Elizabeth Pierson, Chris Peterson, William Rainey, Diana Reese, and Dave Weissman. Kathleen Berge, secretary to L. F. B., was invaluable in coordinating activities between two authors, each frequently in different parts of the world, and for attending to many details of manuscript preparation.

Patricia G. Parker and Thomas A. Waite, *Chapter 10*. This work was supported by funding from the National Science Foundation (grant no. DEB–9322544 to PGP). We thank Tigerin Peare and Julie Rieder for their sea turtle insights from the field and laboratory, and Kim Lundy for her information on Arabian babblers. Jeffreys' multilocus probes were made available by Cellmark Division of ICI. We thank Richard Wagner, Malcolm Schug and Eugene Morton for sharing unpublished data on purple martins.

John A. Endler, *Chapter 14*. I am grateful for very useful comments and suggestions from Rob Bierregaard, Jed Burtt, and the editors.

Katherine Ralls, *Chapter 15*. I thank Richard Buchholz for inviting me to write this essay; my daughter Robin Meadows for her skillful editing; Michael Gilpin for the proverb; Peter Arcese, Richard Buchholz, and Janine Clemmons for comments on the manuscript, and John Eisenberg and Devra Kleiman, my supervisors at the National Zoological Park, for supporting my changing research interests over the last 20 years.

Index